Handbook of Molecular
Force Spectroscopy

Handbook of Molecular Force Spectroscopy

Edited by

Aleksandr Noy

Springer

Aleksandr Noy
Chemistry, Materials and Life Sciences Directorate
Lawerence Livermore National Laboratory
Livermore, CA
USA
noy1@llnl.gov

ISBN-13: 978-0-387-49987-1   e-ISBN-13: 978-0-387-49989-5

Library of Congress Control Number: 2007937164

Printed on acid-free paper.

9 8 7 6 5 4 3 2 1

springer.com

Preface

The concept of force spectroscopy is deceptively simple: if we could attach a pair of handles to two interacting molecules and use those handles to pull the molecules apart, then we could not only obtain a clear and unambiguous value of the bond strength, but also obtain this value with a very direct and straightforward measurement. People have used the "tug test" to measure and compare strength in many forms throughout history and ooccasionally this strength testing could take the entertaining form of a "tug-of-war" contest, such as the Japanese tsunahiki. Tug-of-war even used to be an Olympic sport in the beginning of the twentieth century. We all have certainly played with the force-measuring springs in a science class at school. So what could be simpler than pulling two species apart?

At least that was my impression when I first got exposed to the concept of using tiny springs to probe molecular-scale interactions. At the time, I was a very green and moderately scared first year graduate student at Harvard who, along with the rest of my classmates, was looking for a research project and for a research group to call home for the next several years. The choice process consisted mostly of listening to gossip, going to each group's open house, and then gossiping more. Although the conversations mostly revolved around the free pizza typically served at those functions, a good deal of the professors' research presentations was also discussed. I still remember being very impressed by two events at the open house of Charles Lieber's group. First, instead of the perfectly respectable but boring pizza, the Lieber group served a lavish Chinese takeout buffet, which was met with considerable delight by everyone in attendance. Second, the research presentation mentioned a project where someone would try to attach specific molecules to the needle of an AFM probe and then use specific interactions of those molecules to perform chemically specific imaging at that unprecedentedly small scale. The concept, which later would become known as chemical force microscopy, had a strange appeal to me, which I immediately shared with my classmate sitting on the conference room floor next to me. His response had a noticeable sarcastic overtone: "Good luck doing it, it sounds complicated. ..." Like any good advice, it went unheeded, and after a few days of deliberations I signed on. Thus my decade-long fascination with force spectroscopy and its applications to the study of chemical and biological interactions started. Today I would freely admit that the advice I received from my classmate was quite sound. Over the past decade, it has been indeed fascinating to watch researchers uncover an incredibly rich universe of different physical behaviors that originate from such a conceptually simple setup.

Why do researchers continue to be interested in interaction forces when everything we study in the physics, chemistry, and biophysics courses almost always revolves around interaction energies and interaction potentials? Part of the answer lies in the ubiquitous role the interaction forces play in the majority of condensed phase phenomena. These interactions ultimately shape the dynamics of the molecular behavior on the microscopic scale, and direct probing of interaction forces is important for compiling the full picture of these phenomena. Often these processes, most notably in biological systems, involve spectacular rearrangements and movements of ions, molecules, or whole molecular assemblies driven by mechanical stresses generated by molecular-scale motors. Direct probing of the forces generated by these sophisticated biological machines provides invaluable information about the nature of these processes, and force spectroscopy techniques have been at the forefront of molecular motor studies. Perhaps the most powerful argument for the utility of force spectroscopy techniques is that they provide researchers with a "handle" that they can use to deform the potential energy landscape in the direction of the applied force. Such deformation invariably modifies the kinetics of the molecular bond rupture, and monitoring of the rupture kinetics as a function of the applied force (and force direction) gives us a unique opportunity to study the potential energy landscape of the interactions, often in one direction at a time.

A force spectroscopy measurement almost always involves attaching interacting molecules to a force transducer and then using a mechanical translation device, such as a piezoelectric scanner, to move one of the interacting molecules. In practice, this scheme can be implemented using a large number of very distinct technical approaches. Three of them tend to dominate the force spectroscopy field nowadays. The surface forces apparatus uses ultra-smooth crossing cylinder sheets to probe the interactions between monolayers of interacting species attached to the surfaces of the interacting sheets. Optical and magnetic trapping techniques, which are widely known as "molecular tweezers" techniques, use optical gradients of magnetic fields to trap and move tiny particles or beads. Researchers can use a well-developed arsenal of chemical and biochemical methods to tether different configurations of molecules to the bead surfaces, use the trap to manipulate the beads, and then use highly controlled small forces to study the interaction dynamics. Finally, perhaps the most widespread technique involves using tiny atomic force microscope probes to measure interaction forces between molecules attached to the surfaces of the cantilever tip and the sample.

Each of these measurements addresses several common questions and challenges. First, researchers need to design the experiment to enable probing of a certain specific interactions while discriminating against the non-specific interactions that are always present in real measurements. Second, more often than not, force spectroscopy measurements happen away from equilibrium; therefore researchers need to pay attention to the kinetics of the loading and rupture process and use this information to reconstruct the underlying potential energy landscape of the intermolecular bond. Third, manipulating single molecules on the nanometer scale is rarely precise and researchers are always facing the challenge of estimating properly the number of interacting molecules and relating that to the measured forces.

This book is not intended as a mere survey of the force spectroscopy achievements over nearly two decades of the field's existence, as such surveys are almost always incomplete in an actively developing field. Instead, the intent is to present a series of topics that discuss fundamental concepts and basic methodology used to perform and understand force spectroscopy experiments and illustrate them using examples from current and past research. Thus the ideal audience that we have imagined for this book is a graduate student who is just starting in the force spectroscopy field and is looking to learn the ropes, or a researcher from an adjacent field who wants to get up to speed with force spectroscopy measurements, or simply wants to evaluate the potential benefit of the technique for her research. Our hope is that this audience will be served well by the material presented in this handbook.

D. Leckband starts the volume by describing the basic principles of the surface forces apparatus measurements and their applications for studies of the protein-protein interactions. C. Lieber, A. Noy, and D. Vezenov give a detailed description of chemical force microscopy—the technique for probing intermolecular interactions using AFM tips functionalized with specific chemical functional groups. R. Conroy presents an extensive survey of the force measurements using magnetic and optical tweezers—the technique that in many aspects is complementary to the AFM- and SFA-based measurements.

One of the major advancements in force spectroscopy in the last decade has been the emergence of the kinetic model of the bond strength, which caused a paradigm shift in the interpretation of force spectroscopy experiments and spawned the development of dynamic force spectroscopy. A chapter by P. Williams discusses dynamic force spectroscopy and its applications to the AFM experiments. A contribution by K. Anderson, D. Brockwell, S. Radford, and D. A. Smith describes elegant experiments that use dynamic force spectroscopy to probe protein structure.

Functionalization of the force probes with biological molecules is an extremely important part of any force spectroscopy measurement, and the chapter by C. Blancette, A. Loui, and T. Ratto surveys different approaches to functionalization of the force probes. Attaching biomolecules to the force probes via long flexible polymeric tethers has proven to be an extremely versatile, important, and fruitful approach to such functionalization. The chapter by T. Sulchek, R. Friddle, and A. Noy discusses the implementation of this approach, the models used to interpret the results of these measurements, and their application to studies of the strength of multiple bonds. Development of the approaches to probe equilibrium potential energy landscapes of the interactions remains an important goal of the field, and a chapter by P. Ashby describes the design principles and the setup of the AFM measurements that could allow direct reconstruction of this energy landscape. Finally, the continuing explosive growth of the computing power available to researchers brings molecular modeling to the forefront of force spectroscopy research. D. Patrick presents an overview of the modeling of force spectroscopy experiments, with an emphasis on analyzing chemical force microscopy measurements.

I hope that this book will convey a sense that as a result of the last decade of force spectroscopy development our knowledge of the behavior of a non-covalent chemical bond under an external load is immeasurably richer; yet, at the same time, that we now understand the limitations and the complications of the technique with more clarity. The naïve optimism of the first years of force spectroscopy has been replaced with more realistic expectations rooted in the deep understanding of the physical processes underlying the measurements.

I would like to thank numerous individuals who helped with various stages of this project. First and foremost, I am indebted to the book's contributors for taking the time to summarize their respective areas of research. This book has been partly inspired by the works presented at the symposium on "Nanoscale Probing of Intermolecular Interactions" at the 2005 ACS National Meeting. V.V. Tsukruk has been an early supporter of the idea of this symposium and I thank him for his help and encouragement. David Packer at Springer has been a great editor and I thank him for his patience and for his helping hand. Finally, I thank my wife and my two daughters for their support and patience.

Aleksandr Noy
Lawrence Livermore National Laboratory
Livermore, CA
June 28, 2007

Contents

Contributors

Kirstine L. Anderson, Astbury Centre for Structural Molecular Biology, University of Leeds, Garstang Building, University of Leeds, Leeds, LS2 9JT, United Kingdom, bmbkla@leeds.ac.uk

Paul D. Ashby, Molecular Foundry, Materials Sciences Division, Lawrence Berkeley National Laboratory, Mail Stop 67R2206, 1 Cyclotron Road, Berkeley, California 94720, USA, PDAshby@lbl.gov

Craig D. Blanchette, Chemistry, Materials and Life Sciences Directorate, Lawrence Livermore National Laboratory, L-454, 7000 East Ave, Livermore, CA 94550, USA, blanchette2@llnl.gov

David J. Brockwell, Astbury Centre for Structural Molecular Biology, University of Leeds, Garstang Building, University of Leeds, Leeds, LS2 9JT, United Kingdom, d.j.brockwell@leeds.ac.uk

Richard Conroy, National Institute of Neurological Disorders and Stroke (NINDS), National Institutes of Health LFMI-10 Center Drive Bldg. 10, Rm. 3D17, MSC 1065, Bethesda, MD 20892, USA, conroyri@mail.nih.gov

Raymond W. Friddle, Chemistry, Materials and Life Sciences Directorate, Lawrence Livermore National Laboratory, L-234, 7000 East Ave, Livermore, CA 94550, USA, friddle1@llnl.gov

Deborah E. Leckband, Department of Chemical and Biomolecular Engineering, University of Illinois at Urbana-Champaign, 127 Roger Adams Lab MC-712, Box C-3 600 S. Mathews Ave. Urbana, IL 61801, USA, leckband@uiuc.edu

Charles M. Lieber, Department of Chemistry and Chemical Biology, Harvard University, 12 Oxford Street, Cambridge, MA 02138, USA, cml@cmliris.harvard.edu

Albert Loui, Chemistry, Materials and Life Sciences Directorate, Lawrence Livermore National Laboratory, L-231, 7000 East Ave, Livermore, CA 94550, USA, loui2@llnl.gov

Aleksandr Noy, Chemistry, Materials and Life Sciences Directorate, Lawrence Livermore National Laboratory, L-234, 7000 East Ave, Livermore, CA 94550, USA, noy1@llnl.gov

David L. Patrick, Advanced Materials Science & Engineering. Center, Department of Chemistry, Western Washington University, 516 High St., Bellingham, WA 98225, USA, patrick@chem.wwu.edu

Sheena E. Radford, Astbury Centre for Structural Molecular Biology, University of Leeds, Garstang Building, University of Leeds, Leeds, LS2 9JT, United Kingdom, s.e.radford@leeds.ac.uk

Timothy V. Ratto, Chemistry, Materials and Life Sciences Directorate, Lawrence Livermore National Laboratory, L-231, 7000 East Ave, Livermore, CA 94550, USA, ratto7@llnl.gov

D. Alastair Smith, Chief Executive, Avacta Group plc, York Biocentre, Innovation Way, York Science Park, Heslington, York YO10 5NY, United Kingdom, phydams@ds.leeds.ac.uk

Todd A. Sulchek, Chemistry, Materials and Life Sciences Directorate, Lawrence Livermore National Laboratory, L-231, 7000 East Ave, Livermore, CA 94550, USA, todds@llnl.gov

Dmitry V. Vezenov, Department of Chemistry, Lehigh University, 6 E. Packer Ave., Bethlehem, PA 18015, USA, dvezenov@lehigh.edu

Phil M. Williams, Laboratory of Biophysics and Surface Analysis, School of Pharmacy, University of Nottingham, University Park, Nottingham NG7 2RD, United Kingdom, phil.williams@nottingham.ac.uk

Surface Force Apparatus Measurements of Molecular Forces in Biological Adhesion

Deborah Leckband

Introduction

Adhesion is essential in biology. Intercellular interactions maintain the structural hierarchy of all multicellular organisms across all anatomical length scales. Cells transduce mechanical signals and respond by regulating adhesion, motility, and differentiation. Other adhesive interactions are central to immunity. Pathogenic microorganisms use adhesive interactions with cells in the first steps in infection. Determining the molecular mechanisms underlying these processes is central to understanding the fundamental basis of related diseases and to developing strategies to treating or preventing disease.

Force probe techniques are ideal tools for investigating the mechanisms that control the biological adhesion and the strength of biomolecular linkages. Although the adhesion strength is the central parameter defining function in many biological interactions, adhesion measurements alone are insufficient to determine the relationships between molecular architectures and their mechanical function. In particular, both the range and magnitude of forces are often functionally relevant, particularly in complex environments such as the cell surface. Many adhesion proteins as well as other glycoproteins on cell surfaces are large, and extend several tens of nanometers from the cell membrane. The range of protein interactions, facilitated by these large structures, is often thought to be critical to their function.

In addition, while single-molecule measurements reveal a wealth of information regarding individual molecular linkages, the collective behavior of tens to thousands of proteins or macromolecules typically determines cell-cell or cell matrix adhesive strengths. It is therefore important not only to explore the mechanical strengths of single molecular bonds, but also to determine how populations of bonds govern biological interactions.

The surface force apparatus SFA is uniquely suited to investigations of molecular adhesion because it quantifies both the molecular forces between surfaces and the distances over which the forces act. It also quantifies the adhesion between extended surfaces such as membranes, which are arguably more relevant to cell interactions. The SFA is distinct among force measurement techniques due to its ability to determine absolute intersurface separations to within $0.1\,nm$, while quantifying weak, noncovalent interactions with bond energies on the order of the thermal energy k_BT ($2.48\,kJ/mole$ or $0.59\,kcal/mole$ at room temperature).

This chapter describes the surface force apparatus technique and several of its capabilities. I also describe key examples of how data obtained from SFA measurements generated unique information regarding the relationships between the structures and mechanical functions of proteins that mediate cell adhesion in immunity and the nervous system. These examples focus on the relationships between measured intersurface, i.e., intermembrane potentials and both the protein architectures and the composition of their binding interfaces. We thus investigated the molecular mechanisms of binding and adhesive failure, as well as the regulation of these interactions by post-translational modifications. These examples highlight the importance of both the magnitudes and the ranges of the molecular interactions in bioadhesion. The precise force and distance sensitivity achieved with the SFA are central to these studies.

Surface Force Apparatus

The surface force apparatus (SFA) quantifies the magnitudes and distance dependences of the forces between two macroscopic, curved surfaces as a function of their separation distance. This differs from many force probe techniques that quantify bond strengths or force-extension profiles of macromolecules, because the molecular forces between the interacting surfaces (and the molecules on these surfaces) are measured over absolute separation distances of up to several hundred nanometers with an accuracy of ±0.1 nm.

Absolute Separation Distances are Determined Within ±0.1 nm by Interferometry

In SFA measurements, the absolute separation between two interacting surfaces is determined directly by interferometry.[1] In the instrument, the samples are supported on the surfaces of cleaved, atomically flat ~0.3–0.5 µm thick mica sheets that are fixed to the surfaces of two crossed hemicylindrical, macroscopic silica lenses (Figure 1). The radii of the polished hemicylinder lenses are 1–2 cm.[2, 3] The back surfaces of the mica adjacent to the silica lenses are coated with reflective 50 nm silver films, so that the region between the silver mirrors forms the resonant cavity of a Fabry-Perot interferometer. Light transmitted by the interferometer consists of a series of interference fringes of equal chromatic order (FECO) (Figure 1B), whose wavelengths are determined by the thicknesses and refractive indices of the various films between the two reflective silver surfaces.[1, 4, 5] With a white light source, the transmitted wavelengths change continuously with changes in the surface separation D (central layer thickness).[1, 4] The fringes shift to longer wavelengths with increasing separation, and to shorter wavelengths with decreasing separation.

From these measured wavelength shifts, the distances D are determined within ±0.1 nm. The interferometer has a high coefficient of finesse, owing to the high reflectivity (0.98) of the silver mirrors.[4] This ensures that the interference fringes are narrow and sharp, which is essential for the high precision wavelength measurements needed for subnanometer distance resolution. The standard deviation in wavelength measurements is typically ±0.02 nm. This resolution in $\Delta\lambda$, plus the direct dependence of the transmitted wavelengths on D, enables determinations of the absolute surface separation D within ±0.1 nm.[1]

In force measurements between supported bilayers or protein monolayers interacting across an aqueous gap, the interferometer contains five layers (Figure 2): two mica sheets, two bilayer/protein films (samples), and the aqueous gap between them. These have respective thicknesses Y, Z, and D, and the corresponding refractive indices of the materials are μ_1, μ_2, and μ_3. From the calculated reflection and transmission coefficients of the interferometer, it was shown that for a five-layer interferometer (see Figure 2),[1] the wavelength shifts $\Delta\lambda$ are related to changes in the separation ΔD by

$$\Delta D = \frac{n\Delta\lambda}{2\mu_1} - 2Z \tag{1}$$

for odd order fringes (n odd) and

$$\Delta D\mu_3^2 = \frac{n\Delta\lambda}{2\mu_1} - 2Z\mu_2^2 \text{ for even order fringes (n even).} \tag{2}$$

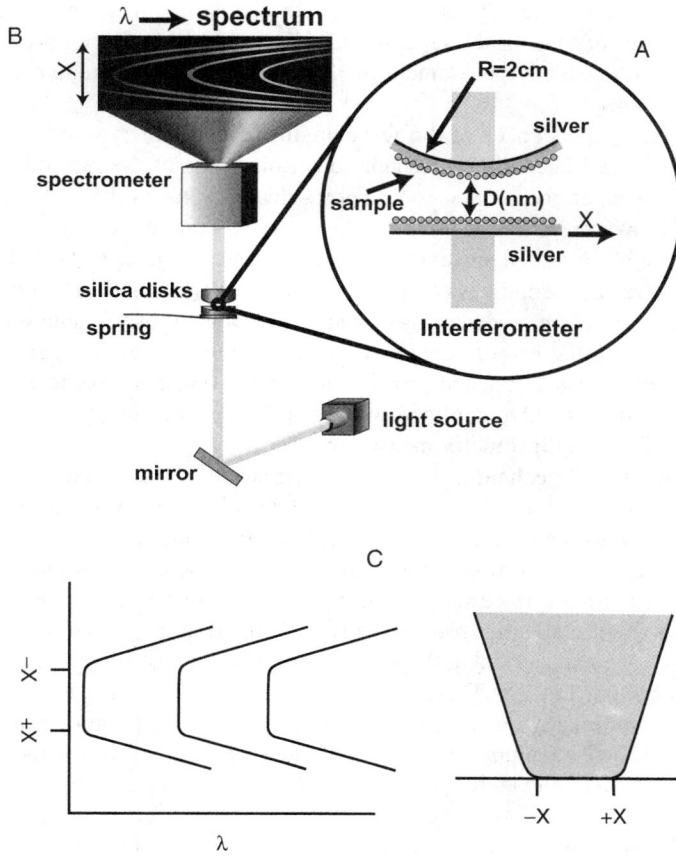

Figure 1. The Surface Force Apparatus. (A) Samples in the SFA are supported on two hemicylindrical lenses oriented at right angles to each other. The equivalent geometry is a sphere interacting with a flat plate. (B) The samples with the reflecting silver mirrors form the resonant cavity of a Fabry-Perot interferometer. White light passed through the samples generates a series of interference fringes. The curvature of the fringes corresponds to the curvature of the contact region between the samples. (C) Example of the distortion in the interference fringes resulting from surface deformation (flattening). The substrate deformation (right) is reflected in the shapes of the fringes (left). The distance from −X to +X is the diameter of the contact area (2x), and is measured directly from the interference fringes. (*See Color Plates*)

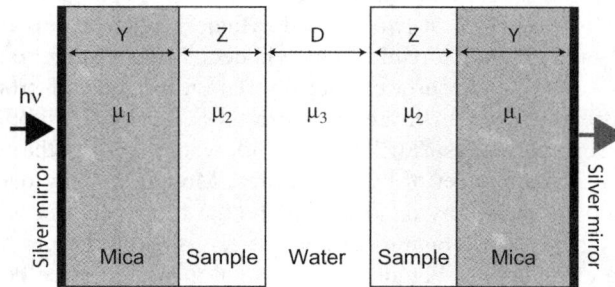

Figure 2. Schematic of the thin films contained in the resonant cavity of the SFA interferometer. In measurements between protein layers, the five films would be mica, samples on the mica surface, and water. The thicknesses of the films are, respectively, Y, Z, and D. The corresponding refractive indices are μ_1, μ_2, and μ_3.

One thus determines ΔD directly from the corresponding, measured values of Δλ. This direct determination of the surface separation differs from other force probe measurements, which only record the relative movement of the cantilever or bead rather than the absolute probe-sample distance.[6]

The different dependences of the wavelengths of odd and even order fringes on both ΔD and the refractive indices further enable determinations of the refractive indices of the samples on the mica surfaces.[1] Equation 1 shows that changes in the odd order fringes only depend on the known refractive index of the mica μ_1. In contrast, changes in the even order fringes depend on all three refractive indices: those of mica, proteins and/or bilayers, and water (Equation 2). The refractive indices of the mica and water are known. The refractive index of the sample μ_2 can therefore be determined from the wavelength shifts measured on even and odd order fringes, and the use of equations 1 and 2.[1, 7] Finally, by comparing the measured value μ_2 to that of a densely packed protein monolayer, one can also determine the macro-molecular surface density. One can therefore use this approach to determine the number of proteins/cm^2, similar to ellipsometry measurements.

The interferometric technique also images the shape of the contact region between the surfaces (Figures 1B and 1C).[8] The SFA substrates (silica lenses) are curved such that the inter-surface distance increases with the radial distance from the center of the contact (Figure 1A).[1, 7] The transmitted wavelength correspondingly increases with increasing distance from the center, and hence with increasing surface separation. Projecting the interference patterns onto an imaging spectrometer (Figure 1B) thus reveals the spatial variation in intersurface distances associated with the local geometry, so that the shape of the interference fringes reflects the shape of the contact area (Figure 1B).[3, 8, 9] There are at least two important consequences of this imaging capability. First, one directly quantifies the local curvature at the point of contact between the surfaces. The radius of curvature scales the magnitude of the intersurface force.[2, 10] Second, it unambiguously establishes that changes in the cantilever position are due to a decrease in D or to deformations of soft materials between the surfaces (Figure 1C).[6] There are several examples of the uses of the imaging interferometer of the SFA in studies of biological materials.[2, 6, 11–14]

SFA Measures the Integrated Force or the Energy Per Area Between Curved Surfaces

The SFA quantifies the total force between two crossed, macroscopic hemicylinders. This geometry effectively integrates the force law between two equivalent flat surfaces to yield the interaction energy per unit area. In other words, the SFA technique quantifies the intersurface potential—or the distance dependence of the interaction energy per area between the surfaces.[3, 6, 15] To appreciate this, consider the differential area dS at a distance Z between the tip of the sphere and the surface (Figure 3). A sphere interacting with a flat plate is geometrically equivalent to two crossed cylinders, if the radius of the sphere $R = (R_1 R_2)^{1/2}$, where R_1 and R_2 are the radii of the cylinders.[15] The geometric average radius of the silica lenses is ~1 cm, so the surface is locally flat on the scale of proteins (or molecular forces). The differential force per area between these locally flat patches on the sphere and flat plate at a separation Z is $dF(Z)/dS = f_m(Z)\rho$, where $f_m(Z)$ is the force per molecular bond at Z and ρ is the number of bonds per area. Moving radially out from the center, the surface separation increases by δ, so the differential force per area will be $df_D(Z + \delta)/dS = f_m(Z + \delta)\rho = f_{flat}(Z + \delta)$ (Figure 3A), where f_{flat} is the force per area between two locally flat surface elements (cf. Figure 3B). The total measured force between the sphere and the opposing plate is the sum of all molecular interactions between the surfaces, or $F_c^T(Z) \approx \sum_{\delta=0}^{\delta=\infty} f_{flat}(Z+\delta)$. To integrate the expression on the right requires a change of variables to account for the geometry,[10, 15] and this introduces a factor of $2\pi R$, where R is the radius

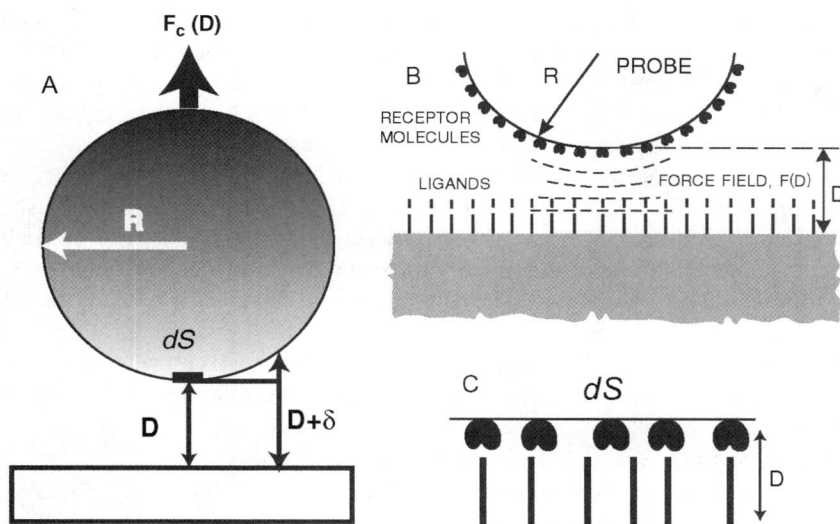

Figure 3. Illustration of the Derjaguin Approximation and the interactions between molecules on opposing macroscopic curved surfaces. (A) Illustration of the integration of the molecular forces between the surfaces to yield the net force between the sphere and flat surface $F_c(D)$. R is the radius of curvature of the sphere, dS is the differential surface element, and D is the minimum distance between the sphere and the plate. (B) Illustration of discrete molecular interactions between a sphere and a flat plate. The distances between the molecules vary as one moves radially from the center of contact. (C) On a molecular length scale, the differential surface elements dS in A and B are locally flat.

of the sphere.[15] Integration gives $F_c^T(D) = 2\pi R \int_{Z=D}^{Z=\infty} f_{flat}(Z)\,dZ = 2\pi R E_A(D)$ where $F_c^T(D)$ is the total force between the sphere and the plate and E_A is the energy per area between two equivalent flat plates.[10, 15] This relationship shows that the force between a sphere and a plate is directly proportional to the energy per area between equivalent flat plates. This is the well-known Derjaguin approximation, which applies when R>>D.[10, 15] Importantly, this shows that the curvature affects only the magnitude of the force through the $2\pi R$ prefactor: $F_c(D) = 2\pi R E_A(D)$. The curvature does not distort the potential—that is, there is no dependence of the shape of the force vs distance profile on R. Between two crossed cylinders, as in SFA measurements, R is the geometric average radius of the hemicylinders, $R = (R_1 R_2)^{1/2} \sim 1$ cm.[10, 15] The range of measured forces D is less than $0.2\,\mu$m. Importantly, the Derjaguin approximation has been tested and validated in over 60 years of research.

It is counterintuitive that, in the SFA measurements, the curved geometry of the disks does not distort the normalized force-distance profiles between the surfaces. For example, protein monolayers are immobilized on the surfaces of opposing hemicylinders. Therefore, at different distances between the tip of the equivalent sphere and plate, the proteins are at different relative distances (Figure 3B). Yet, if the proteins form a distinct bond at a membrane separation D, we measure a single adhesive minimum at D, rather than a smear of attractive interactions or a broad range of attraction. This counterintuitive result is due to the fact that the curved surfaces are locally flat on the protein length scale (Figure 3C). One can easily show, using the chord theorem, that moving out radially from the center line by 2 microns causes a relative shift of only 0.12 nm in the positions of the proteins (or δ in Figure 3A). This locally flat geometry thus enables the experimental integration of the force law. As a result, the maximum attractive force (minimum in the curve) corresponds to the maximum gradient in the intersurface potential, which occurs at a defined distance. The relationship between the normalized force F_c/R and the energy per area between planar surfaces E_A applies generally, even when the molecules interact through complicated force laws with multiple repulsive maxima and attractive minima.

Exact calculations using numerical surface element integration (SEI) showed that the Derjaguin approximation applies even for complicated oscillatory force functions. Sivasankar et al.[16] thus calculated the normalized force F_c/R between two curved surfaces coated with molecules that interact through an oscillatory force law. The calculated net force between the curved surfaces as a function of the distance $F_c^T(D)$ agreed quantitatively with the calculated energy between two equivalent plates, when scaled by $2\pi R$. Sivasankar et al.[16] also showed that the probe curvature only distorts the normalized force curves when $R < 1\,\mu m$. Since $R \sim 1\,cm$ in a typical SFA experiment, the radius does not affect the shape of the potential.

Force Measurement Sensitivity in SFA Measurements

In an SFA experiment, the normalized force sensitivity is $\Delta F/R$ is $\pm 0.1\,mN/m$ or $0.1\,mJ/m^2$, as determined from the deflection of a sensitive leaf spring that supports the lower disk. The absolute force sensitivity ΔF of the leaf springs used in SFA measurements is $\sim 1\,nN$. This is lower than the sensitivity of $\sim 10\,pN$ of cantilevers used in AFM measurements for single-molecule measurements. However, in measurements of the forces between *surfaces*, the measurement sensitivity depends on the normalized force $\Delta F/R$.[6] Therefore, if we compare the sensitivity of two approaches in which the probe radii are $1\,cm$ and $10\,nm$, then the force sensitivity of the method using the smaller probe would have to be $\dfrac{R_1}{R_2} = \dfrac{10^{-2}\,m}{10^{-8}\,m} = 10^6$ greater than that using the larger probe to measure the same force. In order to detect the same force between the curved disks in the SFA with an AFM probe, for example, the force resolution needs to be $\Delta F = 10^{-9}\,N \times 10^{-6} = 10^{-15}\,N$. This is lower than the value of $\Delta F \sim \pm 10\,pN$ achieved with an AFM. Because of the measurement sensitivity of the normalized force and the absolute distance resolution, the SFA is currently the most sensitive technique for quantifying normalized force profiles between surfaces over large distances.

Adhesion Energies

In investigations of adhesion, a key parameter is the work required to separate two surfaces, or the adhesion energy per area.[10, 15, 17] This differs from the forced rupture of a single molecular bond, which is generally a nonequilibrium measurement.[18–24] In measurements between surfaces, the work of adhesion reflects the collective contribution of multiple bonds in parallel. The surface detachment may also occur under a range of pulling conditions from near equilibrium to far-from-equilibrium.

There are important similarities and differences between single bond rupture and adhesion between surfaces. In both cases, the application of an external force lowers the activation energy for unbinding and accelerates the rate of debonding.[19, 22, 25] In the case of single bonds, the force to rupture the bond in 1ms, for example, depends on the activation energy for unbinding and on the pulling rate.[19] In single-molecule studies, the bonds rarely reform, so the system is typically far from equilibrium. Between surfaces, however, the bound and unbound states are in dynamic equilibrium.[25, 26] The surviving bonds hold the surfaces close together, so that some broken bonds can re-form on the time frame of the measurement. Thus the work to separate two surfaces is not simply the product of the number of bonds and a critical rupture force. In addition, adhesion scales with either the kinetic or the thermodynamc properties of the intersurface bonds.[25] Whether the adhesion scales with the activation energy for unbinding or with the Gibbs free energy of the bonds depends on how fast the surfaces are pulled relative to the intrinsic lifetime of the unstressed linkages.[25] Under very slow, near-equilibrium loading conditions, the adhesion energy scales with the Gibbs free energy. Far from equilibrium, the adhesion energy scales with the activation energy for unbinding, as in single bond rupture measurements.[25] While single bond rupture is typically a nonequilibrium process, measurements between surfaces can access both near-equilibrium and far-from-equilibrium loading regimes.

With the SFA, the force measurements are conducted over several minutes, and the system is close to equilibrium during the separation process. The normalized pull-off force between the two surfaces F_{po}/R is directly proportional to the adhesion energy per area E_A between equivalent flat plates. However, the exact relationship depends on whether the surfaces deform when in contact,[15] In many SFA measurements, the mica sheets and epoxy used to fix them to the silica deform and flatten under the influence of the intersurface forces or the external load (cf. Figure 1C). If the surfaces deform, then the Johnson-Kendall-Roberts theory for the adhesion between deformable surfaces relates the pull-off force to the adhesion energy per area by $E_A = 2F_{po}/3\pi R$.[17] However, within the limit of small deformations, the adhesion energy is better described by the Derjaguin-Müller-Toporov theory, in which $E_A = F_{po}/2\pi R$[15].

The pull-off force gives the average adhesion energy per area E_A, but the magnitude also depends on the density of adhesion proteins on the surfaces. The adhesion energies are therefore normalized by the protein or ligand surface density Γ. The adhesion is normalized by the more dilute molecule. The average adhesion energy per bond E_b is then estimated by normalizing the adhesion energy by the protein coverage Γ. Taking into account the Boltzmann distribution between bound and free states, the estimated bond energy is

$$E_b = \frac{E_A\left(1 + \exp\left(-\dfrac{E_b}{k_B T}\right)\right)}{\Gamma}$$. This is an estimate of the average bond energy over a large bond

energy over a large population of molecules (~250,000). Additionally, on fluid membrane surfaces, some proteins may convect to the perimeter of the contact region, possibly increasing the protein density at the edge. Furthermore, any inactive protein in the population will lower the specific activity (adhesion/moles protein). For these reasons, the estimated bond energies are lower bounds. Nevertheless, bond energies estimated from SFA measurements have typically been within a factor of 2 of the Gibbs free energies determined from solution binding measurements.

This comparison with equilibrium bond energies raises an important issue concerning whether the system is truly at equilibrium during the pull-off process, and whether the Gibbs free energy or the activation energy for unbinding scales the adhesion. The most common way to determine whether the system is at equilibrium during separation is to measure the loading-rate dependence of the adhesion. If the system were not at equilibrium, then the measured adhesion would vary with the unloading rate. The scaling of the adhesion with thermodynamic or kinetic parameters would also vary with the pulling rate. To identify the rate-independent, i.e., equilibrium regime, measurements are conducted at decreasing pulling speeds, until the measured adhesion is rate-independent. The critical pulling rate v_c is potentially a more rigorous parameter defining crossover between equilibrium and non-equilibrium pulling[25]. This is defined by the distance between the ground state and the transition state (or the bond length L_b) divided by the intrinsic lifetime of the unstressed bond: $v_c = \dfrac{L_b}{\tau_0}$. In other words, if the time taken to pull the bond to the transition state is much less than the intrinsic lifetime, then the system will be thermally equilibrated during the pulling process. However, because relaxation times and bond lengths are not always known, the empirical approach is more practical.

Sample Requirements and Strategies

Because the SFA measurements reflect the integrated force over large areas ($\sim300\,\mu m^2$), it is essential that the molecules on the surface be uniformly distributed and oriented. Achieving this can be challenging, but one of the more straightforward approaches is to use supported lipid bilayers and engineered epitope tags on the proteins (Figure 4). Several commercial lipid analogs are available with reactive headgroups that form strong physical or covalent bonds with different amino acids. For example, maleimide-functionalized lipid headgroups covalently bind cysteines,[27] whereas nitrilo-tri-acetic acid (NTA) headgroups chelate histidine tags.[28] If these protein tags are at unique sites, then the proteins will bind at these unique tags and orient on the membrane. Histidine-tagged fragments of protein A also form uniformly oriented docking sites for proteins engineered with the Fc domains of immunoglobulin G (IgG).[29] Engineered glycosyl-phospho-inositol (GPI) anchors were used to orient proteins on lipid membranes.[30] Oriented biotinylated protein monolayers also self-assemble on lipid-supported streptavidin monolayers.[31] Fluorescence and AFM imaging can verify the protein distributions on these supported films. The orientations of some histidine-tagged proteins on NTA-lipid membranes were also confirmed by neutron and X-ray reflectivity.[32, 33]

The planar bilayers are supported on the mica sheets, which are supported on the silica lenses. The samples, mounted in the SFA chamber, are bathed in solution at all times.[34] The bilayers are prepared by the Langmuir-Blodgett (LB) deposition of two successive lipid monolayers onto the freshly cleaved mica sheets (Figure 4).[12–14, 16, 27, 28, 31, 35–38] The use of LB films ensures good control of the lipid packing density and composition over large ($1\,cm^2$) sample areas. Asymmetric bilayers in which the first layer is gel phase DPPE or DPPC (Figure 4) give the highest lipid mobility and mobile fraction (>0.9) of lipids in the outer leaflet. Fluorescence recovery after photobleaching measurements showed that the diffusion coefficients of lipids in monolayers supported on gel phase lipids is within a factor of 3–4 of the values measured on spread lipid bilayers on glass (Leckband, unpublished observations).

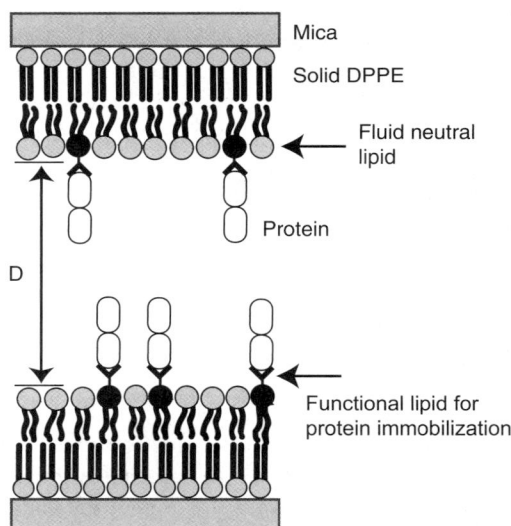

Figure 4. Supported lipid bilayers for immobilizing oriented protein monolayers. Asymmetric bilayers, formed by Langmuir-Blodgett deposition, consist of gel (solid) phase DPPE (or DPPC) and an outer leaflet containing a mixture of lipids with modified headgroups and neutral lipids. The neutral lipids may be in the gel phase or the fluid phase, in order to yield immobile or mobile lipids, respectively. The proteins attach to the functionalized lipids in the outer monolayer. SFA measurements quantify the force between opposing membranes as a function of the distance D between the bilayer surfaces.

To control the protein mobility and protein density on the membrane, the reactive lipid analogs can be mixed with neutral lipids that are either in the fluid or gel phase. The protein anchors are mobile in the former case, but not in the latter. As long as the two lipid components are miscible, the lipid analogs, and hence their attached proteins, are uniformly distributed over the bilayer surface. AFM and fluorescence imaging verified this. The mobility of the attached proteins is particularly important. In SFA measurements, the proteins on opposing membranes are not initially in register, and may need to diffuse laterally to bind the ligands on the opposite membrane. Without lateral mobility, the measured adhesion energy can drop five-fold relative to adhesion between fluid membranes at the same protein densities.[31]

Force Measurements of Protein-Mediated Binding: Structure Function Relationships and the Molecular Basis of Adhesion

The SFA has been used to investigate the interactions of several proteins including streptavidin,[31, 35] cytochrome c, cytochrome b5,[27, 37] antibodies,[36] cadherins,[16, 28, 39–41] the neural cell adhesion molecule,[42, 43] and the immune proteins CD2, CD58, and CD48.[44, 45] Each of these studies exploited the distance resolution of the SFA, in order to quantify the interaction forces, adhesion energies, and the dimensions of protein complexes. The following examples illustrate the unique molecular insights into structure-function relationships that can be obtained from measurements of the distance-dependent forces between membranes decorated with immobilized, oriented proteins.

CD2 Family of Cell Adhesion Proteins in Immunity

Cellular immunity results from interactions between thymus cells (T-cells) and antigen-presenting cells (APCs). Together, several proteins facilitate this important intercellular junction. In particular, binding between the Major Histocompatibility Complex on the APCs and the T-Cell Receptor (TCR) on T-cells triggers an immune reaction. The formation of these intercellular junctions and the association of the MHC and TCR are facilitated by auxiliary proteins such as CD2 and CD58.[46, 47]

CD2 is a member of the immunoglobulin (Ig) superfamily, and is expressed on T-cells. Its ligand CD58 is expressed on APCs. CD2 and its ligands are members of the "CD2 family" of proteins.[47] The structure of CD2 was determined by both X-ray crystallography and by NMR,[46, 48, 49] and the structures of the extracellular domains of the CD2 ligands CD58[50] and rat CD48[51] were also determined. Members of the CD2 protein family have a similar overall architecture, although their chemical compositions differ.[47] The extracellular regions of proteins in this family consist of two tandemly-linked Ig-type domains (D1 and D2) that are bound to cell membranes by hydrophobic anchors (Figure 5A). The linear dimensions of the extracellular regions are ~7.5 nm. Based on the structure of the heterologous complex between the outer D1 domains of CD2 and human CD58,[52] and on biochemical data, the proteins are postulated to bridge cell membranes in a head-to-head configuration as shown in Figure 5B.[46, 50–53]

SFA measurements of the interactions between proteins in the CD2 family determined the dimensions of the heterophilic adhesive complexes, and also quantified the contribution to adhesion from distinct salt-bridges in the CD2-CD58 binding interface.[44, 45]

Dimensions of the CD2-CD58 Complex from Force-Distance Profiles

Site-directed mutagenesis and solution binding studies support a model in which CD2 binds its ligands in a head-to-head orientation (Figure 5B). In this configuration, the CD2-ligand or CD2-CD58 complex is predicted to span a membrane gap distance of ~13.5 nm[46].

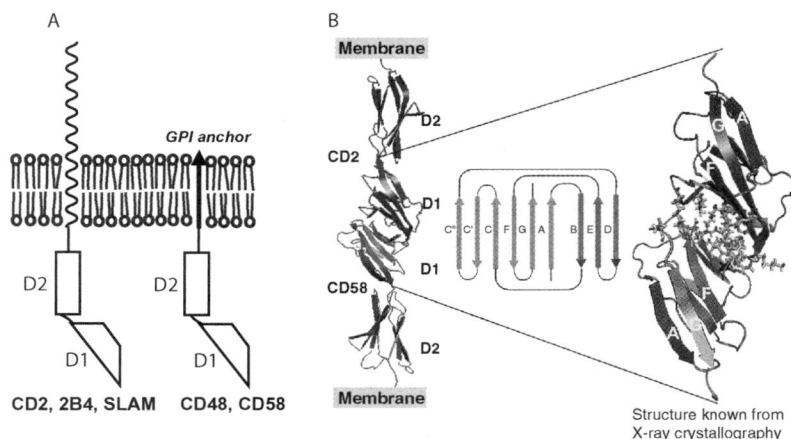

Figure 5. (A) General structure of the CD2 protein family. The proteins are anchored to the membrane via hydrophobic tails. The extracellular region consists of two domains D1 and D2. This figure shows two different hydrophobic anchors observed in this family. (B) Proposed head-to-head binding alignment between CD2 and its ligands. The structure of the complex between the outer D1 domains is shown on the right side. (*See Color Plates*)

It has also been postulated that this distance is functionally important because it closely matches that of the MHC-TCR complex.[46, 54] CD2 may therefore function as both an adhesive protein and as a scaffold to control the intermembrane spacing, and thereby facilitate binding between the MHC and TCR on adjacent cells. Indeed, CD2 molecules engineered with an additional Ig domain spacer impede T-cell activation.[55]

Force measurements between monolayers of oriented CD2 and CD58 extracellular fragments tested this structural model for binding and quantified the adhesion energy.[44, 45] Soluble extracellular domains of both proteins were engineered with hexahistidine tails at the C-terminus of the second Ig domain (D2). This ensured their selective immobilization and proper orientation on NTA-DLGE lipid layers (cf. Figure 4). In the SFA measurements, the distance D is the distance between the surfaces of the supporting membranes (Figure 6, lower panel).

Figure 6 shows the normalized force-distance profiles between the CD2 and CD58 monolayers. Force curves were measured during approach (decreasing D) and separation (increasing D). During approach, the onset of repulsion (F/R > 0) at D<16 nm agrees with the expected range of steric repulsion between end-on oriented proteins. The hysteresis in the force curve upon separation is due to protein-protein adhesion. At the minimum in the curve at 15.3±0.5 nm—that is, the maximum attractive force—the CD2-CD58 bonds failed, and the surfaces jumped out of contact (out arrow). The distance of the adhesive minimum includes a ~1.0 nm contribution from the anchoring NTA tethers on each of the opposing membranes. Thus the end-to-end dimension of the complex is $(15.3\pm0.5)-(2\times1.0) = 13.3\pm0.5$ nm, in quantitative agreement with the predicted dimensions of the complex.[54–56] The magnitude of the adhesion energy per area was 0.38 ± 0.1 mJ/m^2. Normalizing this by the protein density as described above, the estimated CD2-CD58 bond energy was $8.0\pm0.5\,k_B T$ (4.7 kcal/mole at room temperature), where k_B is the Boltzmann constant and T is the absolute temperature.

The accuracy of the intermembrane distance measurements thus enabled the experimental validation of the predicted dimensions of the CD2-CD58 complex. These results further confirmed that the CD2-CD58 complex is structurally matched to the size of the TCR-MHC complex, which the intercellular adhesive junctions must accommodate. The SFA measurements also support both the adhesive and the proposed scaffolding roles of these proteins.

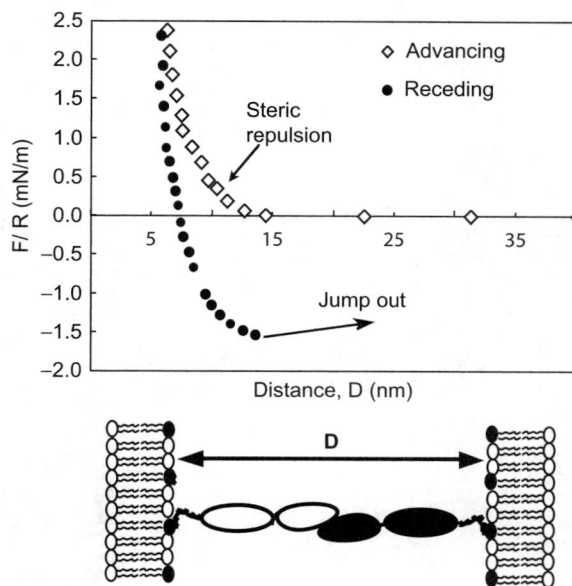

Figure 6. Normalized force versus distance profile between monolayers of human CD2 and CD58. The open triangles show the advancing force profile measured during decreasing distance D, and the filled circles show the receding force profile measured during separation (increasing D). At the minimum in the curve (maximum attractive force), the bonds rupture and the sample surfaces jump out of contact (out arrow). The cartoon in the lower panel shows the sample architecture in the force measurements, and the relative protein alignment at the minimum in the force profile.

Quantifying the Contribution of Salt-Bridges to CD2 Adhesion

The structure of the complex between the outer domains of human CD2 and CD58 also suggests that the adhesive interface is stabilized by several ionic bonds between charged amino acids on the opposing proteins (Figure 5B).[50, 56–58] SFA measurements quantified the contribution of salt bridges at the CD2-CD58 interface to the adhesion energy. Several reports suggest that the complementary alignment of positive and negative charges stabilizes the complex in solution. SFA measurements tested whether the same amino acids form load-bearing contacts that stabilize the bonds under force.

To determine the contributions of these salt bridges to adhesion, a select group of charged amino acids on the CD2 binding face were switched to alanine. This included the four CD2 charge variants: D31A, K41A, K51A, and K91A. SFA measurements then quantified the resulting changes in the adhesion energies between the CD2 mutants (CD2^{mut}) and wild-type CD58.

The amino acids were selected on the basis of steered molecular dynamics (SMD) simulations.[59] In the SMD simulations, the proteins were "pulled" apart by anchoring the C-terminal end of one of the proteins, e.g., CD58, and then imposing a time-varying potential on the C-terminus of the second molecule, e.g., CD2.[18, 60] As the molecules unbind, the sequence in which different contacts between the proteins fail is correlated with the imposed external force on the complex (Figure 7A). Based on these results, several salt-bridges were rank ordered in terms of the magnitude of the external force applied before the bonds broke (Figure 7B). The bond classification is summarized in Table 1. Interestingly, although the CD2 K91 residue did not form a salt bridge in the crystal structure, it did form a transient ionic bond along the unbinding trajectory. The question is whether this residue in fact contributes to the stability of the CD2-CD58 bond.

SFA measurements tested the relative importance of the CD2 side chains of D31, K41, K51, and K91.[45] These proteins were immobilized via engineered C-terminal hexahistidine

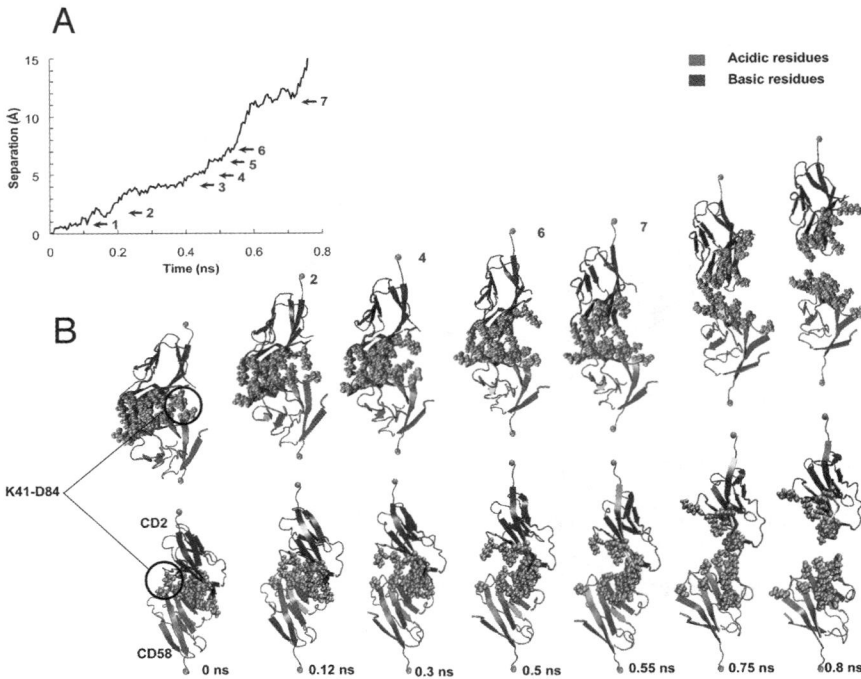

Figure 7. (A) Simulated separation vs. time trajectory as CD2 and CD58 are pulled apart. (B) Snapshots of the complex when discrete salt bridges rupture during protein detachment. The corresponding positions on the separation-time trajectory are indicated in (A). (*See Color Plates*)

Table 1. Comparison of Predicted and Measured Effect of CD2 Charge Mutations.

Protein	Simulations	Measured Average Bond Engery, $K_b t$
Wild Type	—	8.2±0.3
K41A	Not Critical	7.0±0.3
K91A	Not seen in structure	6.5±0.3
D31A	Critical	5.2±0.4
K51A	Critical	3.3±0.2

tails to the NTA-containing membranes. The force-distance profiles were qualitatively the same as those of the wild-type ectodomains, but the magnitudes of the adhesion differed (Table 1). The rank ordering of the salt bridges in terms of their ability to resist force in the simulations also predicted their measured nonadditive contributions to adhesion.

These SFA measurements experimentally confirmed the qualitative trends predicted by steered molecular dynamics simulations (Table 1). They also show that the salt-bridges contribute to the mechanical stability of the CD2-CD58 interface. Interestingly, although K91 only formed a transient salt bridge during the unbinding process, K91A does modestly reduce the adhesion relative to the wild-type CD2-CD58 bond.

There are also species variations between different CD2 ligands. In particular, while the human CD58 can form at least eight salt bridges with human CD2, the rat homolog CD48 forms four fewer salt bridges with rat CD2.[51] SFA measurements also quantified the relative adhesion differences between the human and rat forms. The loss of four salt bridges reduced the adhesion energy eight-fold.[45]

In summary, these SFA studies confirmed that the CD2 proteins bind in a head-to-head orientation that is dimensionally matched to the size of the MHC-TCR complex. They also both confirmed the importance of ion pairs in CD2 adhesion, and showed that their contributions are nonadditive.

The Neural Cell Adhesion Molecule (NCAM)

The neural cell adhesion molecule (NCAM) mediates cell-cell adhesion in the central nervous system by binding to identical NCAM proteins on adjacent cells.[61] This is one of the most abundant adhesion proteins in the brain, and it is linked to long-term memory formation and circadian rhythms.[61] The extracellular region consists of seven tandemly arranged domains, which are structurally similar but differ in their chemical composition (Figure 8).[62] The first five domains are immunoglobulin-type (Ig) domains and are numbered 1–5 (Ig1–5), beginning with the outermost domain. The last two, juxtamembrane, domains are fibronectin type III (FN III) repeats. An additional feature of the extracellular region is a distinct bend in the structure, presumably between the Ig5 domain and the first FN domain, that was identified by electron microscopy.[63, 64] This is also illustrated in Figure 8. The functional significance of the bend has not been determined.

On the basis of structural studies and equilibrium binding measurements, three different models were proposed for the mechanism of homophilic NCAM binding (Figure 9).[65–73] They are based on equilibrium binding studies with NCAM fragments, solution NMR, and X-ray structures of NCAM fragments. In Model 1 (Figure 9A), the NCAM Ig1–5 segments form an overlapping, antiparallel complex. The ability of NCAM fragments to mediate bead aggregation identified Ig3 as the principal adhesive domain, and suggested that Ig3/Ig3 contacts form the strongest NCAM bond.[66] Removing the Ig3 domain eliminates cell adhesion.[71] Despite these findings, isolated Ig3 domains fail to associate in solution.[65]

Model 2 (Figures 9B) is based on surface plasmon resonance binding studies with NCAM fragments[67] and on X-ray and NMR structures of the complex formed between Ig12 fragments.[67, 68, 72] In this case, the Ig12 domains dock in an antiparallel configuration mediated by salt bridges between the two fragments. Cell adhesion tests of the functional relevance of this interaction are nevertheless contradictory.[65]

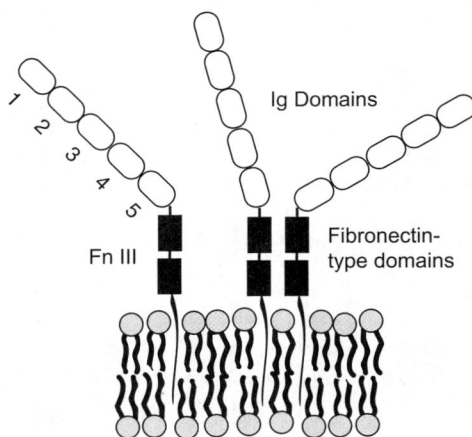

Figure 8. Cartoon of the NCAM structure showing the five immunoglobulin type domains (Ig1–5), the two juxtamembrane fibronectin type III repeats, and the membrane anchor. Three NCAM isoforms found in vivo differ only in the length of the transmembrane anchor.

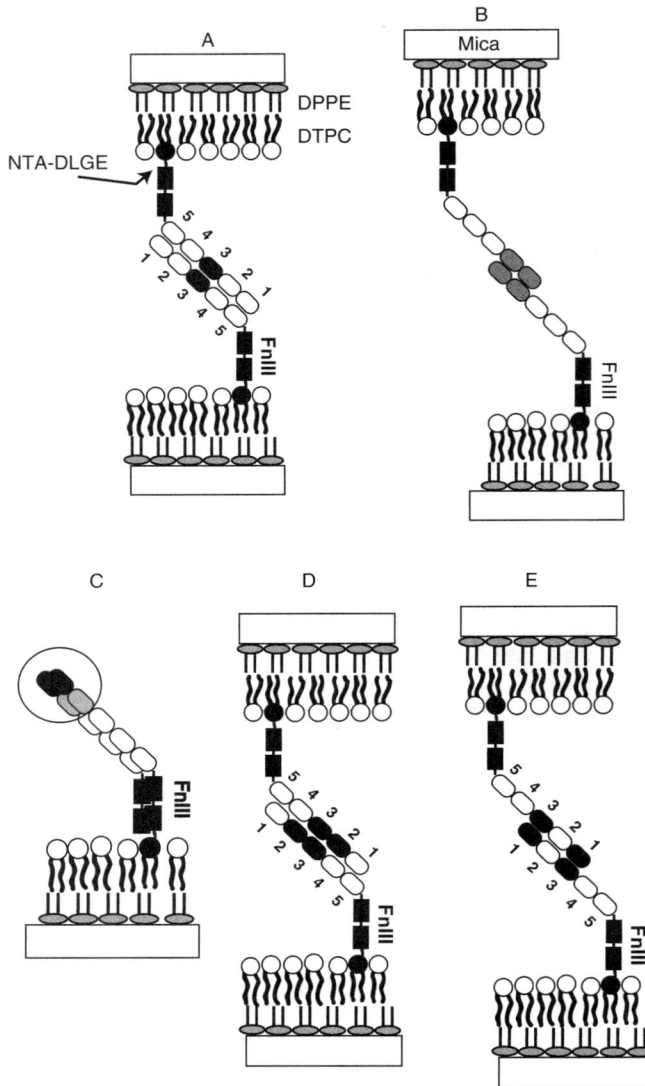

Figure 9. Proposed NCAM binding models based on biochemical and structural data. (A) In Model 1, the proteins bind via interactions between antiparallel Ig3 domains (black). (B) In Model 2, NCAM binds via a double reciprocal bond between Ig12 domains. In Model 3, Ig12 domains form a lateral bond (C), and bridge membranes via interactions between Ig domains 2 and 3 (D) and by a reciprocal Ig1/Ig3 association (E).

Finally, Model 3 (Figures 9C-E) is based on the crystal structure of the Ig1–3 fragment and on the inhibition of NCAM-mediated cell adhesion by short peptides derived from the amino acid sequence at the putative adhesive contacts.[73] In this model, the Ig12 domains form lateral contacts with NCAMs on the same membrane (Figure 9C). The authors also proposed an additional bond between antiparallel Ig1 and Ig3 domains (Figure 9E), as well as an interface between antiparallel Ig2 and Ig3 domains (Figure 9D).

Each model in Figure 9 predicts NCAM binding at a different membrane gap distance. Measurements of the distance-dependence of interactions between NCAM ectodomains could directly test these different models. Because of the bend in the protein, the simplest force measurement to interpret is that between engineered Ig1–5 fragments bound to the membrane via C-terminal hexahistidine tags (Figure 10B). The measured force-distance profiles between these membrane-bound NCAM fragments, and the positions of adhesive minima in particular, enabled us to discriminate between these different binding models and demonstrate the mechanism of NCAM adhesion.

The positions (distances) of the adhesion between the Ig1–5 fragments support both Models 1 and 2, but the data were not consistent with Model 3. Figure 10A shows the normalized force-distance profile between oriented Ig1–5 domains. The length of the Ig1–5 segment is ~19 nm (= 5×3.9 nm), and the protein monolayers interact sterically at distances D<40 nm. In contrast to the CD2 interactions, opposing NCAMs bound at two different membrane distances: namely, 18.0±0.5 nm and 29.0±0.5 nm. This indicates that the proteins form two bonds that span two different membrane gap distances and involve different Ig domains.

These SFA measurements revealed an NCAM-mediated intersurface potential with two attractive minima. In SFA measurements, such multiple minima are detected by controlling the minimum separation D before pulling the surfaces apart. This controls the extent of molecular overlap prior to separation, and allows determination of the relative molecular alignments that generate binding. If the surfaces interact through an oscillatory potential with multiple minima and maxima as in Figure 10A, then the measured force-distance curve will similarly exhibit peaks and valleys corresponding to –dE/dD. At the minima (maximum attractive forces), the bonds rupture, and the surfaces snap out of contact. To identify the different distances at which the proteins bind, we scanned for binding at all separation distances between ~15 nm and 45 nm.

Comparing the adhesion at 18±0.5 nm with the protein dimensions suggests the Ig1–5 domains bind in the overlapping configuration predicted in Model 1 (Figure 10B). Similarly, the second bond at 29.0±0.5 nm agrees with Model 2 (Figure 10B), in which NCAM forms a double-reciprocal bond between Ig1 and Ig2. In contrast to Model 3, the latter finding confirms that the Ig12 domains bind Ig12 domains on an opposed NCAM molecule. The force data are also incompatible with the Ig1-Ig3 interface proposed by Model 3 (Figure 9E). The third adhesive contact postulated in Model 3 (Figure 9C) could not be ruled out from the binding distances alone. However, studies with different NCAM mutants lacking specific Ig domains ruled out the latter possibility.[42] Independent equilibrium binding measurements and single bond rupture measurements with an AFM confirmed the interpretation of the SFA results.[74]

The full-length extracellular domains also bind at the two membrane separation distances 31±0.5 nm and 39±0.5 nm (Figure 10C). These distances correspond to the same Ig1–5 overlap distances shown in Figure 10B, if there is a 138° bend at the hinge between Ig5 and the first fibronectin domain (Figure 10D). This is the maximum bend angle seen in EM data,[64] and the maximum expected extension of the folded proteins under tension.

One might question whether these multiple minima reflect protein interactions, or whether they are an artifact of the measurement technique. It is therefore important to point out that these NCAM results are among several examples of SFA measurements of oscillatory forces—or force profiles exhibiting multiple minima and maxima between surfaces. In other examples, the materials investigated formed ordered layers adjacent to the surfaces, and oscillations in the force-distance curves were due to squeezing out successive molecular layers with decreasing surface separation.[75–85] Although these examples are from materials science and physics, the force profiles nonetheless reflect the structure of materials or fluids between the surfaces and the impact of this architecture on the intersurface potentials. A notable example is the detection of water layering between two mica surfaces.[81, 82] Using the same method described above for detecting multiple, attractive minima between NCAM monolayers, Israelachvili and Pashley[81] identified up to eight water layers in the thin gap between two mica surfaces in water. Other examples include multilayers of cytochrome c and confined polymer melts,[80, 84] in which the periodicity of oscillations corresponded exactly to

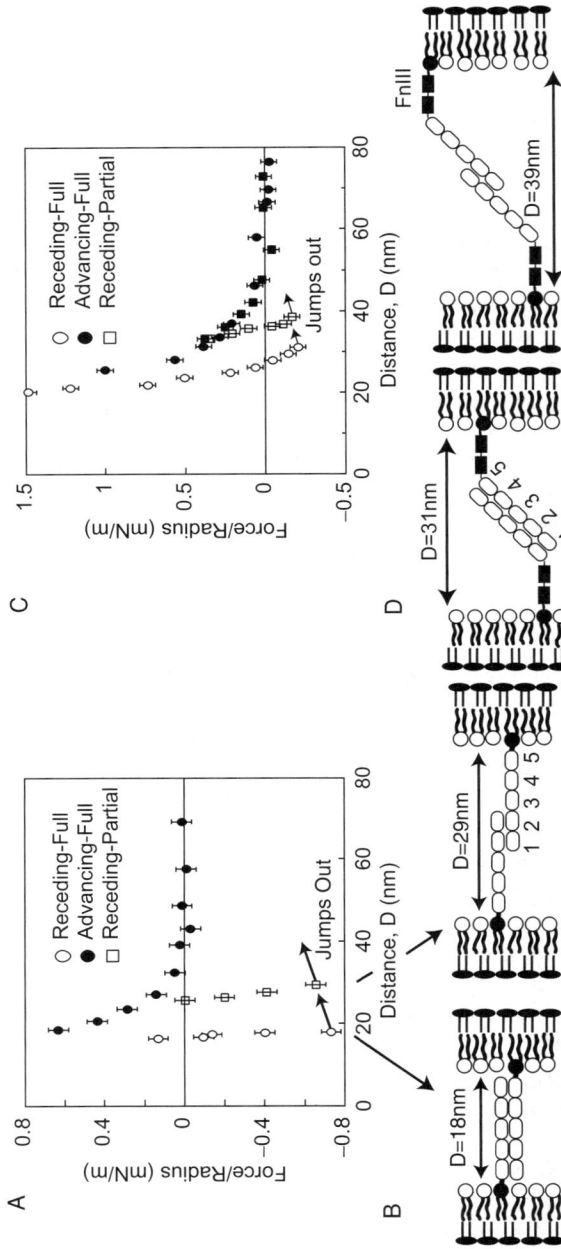

Figure 10. Force measurements between NCAM ectodomains. (A) Force *vs* distance profile between NCAM ectodomains. (B) Proposed Ig1–5 alignments corresponding to the positions of the adhesive minima. (C) Force *vs* distance between identical Ig1–5 fragments. The proteins adhere at 18 nm and 29 nm (out arrows). (C) Force *vs* distance between full NCAM ectodomains. (D) Postulated NCAM alignments responsible for the adhesive minima in (C). The proteins adhere at 31 and 39 nm.

the theoretically predicted dimensions of the successive molecular layers.[86, 87] Lamellar liquid crystals gave similar signatures as successive lamellae were squeezed out with decreasing gap distance.[85] Finally, a different type of adhesion protein, cadherin, also forms multiple bonds[16, 28] that map to different protein domains.[41]

The oscillatory intersurface forces measured with the SFA are due to the distance-dependent attractive and repulsive forces that arise from the molecular architecture of the materials between the surfaces. Importantly, the examples described above differ from NCAM only in the origin of the repulsive and attractive forces. In all cases (multidomain proteins and ordered fluids) the normalized force-distance curves reflect the molecular architecture of the interacting materials and the structure-dependent intersurface potentials.

The NCAM measurements illustrate the importance of both the adhesion and range of interactions when determining the binding mechanisms of these complicated proteins. Traditional biophysical approaches, such as solution binding measurements, cell adhesion measurements, and crystallography, for example, provide important biophysical information regarding molecular associations. However, both the range and magnitude of the interaction forces obtained with the SFA provide unique functional and structural information that would be difficult to deduce from static structures and single parameter measurements such as binding affinities.

Regulating Cell Adhesion by Posttranslational Modification

Another interesting and biologically significant feature of NCAM is that it has both adhesive and antiadhesive forms. The antiadhesive NCAM is modified at the Ig 4 domain with two long, linear polymers of polysialic acid (PSA).[88] The adhesive form is unmodified. The PSA-modification is linked to neural plasticity in the developing brain, to the regulation of circadian rhythms, to tumor progression in some cancers, and to spinal cord regeneration.[89–97]

In vivo, cell surfaces in different tissues or at various stages in development display different amounts and types of carbohydrates. High levels of polysialic acid (PSA) in the brain are associated with the ability to break and form new neural contacts in the early stages of brain development. Electron micrographs also show that PSA increases intercellular spacing.[98]

There are two main hypotheses for how PSA regulates NCAM function, and causes these dramatic physiological changes. One hypothesis is that the grafted PSA polyelectrolytes increase the electrosteric repulsion between adjacent cells.[91, 96, 99] This increased intercellular repulsion would weaken the net intercellular attraction (Figure 11, upper panel). An alternative explanation is that PSA disrupts lateral interactions between proteins on the same cell surface (Figure 11, lower panel), and this would switch off their adhesive activity.[61] The latter mechanism only requires that PSA act laterally, whereas the former requires PSA to extend normal to the membrane over the distance spanned by the NCAM bonds (Figure 11, center panel). In other words, if PSA reduced NCAM adhesion through nonspecific intercellular repulsion, then the steric repulsive force between membranes must be sufficient to overwhelm the NCAM-NCAM bonds at 31 and 39 nm.

Surface force measurements carried out with both the modified and unmodified full NCAM ectodomains directly demonstrated that PSA increases both the range and magnitude of the intermembrane repulsion, and that the range of the repulsion extends slightly beyond the length of the NCAM ectodomains, under physiological conditions (Figure 12).[43] This repulsion abolishes NCAM-mediated intermembrane adhesion at 31 nm and 39 nm. Although the range of the PSA-dependent repulsion was only slightly larger than the steric repulsion between bare NCAM monolayers at physiological salt concentrations, the magnitude of the repulsion was more than eight times greater than the protein-protein attraction. This was more than sufficient to abolish adhesion at both NCAM binding distances (Figure 12). Enzymatically removing PSA by treatment with endoneuraminidase restored NCAM binding.[43]

SFA measurements of the ionic strength dependence of the range and magnitude of the steric repulsion also confirmed that the PSA-dependent abrogation of NCAM adhesion was due to nonspecific steric repulsion. The hydrodynamic radius of the polymer determines the

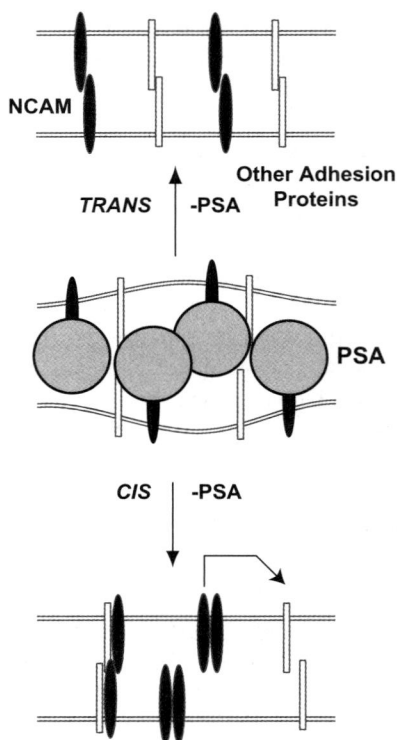

Figure 11. Proposed models for the regulation of intercellular adhesion by PSA. In the steric repulsion model (upper panel), the increased intermembrane repulsion due to PSA overwhelms the attraction between adhesion proteins on opposite membranes, thereby disrupting cell adhesion (center panel). In the second model (lower panel), PSA disrupts the formation of lateral dimers between adhesive proteins on the same cell membrane, thereby inactivating them and disrupting cell adhesion (center panel).

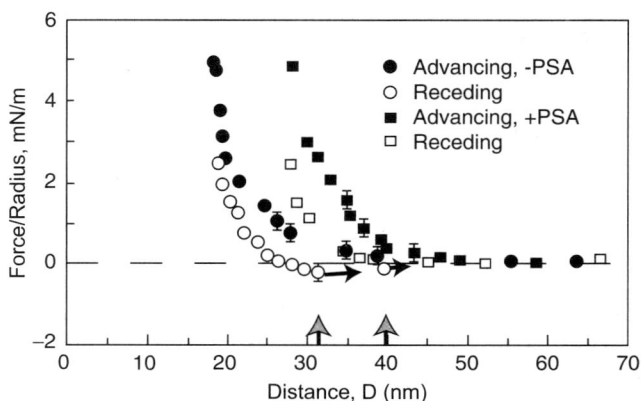

Figure 12. Comparison of force versus distance profiles measured between unmodified (circles) and PSA-modified (squares) NCAM. The presence of PSA substantially increases the magnitude of the repulsion, and there was no detected adhesion between the protein layers. In contrast to the adhesion between bare NCAM (out arrows), there is no adhesion between PSA-NCAM monolayers. The grey arrows indicate the membrane distances at which the bare NCAMs adhere.

range of the repulsion. PSA is a polyelectrolyte, and the hydrodynamic radius of unstructured polyelectrolytes decreases with increasing ionic strength. Consistent with this, in 1M $NaNO_3$, the range and magnitude of the repulsion between PSA-NCAM monolayers decreased substantially.[43] This was attributed to the ionic strength-dependent collapse of the polyelectrolyte chains. The chain collapse was confirmed by X-ray and neutron reflectivity.[32] Concurrent with the reduced steric repulsion, the measured NCAM binding re-emerged at the same distances as between unmodified NCAM ectodomains. This unique ability to quantify the distance dependence of the repulsion and attraction provides direct evidence for the molecular basis of the salt-dependent modulation of adhesion by PSA-expressing cells.[100]

These SFA measurements directly demonstrated, at the molecular level, that PSA regulates NCAM adhesive activity by modulating the range and magnitude of the steric repulsion between cells (cf. Figures 11 and 12). This mechanism contrasts dramatically with the majority of regulatory mechanisms in biology that require specific protein interactions or enzymatic reactions.[101] These force measurements also revealed the molecular origins of a variety of cell behaviors. For example, directly measured variations in both the range and magnitude of the steric repulsion associated with PSA explain the increased intercellular spacing in tissues that express PSA-NCAM.[98] These findings also explain the molecular basis of the ionic strength dependence of adhesion between cells expressing PSA-NCAM.

In summary, this review outlines the important capabilities of the surface force apparatus for investigations of biological adhesion. The forces and distances obtained in these measurements provide unique information regarding the molecular mechanisms and forces governing biological interactions. The specific examples of the CD2 and NCAM proteins described in this chapter clearly demonstrate that the structures of these multi-domain adhesion proteins impact both the magnitude and the range of intermembrane forces. SFA measurements provided unique details of molecular binding mechanisms and their regulation by posttranslational modification, as well as the contribution of specific side chain interactions to adhesion. The SFA is a unique and powerful complement to other biophysical approaches, including single-molecule AFM studies described elsewhere in this text.

References

1. Israelachvili, J., Thin Film Studies Using Multiple-Beam Interferomtry. *J. Coll. Int. Sci.* 1973, 44, 259–272.
2. Israelachvili, J., Adhesion forces between surfaces in liquids and condensable vapours. *Surface Science Reports* 1992, 14, 110–159.
3. Israelachvili, J. N., Adams, G. E., Measurement of Forces between Two Mica Surfaces in Aqueous Electrolyte Solutions in the Range 0–100 nm. *J. Chem. Soc. Faraday Trans. I* 1978, 75, 975–1001.
4. Born, M., Wolf, E., *Principles of Optics*. 6th ed.; Pergamon: Oxford, 1980.
5. Tolansky, S., Applications of multiple-beam interferometry. *Nature* 1951, 167, (4255), 815–6.
6. Leckband, D.; Israelachvili, J., Intermolecular forces in biology. *Q Rev Biophys* 2001, 34, (2), 105–267.
7. Tadmor, R.; Chen, N.; Israelachvili, J. N., Thickness and refractive index measurements using multiple beam interference fringes (FECO). *J Colloid Interface Sci* 2003, 264, (2), 548–53.
8. Tolansky, S.; Omar, M., Evaluation of small radii of curvature using the light-profile microscope. *Nature* 1952, 170, (4331), 758–9.
9. Israelachvili, J., McGuiggan, P., Adhesion and short-range forces between surfaces: New apparatus for surface force measurements. *J. Mater. Res.* 1990, 5, 2223–2231.
10. Hunter, R., *Foundations of Colloid Science*. Oxford University Press: Oxford, 1989; Vol. 1.
11. Helm, C. A.; Israelachvili, J. N., Forces between phospholipid bilayers and relationship to membrane fusion. *Methods Enzymol* 1993, 220, 130–43.
12. Helm, C. A.; Israelachvili, J. N.; McGuiggan, P. M., Role of hydrophobic forces in bilayer adhesion and fusion. *Biochemistry* 1992, 31, (6), 1794–805.
13. Helm, C. A.; Israelachvili, J. N.; McGuiggan, P. M., Molecular mechanisms and forces involved in the adhesion and fusion of amphiphilic bilayers. *Science* 1989, 246, (4932), 919–22.
14. Leckband, D. E., Helm, C. A., Israelachvili, J., Role of Calcium in the Adhesion and Fusion of Bilayers. *Biochemistry* 1993, 32, 1127–1140.
15. Israelachvili, J., *Intermolecular and Surface Forces*. 2 ed.; Academic Press: New York, 1992.
16. Sivasankar, S., Gumbiner, BM, Leckband, D, Direct Measurements of Multiple Adhesive Alignments and Unbinding Trajectories between Cadherin Extracellular Domains. *Biophys. J.* 2001, 80, 1758–1768.

17. Johnson, K. L., Kendall, K., Roberts, A.D., Surface energy and the contact of elastic solids. *Proc. R. Soc. Lond. A.* 1971, 324, 301–313.

18. Balsera, M., Stepaniants, S., Izrailev, S., Oono, Y., Schulten, K., Reconstructing Potential Energy Functions from Simulated Force-Induced Unbinding Processes. *Biophys. J.* 1997, 73, 1281–1287.

19. Evans, E., Ritchie, K., Dynamic Strength of Molecular Adhesion Bonds. *Biophys. J.* 1997, 72, 1541–1555.

20. Dudko, O. K.; Hummer, G.; Szabo, A., Intrinsic rates and activation free energies from single-molecule pulling experiments. *Phys Rev Lett* 2006, 96, (10), 108101.

21. Hummer, G.; Szabo, A., Free energy reconstruction from nonequilibrium single-molecule pulling experiments. *Proc Natl Acad Sci USA* 2001, 98, (7), 3658–61.

22. Hummer, G.; Szabo, A., Kinetics from nonequilibrium single-molecule pulling experiments. *Biophys J* 2003, 85, (1), 5–15.

23. Hummer, G.; Szabo, A., Free energy surfaces from single-molecule force spectroscopy. *Acc Chem Res* 2005, 38, (7), 504–13.

24. Paramore, S.; Ayton, G. S.; Voth, G. A., Extending the fluctuation theorem to describe reaction coordinates. *J Chem Phys* 2007, 126, (5), 051102.

25. Li, F.; Leckband, D., Dynamic strength of molecularly bonded surfaces. *J Chem Phys* 2006, 125, (19), 194702.

26. Vijayendran, R., Hammer, D., and Leckband, D., Simulations of the adhesion between molecularly bonded surfaces in direct force measurements. *J. Chem. Phys.* 1998, 108, 1162–1169.

27. Yeung, C., Purves, T., Kloss, A. A., Kuhl, T. L., Sligar, S., Leckband, D., Cytochrome c Recognition of Immobilized, Orientational Variants of Cytochrome b5: Direct Force and Equilibrium Binding Measurements. *Langmuir* 1999, volume 15, 6829–6836.

28. Sivasankar, S., Brieher, W., Lavrik, N., Gumbiner, B., and Leckband, D., Direct Molecular Force Measurements of Multiple Adhesive Interactions Btween Cadherin Ectodomains. *Proc. Natl. Acad. Sci. USA* 1999, 96, 11820–11824.

29. Johnson, C. P.; Jensen, I. E.; Prakasam, A.; Vijayendran, R.; Leckband, D., Engineered protein A for the orientational control of immobilized proteins. *Bioconjug Chem* 2003, 14, (5), 974–8.

30. Perez, T. D.; Nelson, W. J.; Boxer, S. G.; Kam, L., E-cadherin tethered to micropatterned supported lipid bilayers as a model for cell adhesion. *Langmuir* 2005, 21, (25), 11963–8.

31. Leckband, D., Schmitt, F.-J., Israelachvili, J., Knoll, W., Direct force measurements of specific and nonspecific protein interactions. *Biochemistry* 1994, 33, 4611–4624.

32. Johnson, C. P.; Fragneto, G.; Konovalov, O.; Dubosclard, V.; Legrand, J. F.; Leckband, D. E., Structural studies of the neural-cell-adhesion molecule by X-ray and neutron reflectivity. *Biochemistry* 2005, 44, (2), 546–54.

33. Martel, L., Johnson, C., Boutet, S., Al- Kurdi, R., Konovalov, O., Robinson, I., Leckband, D., Legrand, J. F., X-Ray Reflectivity Investigation of the Structure of Cadherin Monolayers. *J. Phys. IV France* 2002, 12, 365–377.

34. Marra, J., Israelachvili, J., Direct Measurements of Forces between Phosphatidylcholine and Phosphatidylethanolamine bilayers in Aqueous Electrolyte Solutions. *Biochemistry* 1985, 24, 4608–4618.

35. Leckband, D., Müller, W., Schmitt, F.-J., and Ringsdorf, H., Molecular Mechanisms Determining the Strength of Receptor-Mediated Intermembrane Adhesion. *Biophys. J.* 1995, 69, 1162–1169.

36. Leckband, D. E., Kuhl, T. L., Wang, H. K., Müller, W., Ringsdorf, H., 4–4–20 Anti-Fluorescyl IgG Fab' Recognition of Membrane Bound Hapten: Direct Evidence for the Role of Protein and Interfacial Structure. *Biochemistry* 1995, 34, 11467–11478.

37. Yeung, C., Leckband, D., Substrate Alterations of the Apparent Affinities of Immobilized Receptors. *Langmuir* 1998, Kloss, A. A., Lavrik, N., Yeung, C., Leckband, D., Effect of the microenvironment on the recognition of immobilized cytochromes by soluble redox proteins, Langmuir, 16, 3414–3421 submitted.

38. Yu, Z.-W., Calvert, T., Leckband, D., Molecular Forces between Membranes Displaying Neutral Glycosphingolipids: Evidence for Carbohydrate Attraction. *Biochemistry* 1997, 37, 1540–1550.

39. Prakasam, A.; Chien, Y. H.; Maruthamuthu, V.; Leckband, D. E., Calcium site mutations in cadherin: impact on adhesion and evidence of cooperativity. *Biochemistry* 2006, 45, (22), 6930–9.

40. Prakasam, A. K.; Maruthamuthu, V.; Leckband, D. E., Similarities between heterophilic and homophilic cadherin adhesion. *Proc Natl Acad Sci U S A* 2006, 103, (42), 15434–9.

41. Zhu, B.; Chappuis-Flament, S.; Wong, E.; Jensen, I. E.; Gumbiner, B. M.; Leckband, D., Functional analysis of the structural basis of homophilic cadherin adhesion. *Biophys J* 2003, 84, (6), 4033–42.

42. Johnson, C. P.; Fujimoto, I.; Perrin-Tricaud, C.; Rutishauser, U.; Leckband, D., Mechanism of homophilic adhesion by the neural cell adhesion molecule: use of multiple domains and flexibility. *Proc Natl Acad Sci U S A* 2004, 101, (18), 6963–8.

43. Johnson, C. P.; Fujimoto, I.; Rutishauser, U.; Leckband, D. E., Direct evidence that neural cell adhesion molecule (NCAM) polysialylation increases intermembrane repulsion and abrogates adhesion. *J Biol Chem* 2005, 280, (1), 137–45.

44. Bayas, M. V.; Kearney, A.; Avramovic, A.; van der Merwe, P. A.; Leckband, D. E., Impact of salt bridges on the equilibrium binding and adhesion of human CD2 and CD58. *J Biol Chem* 2007, 282, (8), 5589–96.

45. Zhu, B., Davies, E. A., van der Merwe, A., Leckband, D. , Direct measurements of heterotypic adhesion between the cell adhesion proteins CD2 and CD48. *Biochemistry* 2002, 42, 12163–12170.

46. Davis, S. J., vanderMerwe, P. A., The structure and ligand interactions of CD2: implications for T-cell function. *Immunology Today* 1996, 17, 177–187.
47. Davis, S. J., vanderMerwe, P. A., CD2-An Exception to the Immunoglobulin Superfamily Concept. *Science* 1996, 273, 1241–1242.
48. Jones, E. Y., Davis, S. J., Williams, A. F., Harlos, K., Stuart, D. I., Crystal structre at 2.8Å resolution of a soluble form of the cell adhesion molecule CD2. *Natue* 1992, 360, 232–239.
49. Bodian, D. L., Jones, E. Y., Stuart, D. I., Davis, S. J., Crystal structure of the extracellular region of the human cell adhesion molecule CD2 at 2.5 A resolution. *Structure* 1994, 2, 755–766.
50. Ikemizu, S.; Sparks, L. M.; van der Merwe, P. A.; Harlos, K.; Stuart, D. I.; Jones, E. Y.; Davis, S. J., Crystal structure of the CD2-binding domain of CD58 (lymphocyte function-associated antigen 3) at 1.8-A resolution. *Proc Natl Acad Sci USA* 1999, 96, (8), 4289–94.
51. Evans, E. J.; Castro, M. A.; O'Brien, R.; Kearney, A.; Walsh, H.; Sparks, L. M.; Tucknott, M. G.; Davies, E. A.; Carmo, A. M.; van der Merwe, P. A.; Stuart, D. I.; Jones, E. Y.; Ladbury, J. E.; Ikemizu, S.; Davis, S. J., Crystal structure and binding properties of the CD2 and CD244 (2B4)-binding protein, CD48. *J Biol Chem* 2006, 281, (39), 29309–20.
52. Wang, J. H.; Smolyar, A.; Tan, K.; Liu, J. H.; Kim, M.; Sun, Z. Y.; Wagner, G.; Reinherz, E. L., Structure of a heterophilic adhesion complex between the human CD2 and CD58 (LFA-3) counterreceptors. *Cell* 1999, 97, (6), 791–803.
53. McAlister, M. S. B., Mott, H. R., vanderMerwe, P. A., Campbell, I. D., Davis, S. J., and Driscoll, P. C., NMR Analysis of Interacting Soluble Forms of the Cell-Cell Recognition Molecules CD2 and CD48. *Biochemistry* 1996, 35, 5982–5991.
54. Davis, S. J.; Ikemizu, S.; Wild, M. K.; van der Merwe, P. A., CD2 and the nature of protein interactions mediating cell-cell recognition. *Immunol Rev* 1998, 163, 217–36.
55. Davis, S. J.; Ikemizu, S.; Evans, E. J.; Fugger, L.; Bakker, T. R.; van der Merwe, P. A., The nature of molecular recognition by T cells. *Nat Immunol* 2003, 4, (3), 217–24.
56. van der Merwe, P. A.; Davis, S. J., Molecular interactions mediating T cell antigen recognition. *Annu Rev Immunol* 2003, 21, 659–84.
57. Davis, S. J., Davies, E.A., Tucknott, M.G., Jones, E.Y., vanderMerwe, A., The role of charged residues mediating low affinity protein-protein recognition at the cell surface by CD2. *Proc. Natl. Acad. Sci. USA* 1998, 95, 5490–5494.
58. Arulanandam, A. R.; Withka, J. M.; Wyss, D. F.; Wagner, G.; Kister, A.; Pallai, P.; Recny, M. A.; Reinherz, E. L., The CD58 (LFA-3) binding site is a localized and highly charged surface area on the AGFCC'C" face of the human CD2 adhesion domain. *Proc Natl Acad Sci USA* 1993, 90, (24), 11613–7.
59. Bayas, M. V.; Schulten, K.; Leckband, D., Forced detachment of the CD2-CD58 complex. *Biophys J* 2003, 84, (4), 2223–33.
60. Israelev, S., Stepaniants, S., Balsera, M., Oono, Y., Schulten, Molecular Dynamics Study of Unbinding of the Avidin-Biotin Complex. *Biophys. J.* 1997, 72, 1568–1581.
61. Walsh, F., Doherty, P, Neural Cell Adhesion Molecules of the Immunoglobulin Superfamily. *Ann. Rev. Cell. Biol.* 1997, 13, 425–56.
62. Chothia, C., Jones, E. Y., The Molecular Structure of Cell Adhesion Molecules. *Ann. Rev. Biochem.* 1997, 66, 823–862.
63. Becker, J. W., Erickson, H. P., Hoffmann, S., Cunningham, B. A., Edelman, G. M., Topology of cell adhesion molecules. *Proc. Natl. Acad. Sci. USA* 1989, 86, 1088–1092.
64. Hall, A., Rutishauser, U., Visualization of neural cell adhesion molecule by electron microscopy. *J. Cell Biol.* 1987, 104, 1579–86.
65. Atkins, A. R., Chung, J., Songpon, D., Little, E., Edelman, G. M., Wright, P. E., Cunningham, B.A., Dyson, H.J., Solution structure of the third immunoglobulin domain of the neural cell adhesion molecule NCAM: can solution studies define the mechanism of homophilic binding? *J. Mol. Biol.* 2001, 311, 161–172.
66. Cunningham, B. A., Hemperly, J. J., Murray, B. A., Prediger, E. A., Brackenbury, R., Edelman, G. M., Neural cell adhesion molecule: structure, immunoglobulin-like domains, cell surface modulation, and alternative RNA splicing. *Science* 1987, 236, 799–806.
67. Jenson, P., Soroka, V., Thompson, N. K., Ralets, I., Berezin, V., Bock, E., Poulsen, F.M., Structure and interactions of NCAM modules 1 and 2-basic elements in neural cell adhesion. *Nature Structural Biology* 1999, 6, 486–493.
68. Kasper, C., Rasmussen, H., Kastrup, J. S., Ikemizu, S., Jones, R. Y., Berezin, V., Bock, E., Larsen, I. K., Structural basis of cell-cell adhesion by NCAM. *Nature Struct. Biol.* 2000, 7, 389–393.
69. Kiselyov, V., Berezin, V., Maar, T. E., Soroka, V., Edvardsen, K., Schousboe, A., Bock, E., The First Immunoglobulin-like Neural Cell Adhesion Molecule (NCAM) Domain is Involved in Double-reciprocal Interaction with the Second Immunoglobulin-like NCAM Domain and in Heparin Binding. *J. Biol. Chem.* 1997, 272, 10125–10134.
70. Ranheim, T. S., Edelman, G. M., Cunningham, B. A., Homophilic adhesion mediated by the neural cell adhesion molecule involves multiple immunoglobulin domains. *Proc. Natl. Acad. Sci.* 1996, 93, 4071–4075.
71. Rao, Y., Wu, X-F., Gariepy, J., Rutishauser, U., Siu, C.-H., Identification of a Peptide Sequence Involved in Homophilic Binding in the Neural Cell Adhesion Molecule NCAM. *J. Cell Biol.* 1992, 118, 937–949.

72. Soroka, V., Kiryushko, D., Novitskaya, V., Ronn, C. B., Poulson, F. M., Holm, A., Bock, E., Berezin, V., Induction of neuronal differentiation by a peptide corresponding to the homophilic binding site of the second Ig module of NCAM. *J. Biol. Chem.* 2002, 277, 24676–24683.

73. Soroka, V., Kolkova, K., Kastrup, J. S., Diederichs, K., Breed, J., Kiselyov, V. V.,Poulsen, F. M., Poulsen, F. M., Larsen, I. K., Welte, W., Berezin, V., Bock, E., Kasper, C., Structure and Interactions of NCAM Ig1–2–3 Suggest a Novel Zipper Mechanism for Homophilic Adhesion. *Structure* 2003, 10, 1291–1301.

74. Wieland, J. A., Gewirth, A., Leckband, D., Single Molecules Adhesion Measurements Reveal Two Homophilic NCAM Bonds with Mechanically Distinct Properties. *J. Biol. Chem.* 2005, 280, 41037–41046.

75. Christenson, H. K., Horn, R. G., Direct measurement of the force between solid surfaces in a polar liquid. *Chem. Phys. Lett.* 1983, 98, 45–48.

76. Christenson, H. K., Forces between solid surfaces in a binary mixture of non-polar liquids. *Chem. Phys. Lett.* 1985, 118, 455–458.

77. Christenson, H. K., Gruen, D. W. R., Horn, R. G., Israelachvili, J. N., Structuring in liquid alkanes between solid surfaces: force measurements and mean-field theory. *J. Chem. Phys.* 1987, 87, 1834–1841.

78. Heuberger, M., Zach, M., Spencer, N. D., Density fluctuations under confinement: when is a fluid not a fluid? *Science* 2001, 292, 905–908.

79. Horn, R. G., Israelachvili, J. N., Direct measurement of structural forces between two surfaces in a nonpolar liquid. *J. Chem. Phys.* 1981, 75, 1400–1411.

80. Horn, R. G., Israelachvili, J. N., Molecular organization and viscosity of a thin film of molten polymer between two surfaces as probed by force measurements. *Macromolecules* 1988, 21, 2836–2841.

81. Israelachvili, J. N., Pashley, R. M., Molecular layering of water at surfaces and origin of repulsive hydration forces. *Nature* 1983, 306, 249–250.

82. Israelachvili, J. N., Solvation forces and liquid structure, as probed by direct force measurements. *Acc. Chem. Res.* 1987, (20), 415–421.

83. Israelachvili, J. N., Kott, S. J., Liquid structuring at solid interfaces as probed by direct force measurements: the transition from simple to complex liquids and polymer fluids. *J. Chem. Phys.* 1988, 88, 7162–7166.

84. Kekicheff, P., Ducker, W. A., Ninham, B. W., Pilen, M. P., Multilayer adsorption of cytochrome c on mica around isoelectric pH. *Langmuir* 1990, 6, 1704–1708.

85. Petrov, P., Miklavcic, S., Olsson, U., Wennerstrom, H., A confined complex liquid. Oscillatory forces and lamellae formation from an L3 phase. *Langmuir* 1995, 11, 3928–3936.

86. Attard, P., Parker, J. L., Oscillatory solvation forces: A comparison of theory and experiment. *J. Phys. Chem.* 1992, 92, 5086–5093.

87. Frink, L. J., vanSwol, F., A common theoretical basis for surface forces apparatus, osmotic sress, and beam bending measurements of surface forces. *Coll Surf A: Physichochem and Eng Aspects* 2000, 162, 25–36.

88. Nelson, R. W., Bates, P. A., Rutishauser, U., Protein Determinants for Specific Polysialylation of the Neural Cell Adhesion Molecule. *J. Biol. Chem.* 1995, 270, 17171–17179.

89. El Maarouf, A.; Petridis, A. K.; Rutishauser, U., Use of polysialic acid in repair of the central nervous system. *Proc Natl Acad Sci U S A* 2006, 103, (45), 16989–94.

90. Franz, C. K.; Rutishauser, U.; Rafuse, V. F., Polysialylated neural cell adhesion molecule is necessary for selective targeting of regenerating motor neurons. *J Neurosci* 2005, 25, (8), 2081–91.

91. Rutishauser, U., Polysialic acid and the regulation of cell interactions. *Curr. Op. Cell Biol.* 1996, 8, 679–684.

92. Rutishauser, U., Grumet, M., et al., Neural cell adhesion molecule mediates initial interactions between spinal cord neurons and muscle cells in culture. *J. Cell. Biol.* 1983, 97, 145–152.

93. Rutishauser, U.; Landmesser, L., Polysialic acid in the vertebrate nervous system: a promoter of plasticity in cell-cell interactions. *Trends Neurosci* 1996, 19, (10), 422–7.

94. Tang, J.; Rutishauser, U.; Landmesser, L., Polysialic acid regulates growth cone behavior during sorting of motor axons in the plexus region. *Neuron* 1994, 13, (2), 405–14.

95. Rutishauser, U.; Landmesser, L., Polysialic acid on the surface of axons regulates patterns of normal and activity-dependent innervation. *Trends Neurosci* 1991, 14, (12), 528–32.

96. Landmesser, L.; Dahm, L.; Tang, J. C.; Rutishauser, U., Polysialic acid as a regulator of intramuscular nerve branching during embryonic development. *Neuron* 1990, 4, (5), 655–67.

97. Tanaka, F.; Otake, Y.; Nakagawa, T.; Kawano, Y.; Miyahara, R.; Li, M.; Yanagihara, K.; Inui, K.; Oyanagi, H.; Yamada, T.; Nakayama, J.; Fujimoto, I.; Ikenaka, K.; Wada, H., Prognostic significance of polysialic acid expression in resected non-small cell lung cancer. *Cancer Res* 2001, 61, (4), 1666–70.

98. Yang, P. Y., X., Rutishauser, U., Intercellular space is affected by polysialic acid content of NCAM. *J. Cell Biol.* 1992, 116, 1487–1496.

99. Acheson, A., Sunshine, J. L., Rutishauser, U., NCAM Polysialic Acid Can Regulate both Cell-Cell and Cell-Substrate Interactions. *J. Cell Biol.* 1991, 114, 143–153.

100. Yang, P., Major, D., Rutishauser, U., Role of Charge and Hydration in Effects of Polysialic Acid on Molecular Interactions on and between Cell Membranes. *J. Biol. Chem.* 1994, 269, 23039–23044.

101. Alberts, B., Bray, D., Lewis, J., Raff, M., Roberts, K., Watson, J. D., *The Molecular Biology of the Cell.* Garland: NY, 1983.

Force Spectroscopy with Optical and Magnetic Tweezers

Richard Conroy

1 Introduction

Micromanipulation of individual cells and molecules is increasingly important for a wide range of biophysical research because, although ensemble biochemical analysis provides excellent qualitative and quantitative descriptions, it seldom describes phenomena at the molecular level. By observing the force spectroscopy of single molecules, the kinetics, mechanics, and variation of structure, function, and interactions can be fully explored to provide a more complete physiological picture.

The use of electric and magnetic fields for manipulating particles dates back more than a century, with a rich tapestry of applications in separation, filtering and trapping. Recognizing the non-contact advantages of magnetic manipulation, Crick and Hughes probed the physical properties of a cell's cytoplasm more than fifty years ago using magnetic particles [1]. Two decades later, with the development of intense electromagnetic fields from lasers, the manipulation of latex particles with light was experimentally demonstrated by Ashkin in 1970 in his "levitation traps" [2]. Ashkin went on to pioneer optical trapping of both atoms and biomolecules, leading to one of the most successful technology transfers from a physics lab to cell biology.

For many applications, in particular for characterizing biomolecules and their interactions, it is desirable to have a non-contact technique for exerting a force. A non-contact technique allows the behavior of a single molecule under stretching or torsional forces to be measured and manipulated without complicating surface effects or material response limitations. Non-contact techniques also benefit from being easier to multiplex into exerting force on multiple sites of the same molecule or multiple heterogeneous molecules, or to collect parallel statistics on homogeneous copies of the same system. In general they are not limited by access constraints to the interaction volume, and therefore integrate more readily with the desired environmental conditions and other imaging and spectroscopic techniques. For these reasons, and practical reasons such as low cost and biocompatibility, optical and magnetic tweezers have become prominent methods for manipulating and measuring single biological entities and their interactions.

To experience a force in an optical or magnetic field, a molecule must possess either dielectric or magnetic contrast against the surrounding medium. Often the entity under observation does not have favorable intrinsic properties either for imaging or for generating a force, and it is necessary or desirable to label the molecule with a particle or tag to improve contrast. These particles or tags can be multifunctional, acting passively as a position and force sensor and actively as a handle through which a force can be exerted on the attached molecule.

To carry out single-molecule measurements of biological structures and processes requires detection of nanometer displacements and piconewton forces with millisecond resolution. These imaging requirements can be realized using a microscope equipped with CCD cameras and photodiodes, while optical and magnetic tweezers can generate forces in the range 0.1 to 200 pN, making this approach ideal for single-molecule biophysics.

The first half of this chapter will focus on the basic science and the technologies involved in generating these forces using optical and magnetic fields and how force at the piconewton level and displacement at the nanometer scale can be measured. Optical and magnetic tweezers complement other single-molecule manipulation techniques, as detailed in Table 1, by providing extremely sensitive, non-contact manipulation. At one extreme, contact force transducers have high spatial resolution and applied force capabilities. Furthermore, probe tips can be used to provide high resolution chemical and electrostatic information. At another extreme, the global force from fluid flow or an electric field is advantageous in manipulating ensembles of particles using their intrinsic properties. Optical and magnetic tweezers have many of the advantages of either extreme, for example being able to trap multiple particles, and can be readily combined with the other techniques to exploit their relative advantages for the requirements of the system under study.

The energies involved in determining the structure and interactions of biomolecules are carefully balanced to be stable against thermal fluctuations, yet pliable compared to more permanent covalent and ion bonds as illustrated in Figure 1. From a cellular perspective, molecules are ideally reconfigurable using the energy available from nucleotide triphosphate (NTP) hydrolysis, the most common fuel source. Hydrogen bonds (~ 2–$7 k_B T$), hydrophobic interactions (1–$5 k_B T$), and electrostatic forces (0.2–$10 k_B T$) have an energy 1–30 pN nm whereas covalent bonds ($\sim 100 k_B T$) and strong ligand-receptor binding (~ 20–$35 k_B T$) have energies at least an order of magnitude higher, in the range ~ 80–2000 pN nm. Clearly, hydrolysis of individual ATPs ($\sim 20 k_B T$) provides sufficient energy to separate low energy bonds, to reconfigure molecules and drive interactions without disrupting covalent bonds. Optical and magnetic tweezers can produce forces up to 200 pN, which for non-covalent bonds of length of order one nanometer, is sufficient to study

Table 1. Comparison of force spectroscopy techniques.

	Magnetic Tweezers	Optical Tweezers	Electrophoresis	AFM	Micropipette	Fluid Flow
Type	Global/ Point	Point	Global	Point	Point	Global
	Non-Contact	Non-Contact	Non-Contact	Contact	Contact	Non-Contact
Force Range (pN)	0.1–200	0.1–200	0.01–50	10–100000	1–1000	0.1–1000
3D Trap	Yes	Yes	No	Yes	Yes	No
Stiffness (pN nm^{-1})	10^{-6} – 0.1	10^{-6} – 0.1	-	10–10000	0.01–1000	-
Energy Dissipation	No	Yes	Yes	No	No	No
Surface Considerations	No	No	No	Yes	Yes	No
Low Cost	Yes	Yes	Yes	No	Yes	Yes
Parallel	Yes	Yes	Yes	No	No	Yes
Access inside a cell	Yes	Yes	Yes	No	No	No
Self-assembly	Yes	No	Yes	No	No	No

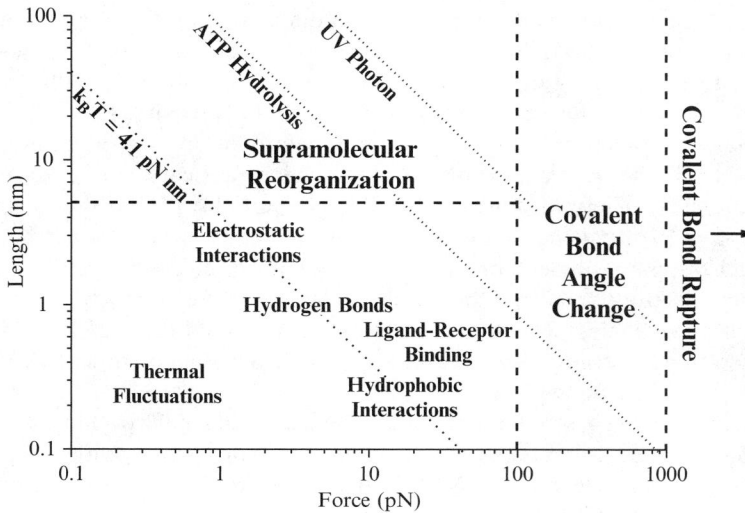

Figure 1. Forces and length scales involved in biomolecular organization. The dotted lines represent the energies associated with thermal noise ($k_B T$), ATP hydrolysis, and a UV photon.

all the processes normally fuelled by NTP hydrolysis. It is worth noting that visible and ultraviolet photons ($\sim$70–150$k_B T$) carry sufficient energy to disrupt nearly all types of bonds if directly absorbed and provide an important limitation to any form of optical imaging. These energy scales provide the setting for the second half of this chapter, which will focus on the force spectroscopy measurements which have been carried out using optical and magnetic tweezers on intracellular biomolecules and their interactions as well as cellular level mechanics and interactions.

2 Optical Tweezers

2.1 Introduction

The effect of light on matter has been known for over four hundred years, dating from Kepler's observation that comet tails always point away from the sun. Indeed, light from the sun can exert a pressure up to $5 \mu N/m^2$ on a totally reflecting surface, ten orders of magnitude less than the force on a cube of the same dimensions due to gravity on the earth's surface. Although resulting in an extremely small force, radiation pressure from sunlight can be significant, for example as the driving force behind solar sails where gravity is negligible. At the beginning of the twentieth century, using thin plates suspended in a evacuated radiometer, Lebedev [3] was the first to experimentally measure the radiation pressure proposed by Maxwell-Bartoli, showing that the pressure for a reflective surface is twice that of an absorbing surface. In 1969 Arthur Ashkin at Bell Laboratories realized that the radiation pressure from an intense laser was sufficient to manipulate dielectric particles [2], demonstrating levitation by balancing radiation pressure with gravity or by using multiple beams. While an axial force on particles in a laser beam was understood in terms of radiation pressure, a radial trapping force was unexpected and it was the 1980s before Ashkin and colleagues showed that a single focused beam could create a three-dimensional optical trap [4]. Following this revelation, demonstrations of the trapping and manipulation of viruses [5] and cells [6] quickly followed, blossoming over the last twenty years into the ubiquitous "optical tweezers."

Optical tweezers are technically a subset of dielectrophoretic (DEP) traps, where an alternating electromagnetic field is used to create a force on objects with dielectric contrast against the surrounding medium, trapping particles of interest at either a maximum (positive trap) or minimum (negative trap) in the field intensity. However, DEP traps typically refer to geometries using two or more electrodes and operating at a frequency below 100 GHz. These traps can be inexpensively microfabricated and used to characterize and trap a wide variety of molecules, cells, and particles [7]. Although terahertz [8] and mid-infrared [9] frequencies are used for spectroscopy, they have not been used for detecting and trapping single biomolecules, providing a clear distinction between the low frequency electrode traps and the purely optical, high frequency traps. Although the higher optical frequencies provide the greatest intensity and the smallest trapping volumes, terahertz and mid-infrared tweezers may have advantages in addressing specific bonds and in cell characterization.

The basis for a dielectric trapping force originates in the polarizability of a particle in an electromagnetic field. From Earnshaw's theorem it is known that a charged particle cannot be held at rest purely by electrostatic fields in free space, yet DEP traps in general and optical tweezers specifically form a stable three-dimensional trap. These traps are stable because the field gradient is three-dimensional, dynamic and not dominated by the scattering force. However, it should be noted that even for simple one-dimensional gradient fields, for example in attractive force magnetic tweezers, stability can be achieved by using a second unrelated force, (e.g., fluid flow) or by modifying the boundary conditions (e.g., at a fluid interface). Nevertheless the unique ability of optical tweezers to form a stable, three-dimensional trap in free space without feedback has been a powerful motivation factor in their adoption beyond specialized physics laboratories.

In recent years, optical tweezers have matured into several commercial products (e.g., Cell Robotics Inc., Arryx Inc., PALM Microlaser Technologies, Elliot Scientific); however, laboratory setups are still at the cutting edge of development, exploiting new methods of light generation, manipulation, and detection, particularly through advances in nanopositioning and optics. In the past decade the range of biologically inspired problems to which optical tweezers have been applied has greatly expanded from cell sorting and classification [10] to intracellular surgery [11] and so-called optical scalpels and scissors [12]. Although not considered here in detail, optical tweezers have also found application in more traditional areas of physics from self-assembly [13] and photolithography [14] to probing for violations of the second law of thermodynamics [15]. There is a rich variety of force interactions which can be probed from colloids [16] to actuators and small turbines driven by light [14], covered in more detail in many of the comprehensive review articles on optical trapping [17, 18, 19]. However, the focus for this section is on the basic science behind how force is generated in an optical trap and how it can be calibrated and used to characterize the force spectroscopy of biomolecules.

2.2 Theory of Radiation Pressure

The interaction between light and matter is a complicated one which is not understood fully for all cases, but informative approximations are available under a number of limits. The origin of a force on matter because of an electromagnetic wave can be understood qualitatively by an electric field exerting a force on charges within a particle, and a magnetic field exerting a force on currents. From Maxwell's equations, an electromagnetic field in a vacuum exerts a force:

$$F_{mech}(r,t) = q[E(r,t) + v(r,t) \times B(r,t)]$$
$$= \int_V [\rho(r,t)E(r,t) + j(r,t) \times B(r,t)]dV \qquad (1)$$

on a single charge, q, moving with velocity $v(r,t)$ in the first expression and a distribution of charges in the second, satisfying the charge conservation law. The conservation of linear momentum in an arbitrary volume V around the charges gives:

$$\frac{d}{dt}\left[P_{field}(r,t)+P_{mech}(r,t)\right]=\int_V \nabla.T\, dV \tag{2}$$

where $F_{mech}=\frac{d}{dt}P_{mech}$ and the field momentum is $P_{field}=\frac{1}{c^2}\int_V [E\times H]\, dV.$

Assuming a particle has the linear relationships $D = \varepsilon E$ and $B = \mu H$, the Maxwell stress tensor, T, can be written as:

$$T=\frac{1}{4\pi}\left[\varepsilon\varepsilon_0 EE-\mu\mu_0 HH-\frac{1}{2}\left(\varepsilon\varepsilon_0 E^2-\mu\mu_0 H^2\right)\delta\right] \tag{3}$$

where δ is the Kronecker delta function. The field momentum is zero when it is averaged over one oscillation period; and applying the Gauss integration law, the time-averaged force becomes:

$$\langle F\rangle=\int_V \nabla\cdot T\, dV=\int_{\delta V}\langle T(r,t)\rangle\cdot n(r)\, da \tag{4}$$

where δV is the surface of V, $n(r)$ is the unit element perpendicular to the surface, and da is a surface element. This equation, using a generalized Maxwell stress tensor, is generally applicable and is only constrained by assuming the particle is rigid.

The radiation pressure, P, can be found from integrating Maxwell's stress tensor on an infinite planar surface A, perpendicular to the Poynting vector direction z:

$$Pn_z=\frac{1}{A}\int_A \langle T(r,t)\rangle\cdot n_z\, da \tag{5}$$

Assuming an incident plane wave $\left(I_0=\frac{1}{2}c\varepsilon_0 E_0^2\right)$ interacting with a particle with complex reflection coefficient R, the electric field outside the particle can be written as a superposition of two counter-propagating waves:

$$E(r,t)=E_0\,\mathrm{Re}\left[\left(e^{ikz}+Re^{-ikz}\right)e^{-i\omega t}\right]n_x \tag{6}$$

and using Maxwell's equations, the magnetic field is:

$$H(r,t)=\sqrt{\frac{\varepsilon_0}{\mu_0}}E_0\mathrm{Re}\left[\left(e^{ikz}-Re^{-ikz}\right)e^{-i\omega t}\right]n_y \tag{7}$$

Under these conditions the Maxwell stress tensor reduces to:

$$\langle T(r,t)\rangle\cdot n_z=-\frac{1}{2}\left\langle\varepsilon_0 E^2+\mu_0 H^2\right\rangle n_z=\frac{\varepsilon_0}{2}E^2\left[1+|R|^2\right]n_z \tag{8}$$

And the radiation pressure on the particle can be expressed as:

$$P=\frac{I_0}{c}\left[1+|R|^2\right] \tag{9}$$

As observed experimentally by Lebedev, the pressure on a perfectly absorbing body ($R = 0$) is half of that for a perfectly reflecting particle ($R = 1$). While many objects have irregular shapes and nonlinear response and are illuminated by a complex spatial, spectral, and temporal light source, this approach of simple integration of the Maxwell stress tensor provides a good approximation for computing the forces on dielectric spheres trapped by optical tweezers.

It is intuitively helpful to consider the two component forces in an optical trap, the gradient force and the scattering force, independently. Generally both forces are present in a light beam but the scattering force dominates, exerting a force in the direction of propagation of the light. However, by tightly focusing a near diffraction limited light source, the gradient force created is sufficient to overcome the scattering force and form a trap at the region of highest intensity.

More in-depth analysis of the approximations and limitations in modeling optical traps have been discussed extensively in a number of publications [20, 21, 22], though to gain an insight into their basic operation and optimization we will consider the two component forces (gradient and scattering force) and the two scaling limits (Rayleigh and ray optic regimes).

2.2.1 The Gradient Force

Light incident on a particle creates a dielectric response, due to the polarizability of the constituent atoms or ions. For one of these atoms or ions in a monochromatic, linearly polarized, continuous light field, E, the time-averaged induced dipole moment is:

$$p = \alpha E \tag{10}$$

where $\alpha = \alpha' + i\alpha''$ is the relative complex polarizability of the particle to the surrounding medium. The interaction of the induced dipole with the electric field of the light creates an electrostatic potential:

$$U = -p \cdot E \tag{11}$$

Thus in a light field with a spatially varying intensity, there is a gradient force:

$$F_{grad} = -\nabla U = -p.\nabla E = -\alpha (E \cdot \nabla) E \tag{12}$$

For a small particle of radius r_p, this leads to the force relation [4]:

$$F_{grad} = -\frac{n_m^3 r_p^3}{2} \left(\frac{n_c^2 - 1}{n_c^2 - 2} \right) \nabla E^2 \tag{13}$$

Thus the gradient force is linearly dependent on the spatial variation of the intensity of the light field and on the dielectric contrast of the particle to be trapped relative to the surrounding media, which can be described by the Clausius-Mossotti relation. For particles with a refractive index higher than the surrounding medium, the gradient force acts toward the point of highest intensity, that is to say the focal point of a diffraction-limited beam in optical tweezers. Conversely, particles with a lower refractive index can be trapped at a minimum in the light field intensity.

The strength of the restoring gradient force in an optical trap of radius r can be characterized as a Hookean spring with stiffness, κ, where the force is linearly proportional to small displacements ($d < r/2$):

$$F = -\kappa \cdot r \tag{14}$$

and the trap period, a measure of correlation time, is:

$$\tau_0 = \frac{6\pi \eta r_p}{\kappa} \tag{15}$$

where η is the viscosity of the surrounding medium. A schematic of the axial and radial potentials and their resulting stiffnesses is shown in Figure 2.

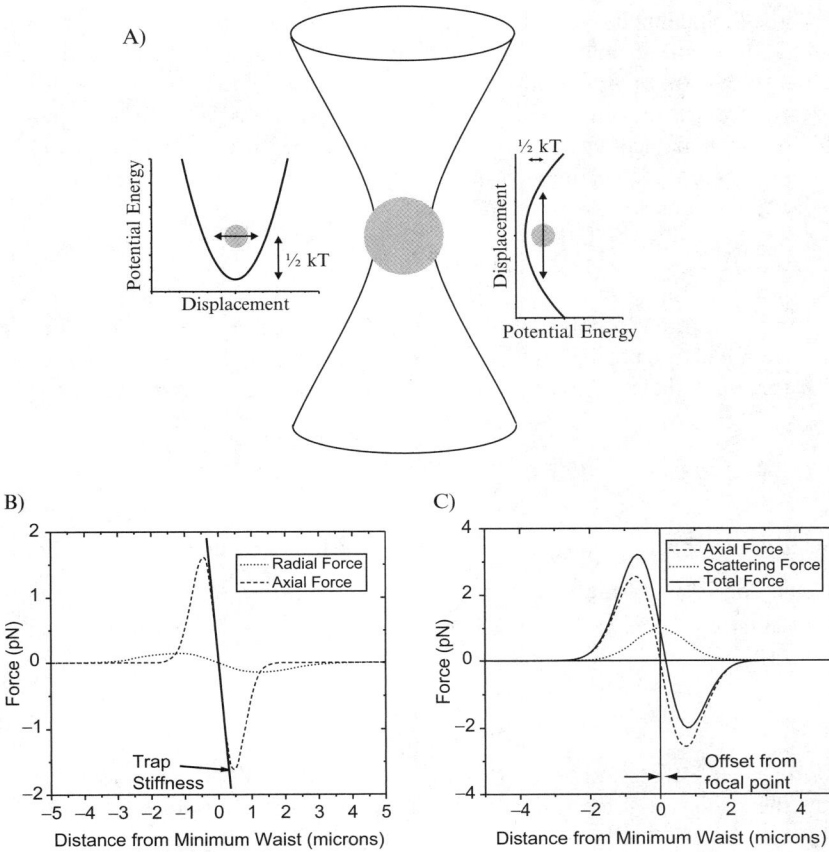

Figure 2. (A) The axial and radial trapping potentials of a bead in an optical trap lead to (B) differing stiffnesses and extents. (C) With the addition of the scattering force, the trap center is offset from the focal point.

Techniques for measuring the stiffness of an optical trap are described later, but for a 1 μm diameter polystyrene bead in a typical optical tweezers setup, the stiffness can be varied easily in the range 10^{-6}–0.1 pN/nm by adjusting the laser power from 10–1000 mW. These characteristics complement the stiffness of physical cantilevers such as AFM tips (10–10^4 pN/nm), which cannot be as easily tuned after fabrication. The trap stiffness is important in determining the minimum force which can be measured through displacement detection and sets an upper limit for the maximum useful sampling rate through the trap frequency. Although not immediately obvious from these simple expressions, the trap stiffness is greatest when the particle to be trapped is the same size as the beam waist; as particle size decreases, the restoring force decreases rapidly, but decreases only modestly when the particle size increases.

2.2.2 The Scattering Force

The second force component in an optical trap arises from the scattering of light and is a consequence of photons having momentum. This force acts in the direction of propagation of the light and is dependent on the light intensity rather than the gradient. The momentum of a single photon of energy E is:

$$p = \hbar k = \frac{h}{\lambda} = \frac{E n_m}{c} \tag{16}$$

A beam of incident photons can be scattered from the particle, resulting in two impulses: one along the direction of light propagation, and the other opposite the direction of the scattered photon. For isotropic scattering, dependent on the size of the particle, the latter impulse has no preferred direction and results in a net force in the direction of light propagation.

The change in momentum, or force, of a particle can be calculated by considering the photon flux impinging on and leaving an object under the conservation of momentum:

$$F_{scat} = \frac{n}{c} \iint (S_{in} - S_{out})dA = \frac{n_m \sigma \langle S \rangle}{c} \tag{17}$$

where n_m is the refractive index of the surrounding medium, $<S>$ is the time-averaged Poynting vector, c is the speed of light, and σ is the particle's optical cross section. In the case of a small, spherical, dielectric particle, the Rayleigh scattering cross-section is:

$$\sigma = \frac{8}{3}\pi \left(\frac{2\pi n_m}{\lambda}\right)^4 r_p^6 \left(\frac{n_c^2 - 1}{n_c^2 + 2}\right)^2 \tag{18}$$

where r_p is the particle radius, $n_c = \frac{n_p}{n_m}$ is the refractive index contrast between the particle (n_p) and the medium (n_m), and $k = \frac{2\pi}{\lambda}$ is the wave vector of the trapping light. The scattering force on a Rayleigh particle can then be written in terms of the light intensity I_0 [4]:

$$F_{scat} = \frac{128\pi^5 r_p^6}{3\lambda^4}\left(\frac{n_c^2 - 1}{n_c^2 + 2}\right)^2 \frac{n_m I_0}{c} \tag{19}$$

Thus the scattering force is dependent on the photon flux or light intensity, the wavelength of the trapping light, the particle size, and its refractive index contrast against the liquid in which it is immersed.

For larger particles ($r_p \gg \lambda$), the scattering cross-section can be expressed as $\sigma = Q_{scat}\pi r_p^2$ where Q_{scat} approaches the limit of 2. However, for intermediate sizes, an accurate force estimate needs to be numerically evaluated using Mie theory [23], in part because the scattering of incident photons is no longer isotropic.

To maximize the gradient force, the particle's radius should be comparable to the wavelength of the trapping laser and its associated minimum focal spot size and consequently is most appropriately described by the intermediate, Mie regime. The need to numerically solve Mie scattering theory is one of the complications in developing a simple model for optical tweezers and makes direct comparison of the gradient force and scattering force difficult. However, one variable which can be controlled and optimized is the refractive index contrast between the trapped particle and the surrounding medium. The optimal refractive index contrast is 1.2–1.3, which maximizes the gradient force with respect to the scattering force to $F_{max} = \frac{0.49 n_m P}{c}$ for the incident optical power P. Conveniently, polystyrene beads in water have a refractive index contrast of $n_c = 1.59/1.33 = 1.2$, close to optimal, with a potential maximum force of $F_{max} = 2.2\,pN/mW$, though optical tweezers generally operate at around 2/3 of this value [24].

Adding the two force components results in the equilibrium position for an optical trap being displaced a distance proportional to the light intensity from the minimum beam waist in the direction of the light propagation, typically 100–500 nm, and illustrated in Figure 2c. This distance can be found experimentally by translating a trapped bead into a surface and measuring the displacement of the bead in the trap when it is in the focal plane of the surface, or comparing it to a bead previously fixed to the surface.

2.2.3 Rayleigh Regime (r << λ)

In the two particle size limits, the Rayleigh regime ($r \ll \lambda$) and the ray optics regime ($r \gg \lambda$), a theoretical treatment for calculating the radiation pressure is relatively straightforward and provides a number of useful insights.

In the Rayleigh regime, particles can be treated as a collection of dipoles polarized by the envelope of the light field forming the trap, with the phase of the field being approximately constant throughout the particle. In the previous sections, equations were presented for the gradient and scattering forces on small dielectric particles; however, in practice, it is difficult to exert sufficient force to trap a dielectric particle below 100 nm in size with current optics and laser limitations. As particle size increases, the difference between Rayleigh and Mie scattering becomes measurable for particles larger than 200 nm for visible trapping fields [25], and Rayleigh approximations break down for most trappable objects.

However, exploiting a nonlinearity such as a plasma resonance, ionic resonance, or intensity dependent refractive index, using a microstructured meta-material or reduction of homogeneous and inhomogeneous broadening can enhance the dielectric contrast to trap particles down to 5 nm in size [26]. Alternatively, the medium surrounding the particle can be modified to minimize Brownian motion to the extreme of trapping and cooling small numbers of atoms in an ultrahigh vacuum chamber [27]. In general, however, the complications and limitations associated with these approaches mean that optical traps rarely operate in a pure Rayleigh regime, and predictions can be inaccurate without experimental validation.

2.2.4 Ray Optics Regime (r >> λ)

In the other limiting case, where the size of the particle to be trapped is much larger than the wavelength of light, and has a small refractive index contrast with the surrounding medium, the component forces can be modeled using ray optics. An incident monochromatic light beam can be decomposed into individual rays with appropriate intensity, momentum, and direction. In a uniform, nondispersive media these rays propagate in a straight line and can be described by geometric optics. For a uniform dielectric sphere the optical forces, including the scattering component, can be calculated directly from ray optics [20]:

$$F_{scat} = \frac{n_m P}{c} \left(1 + R\cos(2\theta_R) - \frac{T_F^2 \left(\cos(2\theta_R - 2\theta_T) + R\cos(2\theta_R) \right)}{1 + R^2 + 2R\cos(2\theta_T)} \right)$$

$$F_{grad} = \frac{n_m P}{c} \left(R\sin(2\theta_R) - \frac{T_F^2 \left(\sin(2\theta_R - 2\theta_T) + R\cos(2\theta_R) \right)}{1 + R^2 + 2R\cos(2\theta_T)} \right) \tag{20}$$

where R & T_F are the Fresnel coefficients, and θ_R and θ_T are the angles for reflection and transmission of the incident rays. Figure 3 schematically illustrates the origins of the axial and radial forces due to diffraction and how the component forces add together. For non-spherical and complex particles approximations can be computed [28]. The ray optics regime is increasingly accurate for dielectric particles of radius $r_p > \dfrac{5 n_c \lambda}{\pi n_m}$, though for these larger particles the radial trapping force diminishes. However, increasing the focal spot size to compensate would decrease the axial trapping force.

As mentioned earlier, trapping efficiency is highest for objects which are approximately a wavelength in size and therefore fall in the intermediate regime between the Rayleigh and ray optics regimes. Early approaches to analytically modeling this intermediate regime used a generalized Lorenz-Mie approach [29], though recently there has been progress through the extension of Rayleigh theory to larger particles [30]. Forces from both models compare well

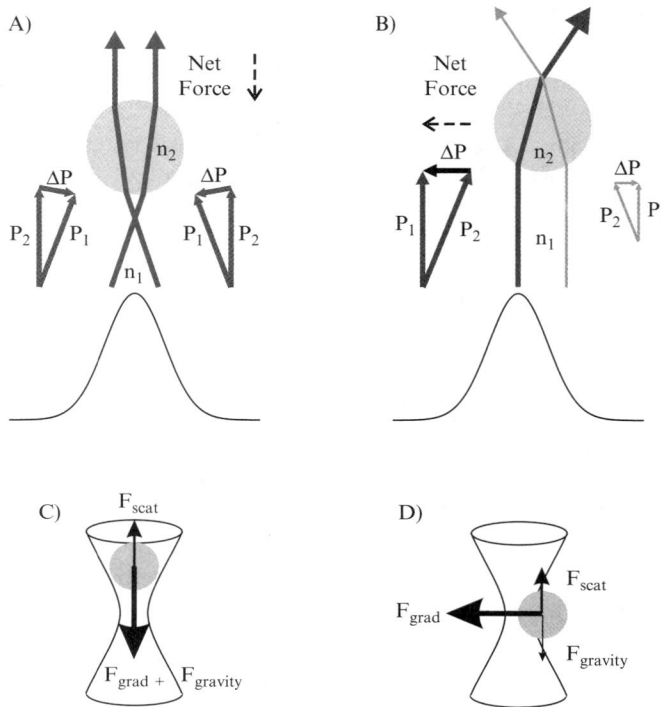

Figure 3. Schematic of the optical forces in the ray-optics regime. Summing the rays gives an (A) axial force due to vertical displacement from trap center; (B) radial force due to lateral displacement from trap center. Taking into account gravity and scattering, (C) the axial and (D) radial gradient force must be the dominant component to form an optical trap.

and have helped in trap calibration through the prediction of the far-field interference pattern resulting from a trapped object; potentially they will help in trap design and optimization.

These theoretical models generally assume a continuous, diffraction limited monochromatic beam focused by a high numerical aperture lens to trap a rigid dielectric sphere with a refractive index higher than the surrounding medium. All of these assumptions can be broken through choice of trap geometry, light source, and particle to be trapped. For the remainder of this section we will consider some of the trap designs, light sources, and particles which have been or can be used.

2.3 Types of Optical Traps

The standard optical trap uses a single high numerical aperture lens, typically a 100x objective with an NA of ~1.3, to focus a near infrared laser to a diffraction limited spot and trap a polystyrene sphere approximately 1 μm in size. Imaging of the bead can be done through the same lens, opening one side of the sample, to manipulation or examination using another technique. Access to higher numerical apertures are limited by lens design and total internal reflection, while the spot size is limited by diffraction, and the maximum power is limited by damage thresholds. With the trap center at the maximum intensity, the maximum power is also limited by considerations of heating and damage to the trapped object and optics. Therefore, irrespective of design, it is unlikely that a conventional single beam optical trap will ever offer forces higher than a nanonewton because of these limitations.

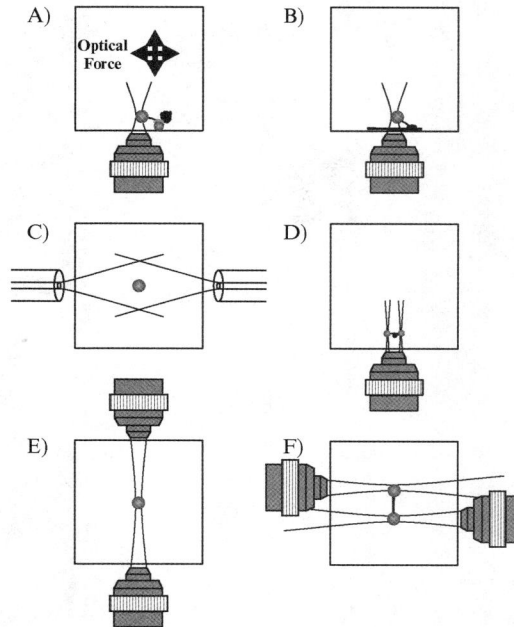

Figure 4. Optical trap designs: (A) immobilized enzyme with substrate tether; (B) immobilized substrate and tethered enzyme; (C) optical divergence trap using optical fibers; (D) multiple traps using the same objective; (E) dual objective trap with one or both objectives used to trap a single particle; (F) dual objectives in a horizontal configuration trapping independent particles.

A second major disadvantage of the standard design is the combined use of the objective for controlling force on the bead and imaging, which, while saving space, limits the direction in which forces can be applied and the range over which particles can be translated and imaged. Finally, light from a single beam can generally only trap a single bead, limiting the rate at which statistics can be collected and systems can be manipulated. Therefore, while the single beam design is an ideal and cost effective introduction to optical trapping, a number of other approaches have become increasingly popular. Figure 4 illustrates a number of common optical trap arrangements, and Figure 5 illustrates a number of common modifications to improve imaging or introduce other techniques.

2.3.1 Single Beam Optical Trap

The most common form of optical tweezers, the single beam optical trap, has matured into a turnkey package commercialized by a number of companies as a non-contact micromanipulator for cell biologists. These systems provide a rapid and convenient method for nonspecialists to work with optical traps.

For manipulation there are two possibilities, movement of the trapping light or movement of the specimen. In general it is preferable to translate the specimen using a stage to avoid changing the optical beam path, even though optical deflection at high speeds can be more readily precise and reproducible.

A single beam optical trap typically has a radius of 250 nm and an intensity-dependent stiffness of 1–500 pN/μm, generating a maximum force in the range 0.2–100 pN. This enables single beads to be moved with velocities up to 1 mm/s, dependent on the trap stiffness. Using the methods described later, displacements down to 1 nm can be measured, corresponding to applied forces down to 100 aN [31].

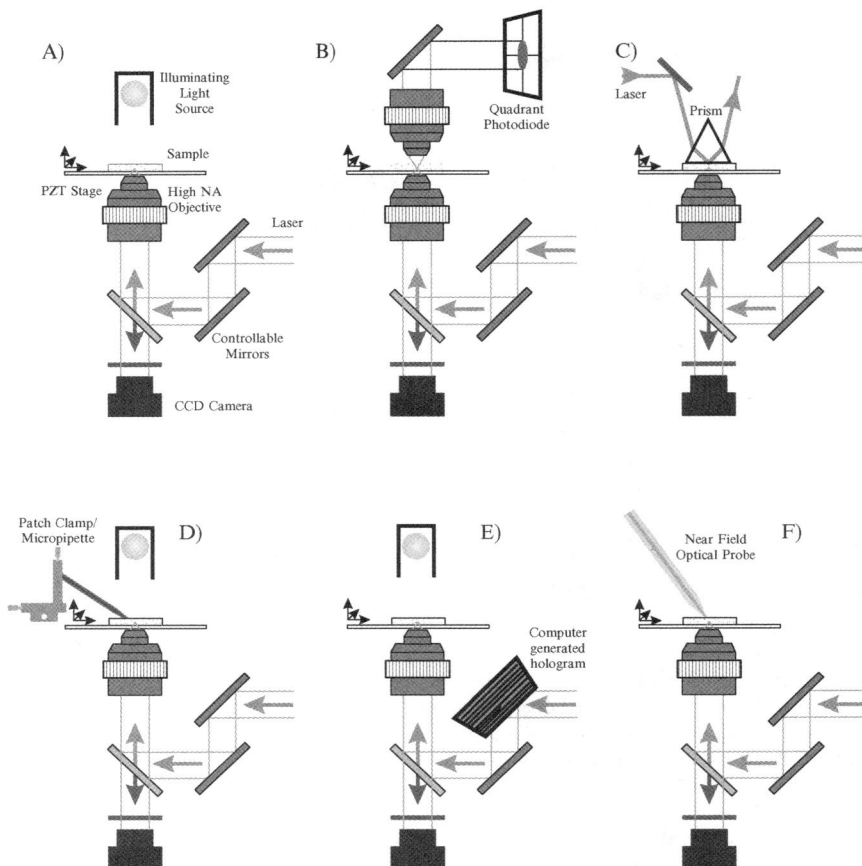

Figure 5. Common extensions to optical traps: (A) steerable mirrors to move trap position; (B) imaging of trapped particle onto a quadrant photodiode; (C) evanescent field excitation of trapped particles near a surface; (D) incorporation of a micropipette or patch-clamp; (E) dynamic trap configurations with a spatial light modulator; (F) excitation and imaging with a near-field optical probe. (*See Color Plates*)

Without feedback about the particle's position with respect to the trap center, single beam traps operate in the constant extension regime, providing a loading rate dependent on the speed at which the trap center is translated. A single beam trap also requires that the system under study be tethered to a solid surface to provide an anchor against which the force can be applied. Ideally, working in such close proximity to a surface is not desirable because it increases the risk of non-specific interactions and modifies parameters such as the drag coefficient because of the stationary boundary. For many applications, however, these limitations do not restrict the results which can be obtained; but to extend the range of operation, other approaches can have more favorable characteristics.

2.3.2 Multiple Beam Optical Traps

In the 1970s the second generation of stable optical traps used a second counter-propagating beam to counteract the scattering force from a single beam [2]; a modern incarnation is composed of two optical fiber tips a short distance apart and is known as an "optical stretcher" or optical fiber divergence trap [32]. Any trapped elastic object will be stretched along the axis between the fibers as a result of the asymmetry of the applied forces, and can result in deformation forces of 0.1–0.3 pN/mW for cells tens of microns in size. Trapping can be achieved with relatively modest powers of tens of milliwatts and stretching achieved without the need for labeling, though large trapping volume and whole cell interactions are traded against trap strength, three-dimensional manipulation, addressing of submicron voxels, and position detection. In combination with microfluidics, an optical stretcher has the potential for high throughput screening [33], though the clinical impact remains to be seen. The use of multicore fibers, multiplexed fibers, and microfabricated optical waveguides should also provide increased flexibility for these traps. The extension of these fibers to subwavelength scales and the near field regime is discussed later in this section.

Closer in design to the single-beam setup, a second objective can be used to focus light counter-propagating and overlapping with the first objective. For identical objectives and illumination, the scattering forces will cancel, permitting the use of longer focal length objectives, and creating higher trapping forces than with the same intensity through a single objective. Alternatively, the objectives can be misaligned to create two independently controllable traps. The primary disadvantages are technical, in the alignment of the counter-propagating beams and loss of a large solid angle of access by the proximity of the second objective to the sample. The higherforces afforded by dual-beam optical tweezers have been used to over-stretch DNA [34, 35], though in general it is easier to scale the laser source to access higher trapping forces than to use a second objective.

For two or more independent traps using a single objective, a single beam can be easily split multiple times by a polarizing beam splitter to form two or more continuous and independent beams. This can be an attractive approach if a high power laser is available; however, independent manipulation of individual beams is non-trivial. An alternative to spatial separation of the beam is to "time-share" it between multiple points, by moving the light rapidly in the back focal plane of the objective using a deflector [36]. As long as the dwell time at each trap is sufficient to give a strong enough restoring force, the scanning rate is faster than the trap period and the beam can exactly repositioned; then multiple particles can be trapped as arbitrary locations in the same plane [37, 38]. Increasing the number of traps decreases the duty cycle for each trap, and to maintain the same trap stiffness the intensity must be increased. For a large number of traps, the fly time becomes a significant percentage of the duty cycle and wastage of the trapping power. The beam can be deflected at large angles using a pair of scanning galvanometer mirrors at rates of up to a kilohertz, but with limited reproducibility. Acousto-optic and electro-optic modulators, although more expensive and less efficient, can sweep smaller angles at up to megahertz frequencies with higher accuracy and stability. Limitations of this time domain modulation approach are that long term thermal stability is required for nanometer reproducibility, programming multiple trap trajectories can be complicated, and the high intensities required to maintain trap stiffness can lead to sample damage.

2.3.3 Holographic Optical Traps

One of the most significant advances in optical trap design recently has been the use of spatial light modulators (SLMs) to create dynamic, holographic traps [39]. The kinoforms required to manipulate the phase of an incident optical beam to create the traps can be computer generated and optimized from the inverse Fourier transform of the required image [40]. These kinoforms are then written to a spatial light modulator to imprint the new transverse phase mask, which results in the desired pattern of constructive and destructive interference in the focal plane of the objective [41]. Spatial light modulators working in reflection mode have higher efficiency compared to diffractive optics such as volume holograms because there are no diffracted orders; and they perform better than galvanometers in creating multiple traps, though currently they have a slow frame rate. Intrinsically they work in a narrow wavelength range and are limited to creating optical features defined by the degree of constructive and destructive interference available, with efficiency falling rapidly for features below half a wavelength in size. The demand for projection systems has helped drive the development of SLMs which now have megapixel arrays with >100:1 contrast ratios and frame rates of up to 100 Hz [42], and these undoubtedly will improve.

Computing power limits trap geometries to predetermined configurations, though for many experiments this is not a significant limitation. Currently the spatial resolution of the available light modulators limits their performance and wider usage; though potentially they could allow parallel operation and faster data collection than that of the single trap geometries, but will never reach the update rates available from acousto-optic or electro-optic modulators. The future is bright for SLMs because the ability to manipulate multiple arbitrary particles using diverse transverse modes in three dimensions would be of benefit in many applications, such as intracellular manipulation and surgery, if the resolution, rate, and contrast can be improved.

2.3.4 Near-Field Optical Traps

The evanescent field created by total internal reflection at a surface [43] or at the metal-coated end of a fiber tapered to sub-wavelength dimensions [44] can be used to create a near field optical trap. These traps are generally only two-dimensional and have limited axial displacements, typically only a few tens of nanometers within 100 nm of a surface. However they can potentially describe sub-wavelength features, are not limited by the transmission of the surrounding medium, and do not require a high-NA objective. Using this approach, interference patterns have been used to manipulate and sort particles and cells on a surface [45]. However, the limitations of working near a surface and the difficulty of forming a three-dimensional trap have restricted the wider application of near-field techniques.

2.4 Types of Optical Beams

The spatial-temporal distribution of the electric field around the focal point of an optical trap is to a large extent controlled by the coherent properties of the light used. For example, a decrease in the spectral coherence of the light, corresponding to an increasing line width, will result in lower trapping efficiency due to increasing chromatic aberration. In the case of decreasing temporal coherence, pulsing of the light source will increase the peak intensity in the focal region while maintaining a substantially lower average power, which can be used to exploit nonlinear effects and minimize thermal effects. These are some of the many variables to manipulate in a laser light source relating to the spatial, temporal, and spectral coherence. One of the simplest examples is to switch the normal linear polarization used to circular polarization. As Beth observed in 1936, circularly polarized light carries angular momentum, which can generate torque on a birefringent particle [46]. However, it has been manipulation of the spatial coherence of the light which has had the most impact on optical trapping.

Spatial coherence, classically demonstrated in Young's double-slit experiment, is the ability for one spatial position of a wavefront to interfere with another. As with the other forms of coherence, the spatial coherence of a laser is determined by the type of optical cavity and gain medium used. Typically in optical tweezers, the laser cavity is a stable geometry, which produces the lowest order Hermite-Gaussian (TEM_{00}) transverse electromagnetic mode resulting in a diffraction-limited beam, giving the smallest waist and highest intensity at the focal point. However, there are a wide range of other coherent transverse modes, including higher-order Hermite-Gaussian modes, Laguerre-Gaussian modes, and Bessel modes.

2.4.1 Hermite Gaussian Beams

In rectangular coordinates, the individual solutions of the paraxial wave equation subject to boundary conditions, in the case of a laser boundaries defined by its mirrors, are described by Hermite-Gaussian modes. The spatial wavefront of the electric field of these transverse electromagnetic modes is described by two numbers corresponding to the number

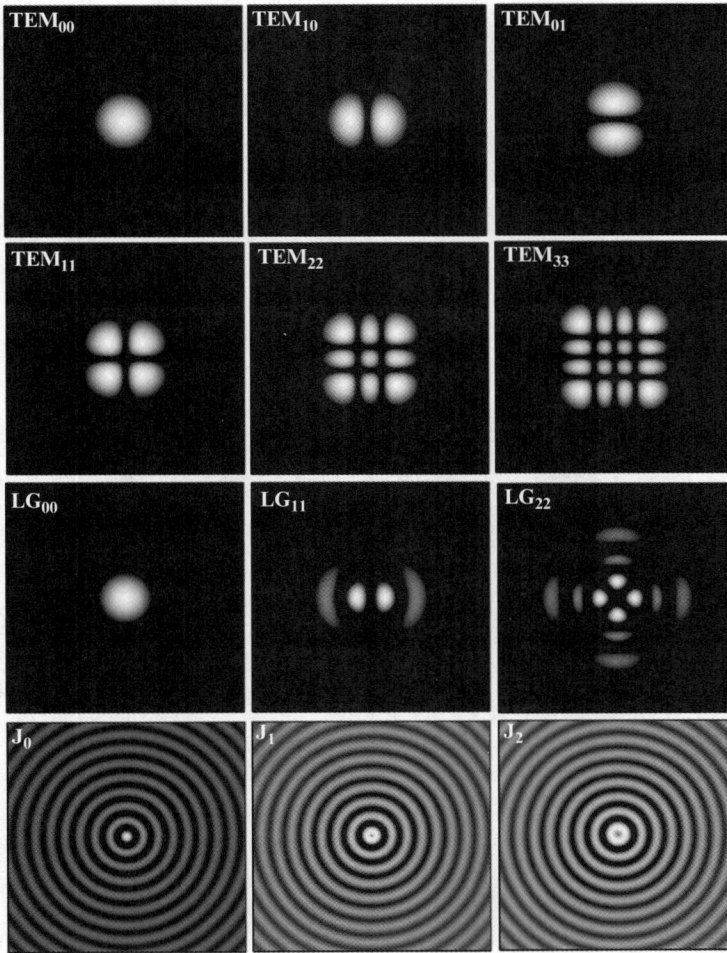

Figure 6. Intensity profiles for the Hermite-Gaussian (TEM$_{x,y}$), the Laguerre-Gaussian (LG$_{m,n}$), and the Bessel function (J$_m$) transverse electromagnetic modes.

of minima in each of two orthogonal directions to the direction of propagation. A (transverse electromagnetic) TEM$_{00}$ beam is the lowest order Hermite-Gaussian mode with a single high-intensity lobe and the smallest divergence at the "diffraction limit," while a TEM$_{33}$ mode looks like a tic-tac-toe board and diverges significantly faster. A number of TEM modes are illustrated in Figure 6. The majority of optical traps use a diffraction limited beam in order to maximize intensity at the focal point of a single trap; however, it is also possible to trap using higher order modes from a laser [47], which offers a number of advantages including increased axial trapping force, as well as the ability to trap at both multiple maxima and local field minima. High order modes can be generated by tilting one mirror of the laser cavity or by using a computer generated hologram; however, to a large extent holographic optical traps have superseded static designs because they can be dynamically tuned and are not limited to only Hermite-Gaussian modes.

2.4.2 Laguerre-Gaussian Beams

For a circularly symmetric aperture, the paraxial wave equation can also be solved in polar coordinates, yielding Laguerre-Guassian modes. The helical or corkscrew topology of

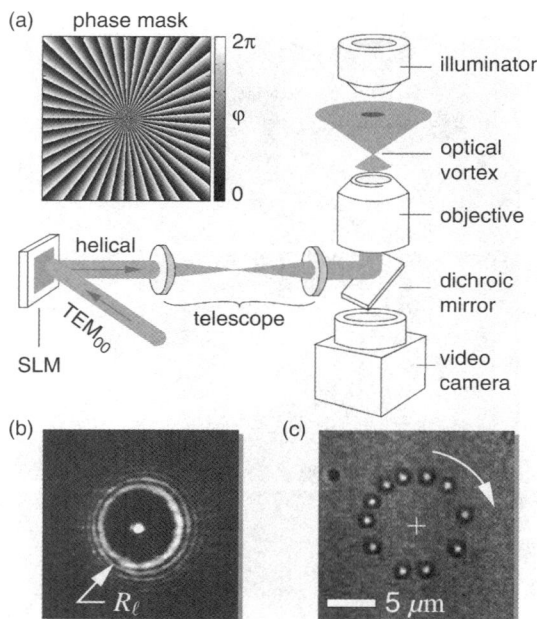

Figure 7. (A) Generation of a Laguerre-Gaussian mode from a Hermite-Gaussian TEM$_{00}$ beam using a computer generated phase mask on a spatial light modulator, (B) producing a radially symmetric intensity profile, (C) which can be used for the controlled rotation of trapped objects [Reprinted Figure with permission of Ref. 309 by the American Physical Society]. (*See Color Plates*)

the phase results in destructive interference along the optical axis, producing a ringed intensity structure, illustrated in Figure 7. For higher order modes the lack of axial light is helpful for trapping scattering, reflecting, or absorbing particles [48, 49], particles with a refractive index lower than the surrounding liquid [50], and large particles [51].

What distinguishes these optical beams is that they can impart orbital angular moment [52], generating torque on a trapped dielectric particle. This radially induced motion has led to these traps being referred to as "optical vortices" [53] and "optical spanners" [54]. Interfering with these beams, for example, with a plane wave to create a multiple armed vortex, can produce many new and novel intensity profiles and consequently trap structures. By superimposing two Laguerre-Gaussian modes that are phased so that they destructively interfere, an optical bottle trap can be formed, where the dark central region is surrounded completely by regions of higher intensity, providing three-dimensional structure [55]. Controlled phase changes of any of these interference patterns will result in rotation of the high intensity regions and objects trapped there.

Rotational control of optically trapped particles has however not been exploited to the same extent as with magnetically trapped particles, in part because it is difficult to impart a defined number of turns to the particle. In most cases a constant torque is applied while the particle is illuminated by the trapping light, making precise control of, for example, the supercoiling of DNA difficult. There are many areas, such as in mixing within microfluidic channels, where this would not be a limitation.

2.4.3 Bessel Beams

Diffraction limits the range over which an optical intensity or spatial light profile can be maintained. Therefore it was a surprise when, less than twenty years ago, it was noted that Bessel functions provide wavelength scale transverse features which do not change along the

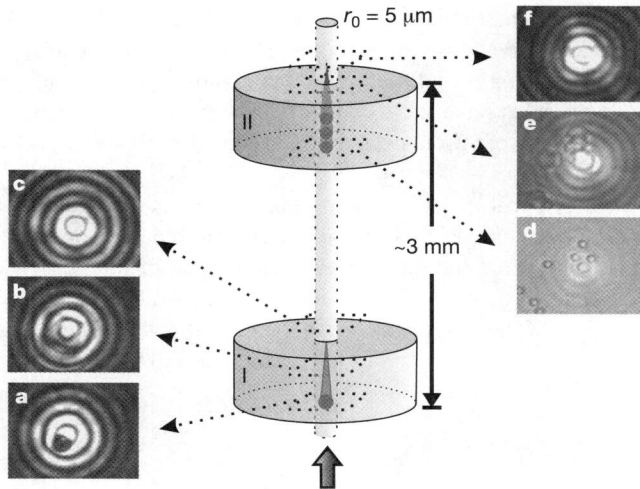

Figure 8. A diffractionless Bessel beam can be used to trap beads in two chambers separated by 3 mm [Reprinted from Ref. 58 by permission of Macmillan Publishers Ltd].

beam axis. Approximating with quasi-Bessel beams, this non-diffracting quality was used to optically manipulate particles in two separate chambers three millimeters apart in 2001 (Figure 8) [56]. Bessel beams do not form three-dimensional traps because there is no axial gradient, but they can trap both high and low refractive index particles simultaneously, as well as multiple particles along the optical axis, and do not require an objective [57]. The waveform is also to some extent self-healing and can be used to trap multiple particles along the axis of the beam, providing the ability to manipulate three-dimensional structures [58]. Again, the implications of using these beams for force spectroscopy of biological samples has not yet been fully explored, though potentially they open new regimes in which particles can be manipulated by light.

2.5 Types of Particles

Optical tweezers are capable of trapping particles composed of a wide range of materials, shapes, and sizes. The primary characteristics sought in a good particle have been low scatter and low absorption, with a surface which can be chemically modified. Traditionally micron-sized dielectric spheres have been trapped at the center of a diffraction limited beam; however, using higher order beams with a dark center, it is possible to trap reflecting and absorbing [49] particles as well as low index particles [50]. The size range of objects which can be trapped spans more than three orders of magnitude, from small metallic particles (5 nm) [26] to large (>10 μm) beads [51] and cells. Core-shell or variable index particles, while potentially difficult to make, may offer a number of advantages both from a trapping and a functional perspective.

The composition of particles has not been as widely exploited, primarily limited to dielectrics and metals [59]. Some metals, for example gold, have a plasmon resonance in the visible spectrum and attractive thiol chemistry which can be exploited. In addition, particles containing light and environmentally sensitive compounds offer the possibility of local reporting and delivery—for example, using controlled hyperthermia to melt a low-melting point polymer containing a drug. Optical tweezing of carbon nanotubes has been demonstrated [60], reinvigorating the possibility of self-assembly and the use of carbon chemistry.

Nanocrystalline particles, in particular those containing rare earth ions, could be used both for enhancing the dielectric contrast, providing bright, narrowband, continuous fluorescence, and for providing localized heating. In addition, the quantization of the band gap and nonlinear phenomena of small particles have yet to be fully exploited in optical traps—for example, using two-photon excitation of resonances and stimulated emission/depletion to determine particle location within a trap.

This wide range of particles offers the possibility of optical tweezers as a unique and versatile tool for the three-dimensional, non-contact manipulation of microscopic objects. The ability to dynamically tune trap properties and the ease of integration with other techniques has led to them being applied to a wide range of problems in force spectroscopy. Before considering some of these applications, we will first consider the closely related field of magnetic tweezers.

3 Magnetic Tweezers

3.1 Introduction

The use of magnetic particles for biophysical measurements dates back more than fifty years. In 1950 Crick and Hughes [1] used magnetic particles for a study of the viscoelastic properties of the cytoplasm, pioneering the field of magnetic twisting cytometry and cell rheology. Microrheology investigations have since been carried out on a wide range of biopolymer networks, cells, and tissue types to provide insight into cytoskeletal and extracellular organization and the dynamics of biochemical processes [61]. The use of magnetic particles to investigate the force spectroscopy of single biomolecules has been a natural extension of these techniques, offering an alternative to optical tweezers.

Commercially, magnetic particles have come to prominence as an effective, generic technique for separating and purifying target cells and biomolecules [62]. By labeling superparamagnetic particles with chemical or biological species that selectively bind a target analyte in a reversible reaction, repeated separation and concentration can be achieved using a simple, inexpensive magnet. Superparamagnetic particles are used because their response is several orders of magnitude higher than diamagnetic or paramagnetic particles, but they have no remnant field, so they can easily be dispersed. There are also fewer restrictions on their size, transparency, and composition because these properties are not critical to their operation. So this enables size to be optimized, whether high force or large surface area is required, while their response to a magnetic field can be tuned by composition without influencing the optical properties or biochemical tagging.

This flexibility in particle size and functionality has led to magnetic particles being used in vivo—for example, as MRI contrast enhancement agents with submillimeter voxel resolution [63,64], and in hyperthermia treatments [65] for targeted energy delivery. These characteristics have not been widely exploited to date in force spectroscopy studies; however, directed energy transfer and localized magnetic resonance imaging using magnetic cantilevers may prove useful in single-molecule studies. One of the other unique characteristics of magnetic beads is that they can exert an attractive or repulsive force on nearby objects. The decay in the magnetic field from the surface of a small particle is very localized, decreasing as a function of the particle's radius, making it ideal for studying nanometer phenomena in a small volume. The resulting high field gradient and hence force can also be used to self-assemble beads into lattices [66], providing sensitive detection and characterization of biomolecules as well as a platform for directed cell growth.

The force due to a magnetic field is analogous to the optical gradient force, with a potential energy due to the response of the particle to an external field. A magnetic field is distinct from an optical field, however, in that a maximum cannot be created in free space and cannot be manipulated other than by shaping the magnetic elements. This results in the ability to create a uniform force over a large area, enabling many systems to be probed at once, resulting in fast data collection [67]. The orientation of the field can also be manipulated independent of the gradient, permitting controlled rotation, attractive for studying phenomena such as supercoiling. However, these properties also make it difficult to form a stable trap without feedback, to manipulate single particles, and to change dynamically the trap characteristics.

"Magnetic tweezers" is an umbrella term for a number of different techniques, only a subset of which form a stable three-dimensional trap, while others rely on the fact that one end of the system under study is tethered to a surface. A stable trap can be formed either using feedback or by balancing the magnetic force with another force—for example, fluid flow. However, for the majority of experiments described in the second half of this chapter, a simple magnetic gradient is used to create a constant force experiment.

This section will mirror the discussion of optical tweezers, with an introduction to the origin of the magnetic force, followed by a discussion of magnetic tweezer design and the particles which can be manipulated. From a practical perspective, the decoupling of the imaging path from the force generation axis is advantageous for both tweezer design and particle selection. Magnetic fields also have the advantage that they are less likely to damage sensitive biological specimens. However, these advantages also place more reliance on the addition and placement of a responsive bead, through which a force can be applied and the results studied.

3.2 Theory of Magnetic Force

Magnetism plays an important role in many aspects of our daily life, from computer storage to electric motors. Qualitatively, materials can be described by their response to an applied field, as illustrated in Figure 9. If a material is placed in a magnetic field of strength H, the individual atomic magnetic moments in the material contribute to its overall magnetic induction response:

$$B = \mu_0(H + M) \tag{21}$$

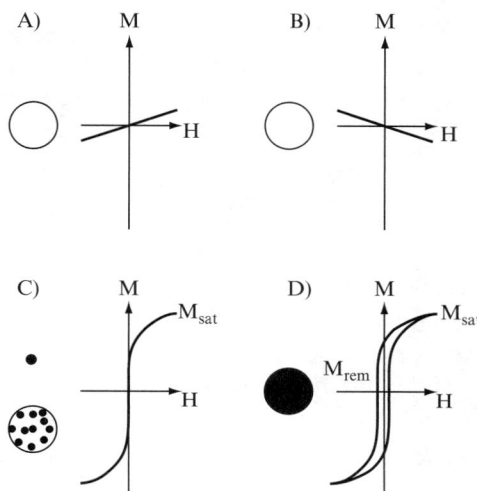

Figure 9. Magnetic response of different materials: (A) paramagnetic, (B) diamagnetic, (C) superparamagnetic, and (D) ferromagnetic.

where B is the magnetic induction, H is the external magnetic field strength, M is the magnetization and μ_0 is the permeability of free space. The magnetization of the material can be classified in terms of its magnetic susceptibility:

$$M = \chi_m H \tag{22}$$

where χ_m is the dimensionless magnetic susceptibility. In paramagnetic materials, χ_m is small and positive ($\chi_m = 10^{-6} - 10^{-1}$), locally strengthening the magnetic field by weakly aligning the magnetic dipoles present in the material with the external field. In contrast, the local field is weakened by the presence of a diamagnetic material, with the magnetic dipoles aligning anti-parallel to the external field ($\chi_m = -10^{-6} - -10^{-3}$). If the magnetic susceptibility is nonlinear and the dipoles can spontaneously align without an applied field, then the material is ferromagnetic ($\chi_m = 10^1 - 10^3$). Superparamagnetism occurs when a ferromagnetic particle is smaller than the domain size required for spontaneous alignment (typically 10–100 nm), because thermal fluctuations cause moment reversals on time scales much shorter than the experimental time frame, resulting in a net zero field in the absence of an external field. As the strength of an external magnetic field increases, the number of aligned dipoles within a superparamagnetic material increases up to some saturation level, giving a characteristic sigmoidal response without the hysteresis of ferromagnetism.

Analogous to optical tweezers, the potential of a single magnetic dipole m in a constant magnetic field B is given by:

$$U_m = -m.B \tag{23}$$

A gradient in the magnetic field will result in a potential gradient and a force on the magnetic dipole:

$$F_m = (m.\nabla)B \tag{24}$$

Thus, because the magnetic dipoles in a particle respond to an external field, the force on the particle is dependent on the number and type of dipoles present. The force on a superparamagnetic bead in a magnetic field gradient is:

$$F_m = M(B)\nabla B \tag{25}$$

where $M(B) = \dfrac{m}{V}$ is the external field dependent volumetric magnetization of the particle, V is the volume of the bead, and ∇B is the magnetic gradient. In a liquid medium, the susceptibility contrast of the particle relative to the water can be expressed as $\Delta\chi = \chi_{particle} - \chi_{liquid}$, which in turn gives $M = \Delta\chi H$, and the force can be expressed as:

$$F_m = \frac{V\Delta\chi}{\mu_0}(B.\nabla)B \tag{26}$$

If there are no currents or time-varying fields, then we can apply Maxwell's equation, $\nabla \times B = 0$ to the above expression:

$$F_m = V\Delta\chi\nabla\left(\frac{B^2}{2\mu_0}\right) = V\Delta\chi\nabla\left(\tfrac{1}{2}B.H\right) \tag{27}$$

where the force is related to the differential of the magnetostatic field energy density. From this equation a particle can be a high field seeker or a low field seeker, dependent on the liquid in which it is immersed, and the force is dependent on the volume magnetization. It is no surprise that equation 27 is comparable to equation 12, which described the gradient force for an electrostatic potential, because they are intertwined in electromagnetic theory.

In both cases the force is dependent on the particle volume (assuming homogeneity), the field gradient, and the contrast of the particle against the surrounding medium—the only differences really arising from the different ways in which the fields are generated.

A magnetic field falls as the inverse of the distance from the surface of a large bar magnet with corrections for the size of the magnet, the rate increasing as the magnet becomes more point-like. For the large magnet, the magnetic field gradient and the force decrease are the inverse square of the distance. One complexity of magnetic tweezers arises because the magnetization of a superparamagnetic particle is dependent on the applied field below saturation. This field and hence position dependence makes extrapolation for calibration and dynamic control of particles non-trivial for three-dimensional traps.

As with optical tweezers, magnetic particles are typically observed with a microscope, to maximize temporal, spatial, and spectral resolution. Alternatively, giant magnetoresistive sensors [68] (illustrated in Figure 10) and miniaturized Hall sensors [69] can also been used

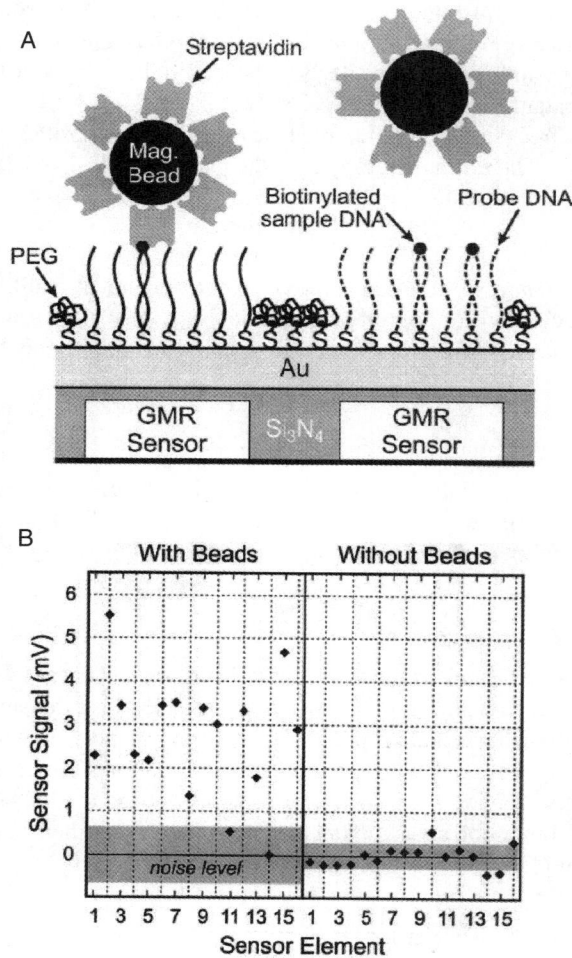

Figure 10. A) Schematic of the bead-array counter (BARC). Thiolated DNA probes specifically bind to complementary DNA strands on magnetic beads, the presence of which can be detected by a GMR sensor. The impact of non-specific binding can be minimized by applying a magnetic force before detection to remove excess beads. B) Binding events can be detected by electrical readout with the signal intensity dependent on the number and distance of the beads from the detector [reprinted from Ref. 310 by permission of Elsevier]

to detect the presence and orientation of beads above their surface, introducing the potential for low cost detection of multiple samples. Magnetic fields already have applications in micro-electromechanical systems and biomedical applications [70], for pumping, detection, filtering, and concentrating; and along with the diverse spectrum on functionalized superparamagnetic particles available, they will play an increasingly significant role in the concept of lab-on-a-chip analysis [71].

3.2.1 Torque

In addition to creating a force, a magnetic field can also be used to exert a torque on a particle. When the magnetic moment of a particle and the external field are not parallel, then a torque

$$\Gamma_m = M \times B = mB \sin(\vartheta) \tag{28}$$

is exerted on the particle, where θ is the angle between the external field and the magnetic moment. Torques of several hundred pN nm can be generated by rotating the magnetic field, either by physically rotating the magnet or using phased electromagnets. In contrast to the constant torque induced by circularly polarized light in optical tweezers, magnetic tweezers induce constant twist.

In practice, for free particles at low frequencies and low viscosities, the torque results in rotation of the particle. There is a drag torque on the particle [72]

$$\Gamma_{drag} = 8\pi\eta r^3 \omega \tag{29}$$

and the rotation frequency, ω, can be evaluated under dynamical equilibrium with rates of tens of hertz easily achieved for micron-sized particles.

For particles attached to a tether, the torsion stored in the tether is [73]

$$\Gamma_{tether} = \frac{C\Omega}{L} \tag{30}$$

where C is the torsional modulus, Ω is the twist angle, and L is the contour length of the tether. If the tether is twisted sufficiently far it will undergo a buckling transition [74] and plectonemes will be formed, with the transition characterized by:

$$\Gamma_b = \sqrt{2L_p k_B TF} \tag{31}$$

where L_p is the persistence length of the tether and F is the applied linear force.

The winding and unwinding of DNA has been explored extensively using magnetic tweezers [75], though the torsional properties of few other biopolymers have been characterized. This is surprising, because twist can induce conformational changes in the substrate to mimic protein activity. For example, structural proteins rely on torsion as part of their rigidity, and diverse rotational molecular motors create torque during their operation, providing a wealth of studies uniquely suited to magnetic tweezers.

3.3 Types of Magnetic Tweezers

There is greater flexibility in the construction of magnetic tweezers than in that of optical tweezers because the microscope objective is no longer required as part of the force transducer setup. Figure 11 illustrates some of the designs used in magnetic tweezer experiments. Instead, a magnetic element must be placed in close physical proximity to the beads

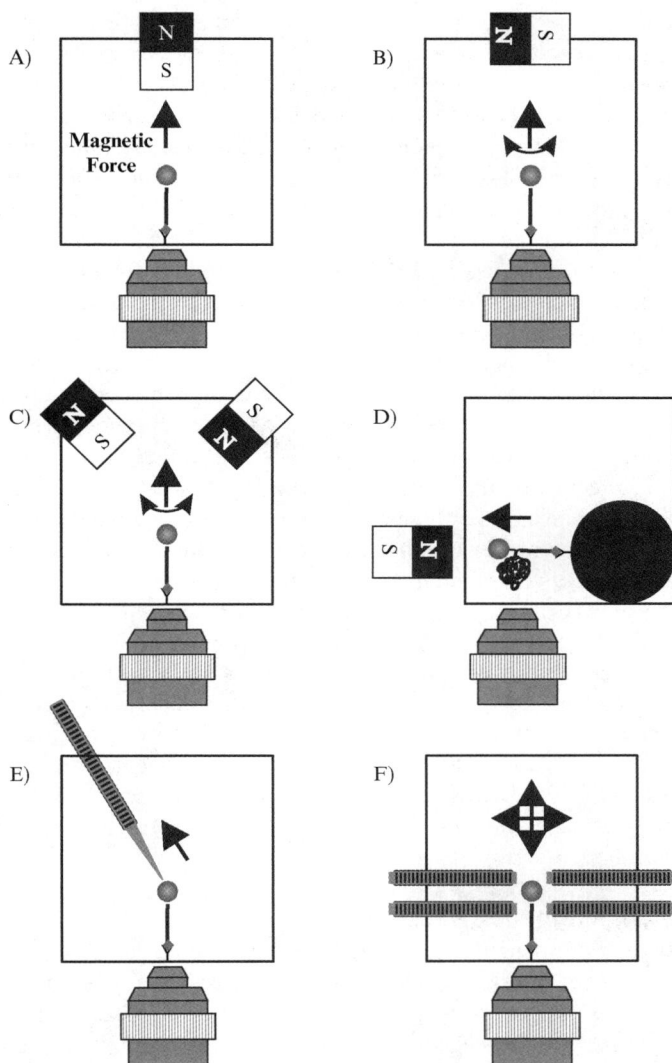

Figure 11. Magnetic tweezer configurations. (A) simple attractive force in parallel to optical axis; (B) magnetic poles perpendicular to optical axis; (C) multiple poles symmetric around optical axis providing rotation; (D) attractive force perpendicular to optical; (E) electromagnet needle; (F) multi-pole electromagnets with feedback providing a 3D magnetic trap.

in order to generate a force, introducing different constraints. The magnetic elements can be either permanent magnets, where their distance to the beads is adjusted to generate different forces; or electromagnets, where the current is adjusted to change to force. The size of the magnet also needs to be considered, with small magnets producing steeper gradients, but which need to be placed closer to the beads; and larger magnets producing shallower gradients, but which can be used at a greater distance and over a larger field. As a rule of thumb, to generate a high force on a superparamagnetic bead, the distance from a permanent magnet must be less than the width of the magnet for sufficient field gradient and magnitude.

A practical consideration is the orientation of the magnetic field and the gradient with respect to the optical axis of the microscope. If the gradient is perpendicular to the optical axis, then a single magnet can be used in the plane of the sample to give a simple attractive force, and displacements can be observed in one direction of the focal plane while the force is calibrated in the other. If the gradient is parallel to the optical axis, then displacements will result in defocusing of the beads, while motion in the focal plane will be due to Brownian motion. Rotation of the beads can also be more easily detected with the gradient parallel to the optical axis, though this requires a ring magnet to avoid blocking the optical illumination path. In addition, for rotation of a tethered particle, the field gradient should be parallel to the optical axis and rotation axis to generate torque or rotation at constant force in the image plane. For self-assembly of beads at a surface, the direction of the field should be perpendicular to the surface and usually parallel with the optical axis, while the generation of columns within the field of view is best done with the field direction perpendicular to the axis.

In order to create a three-dimensional magnetic trap, as opposed to an attractive gradient, at least six magnetic elements are required, as well as a feedback system to control the force toward each. Practically, this can only be achieved using electromagnets and optical feedback and by compromising the force available from a single magnet. Multipole arrangements do however have a number of other advantages, including rotation using phased currents [76], and integration with other approaches such as microfluidics.

As with optical tweezers, magnetic tweezers have been combined with a number of other techniques, most notably glass microneedles [77]. Surprisingly, the widespread use of overlapping optical and magnetic traps in cold atom research has yet to carry through to single-molecule studies. Superparamagnetic beads are also useful as a calibration tool—for example, to calibrate the stiffness of a microneedle or other cantilever by applying a known force.

A number of general papers on the construction and characterization of magnetic tweezers have been published [78, 79], discussing the advantages, disadvantages, and trade-offs involved. Here we will briefly consider the choice of magnet system and magnetic particles to highlight some of the possible approaches.

3.3.1 *Permanent Magnetic Tweezers*

A wide range of permanent magnets are available, which can be used to create strong magnetic field gradients, most notably those made from neodymium, iron, and boron (NIB, $Nd_2Fe_{14}B$). These rare earth magnets have the highest residual magnetic flux density (>12,000 G) of any permanent magnet and can be machined or sintered into any required shape. A single magnet can be used to create an attractive force of up to 200 pN with a resolution as low as 10 fN [80]. NIB magnets have the highest quality factor for a magnet, implying that a required magnetic flux can be obtained with a smaller volume of the material. The other permanent magnetic materials with attractive properties for magnetic tweezers are the alnico family of alloys of aluminum, nickel, and cobalt with iron. Some of these alloys have higher remnant fields than NIB magnets and a coercive field of less than 0.1 T, which may be beneficial for bi-stable switching.

The high surface flux densities of permanent magnets can be exploited by placing small, soft ferromagnetic structures in close proximity to enhance the local field. For example, adding a stainless steel needle to the surface of a NIB magnet can enhance the force to more than 1 nN in the region of the needle tip [81]. An alternative approach is to use the gradient created perpendicular and close to the corners of two opposing magnets in close proximity to each other. In this way, the field gradient again can be enhanced above 1 T/mm and used to levitate and trap paramagnetic particles against gravity or flow [82]. Emptying the trap can be done by increasing fluid flow to exceed the upward repulsive force of the magnets, illustrating one way in which the fixed force limitation of permanent magnets can be overcome.

3.3.2 Electromagnet Tweezers

Electromagnets have the benefit of precise control of the magnetic field and the ability to easily switch the magnitude and direction of the field as well as the number of poles energized. Permanent magnets need to be mechanically moved in order to exert temporal control of the field at a point in space, making rapid changes and vibration-free rotation difficult to achieve. However, heat dissipation, magnetic hysteresis, and lower field magnitudes have hampered the more widespread use of electromagnets.

The magnetic field gradient available at the end of an electromagnet's core is primarily determined by its composition and geometry. Ideally, the core is composed of a soft ferromagnetic material with no residual magnetic field and a large magnetic susceptibility, to maximize the magnetic flux density for a given solenoid current. If the field generated by the solenoid can be made sufficiently high, then iron is the ideal choice as the core material, with a saturation field of up to 2 T, though with a low relative permeability ($\mu_r = 7,000$), it is less attractive at low fields. Mu-metal and the other nickel iron alloys of the same family are better choices for a low field solenoid, with a higher permeability ($\mu_r = 300,000$) but a lower saturation field (0.77 T). High purity, single crystal, orientated iron would outperform both these materials, but would not be stable or practical to use in these experiments. Independent of the choice of material, the core piece is at risk of heating and expanding and may have a limited frequency response, which needs to be considered in any design.

The gradient can also be enhanced by shaping the core into a point with a small radius of curvature, most commonly a needle shape. Simple, single core devices used to create an attractive gradient can be mechanically engineered to create forces up to 1 nN [77], and more recently up to 50 nN [83], by electropolishing. However, it should be noted that there is a trade-off between the enhancement gained from shaping the core and the decrease in spatial extent of the desirable gradient. Working with a radius of curvature less than 100 μm requires consideration of access and thermal expansion, but can also be advantageous in being able to address a single bead in a field and manipulating it. Although not extensively employed to date, these devices have the potential to create local repetitive stain by low frequency amplitude modulation and to create localized heating around single beads by high frequency modulation.

Microfabricated magnetic traps, for integration with microfluidics, have also been constructed [84]. This initial demonstration showed that in an external magnetic field, 1 × 4 μm magnetic pads could trap individual beads from up to 5 μm away with a force of up to 100 pN.

Microfabricated designs allow complex material and topological constructions such as the use of field enhancing soft ferromagnetic structures to generate gradients up to 2 T/cm from micron-sized wires [85]. The flexibility in creating arbitrary trap shapes and distributions on a surface using many different material characteristics, for example in the form of electrically-driven magnetic spin valves capturing different magnetically labeled species, is potentially very powerful but has not been fully realized. The difficulty in implementing a strong, fully three-dimensional magnetic trap is an unresolved problem, and perhaps best addressed using microfabrication [86] if there is sufficient need.

Macroscopic three-dimensional magnetic traps, or true magnetic tweezers, have been built using Hall probes to measure the magnetic fields of an octopole arrangement of electromagnets symmetric around the trap center [87]. With a distance of 7.5 mm between opposing 5.5 mm diameter poles, the maximum force on a 2.8 μm diameter beads was 1 pN for a surface field of 0.05 T and could generate a torque of 0.01 pNm. Haber and Wirtz reported a two-coil electromagnet configuration with water cooling perpendicular to the optical axis, which produced a gradient of up to 1 T/cm for 1.5 cm separation and forces up to 10 pN [78]. More recently a hexapole arrangement above the sample plane produced a trap with a stiffness of

$0.1\,pN/\mu m$ and a maximum force of 5 pN, moving beads at up to $10\,\mu m/s$ [79]. To ensure linear feedback in these magnetic traps, a square root function needs to be applied to the error signal, but is complicated by the field-dependent magnetization of the bead below saturation, making displacement and force correlation difficult.

Dynamic electromagnetic traps have a number of advantages over optical tweezers, including a much larger trapping volume preventing loss of the trapped particle because of transients, and a greater range of motion, not limited by the working distance of a high numerical aperture lens; though these are gained at the expense of complexity in operation and lower trapping forces.

3.3.3 Other Magnetic Force Devices

A range of other devices which utilize a magnetic force have been developed to manipulate magnetic particles for biological applications. As mentioned at the beginning of this section, active microrheology, manipulating small magnetic particles to measure viscoelastic response [88], has been used extensively to characterize the viscoelastic properties of living fibroblast [89], macrophage [90], and endothelial [91] cells. A range of magnetic separators have also been developed, including the use of packed columns, multipole arrangements [92], and fluid flow fractionation [93]. Methods for preparing [94], tracking, and selecting [95] immunomagnetic particles and their targets have complemented these separators. Red blood cells, cancer cells [96], bacteria [97], and Golgi vesicles [98] have also been tracked and separated using magnetic particles. The negative susceptibility contrast, between diamagnetic particles and a liquid with a positive susceptibility, has also been exploited to form a weak-field–seeking trap [99]. A stable trap for particles can also be formed by exploiting gravity [100], using a ferrofluid [101], or using additional DC/AC electric fields [102], illustrating the wide range of possible magnetic trap setups.

3.4 Types of Magnetic Particles

The choice of particle composition and size is determined by the application and the force required. Primarily, superparamagnetic particles, and to a lesser extent ferrofluids, are used because of their large positive susceptibilities and zero residual fields. Large (> 500 nm) particles can easily be tracked with light microscopy or magnetic resonance imaging and used to create multiple linkages or for distributing the force over a larger area. Smaller particles (< 50 nm) can address single molecules in vivo and be fluorescently labeled or enhanced optically for imaging with optical or electron microscopy.

The force generated by a particle increases with iron content, but does not necessarily correspond directly to particle volume because of density constraints to keep the particles superparamagnetic and the randomness involved in their synthesis. The magnetic moment and frequency-dependent susceptibility of beads can be measured using a SQUID magnetometer/ susceptometer, and the magnetic field using a Hall probe. From these measurements a sample often shows significant variation because of the random number of iron oxide grains which are incoporated in each bead. Ferrofluids, composed of single iron oxide particles, have a more uniform distribution, though they are difficult to manipulate at the single-molecule level.

The success of functionalizing superparamagnetic beads for separation has stymied development of other composite magnetic particles. Although depositing a shell of magnetic material onto a polystyrene or quantum dot core would produce a more uniform sample of beads, most metals are too refractive or toxic to be presented on the surface. Pure cobalt nanoparticles have been made; and the pigments used in magnetic recording devices such as nanoCAP and barium ferrite have attractive properties, though it remains to be seen whether they can have a biological impact. Gold or silver coatings may be an alternative to embedding in a polymer matrix and can improve optical detection of small particles; however alloy formation may be a problem. Organic magnetic particles can be embedded in micelles and vesicles [103], and iron-sequestering proteins like ferritin have the potential to be engineered into molecular magnets [104]; though neither has been developed to the stage of being used by the wider community. The all-round flexibility of superparamagnetic polystyrene beads is difficult to beat, though changing applications may drive the marketplace in new directions utilizing these alternative technologies.

3.4.1 Superparamagnetic Particles

A large number of companies (e.g., Dynal Inc., Bangs Laboratories Inc., Duke Scientific, Miltenyi Biotech) offer 0.1–100 μm diameter superparamagnetic beads with a wide range of chemically or biologically labeled surfaces that have traditionally been used for magnetic separation. These beads are typically composed of ~10–20 nm iron oxide (both magnetite and maghemite) particles embedded in a porous polymer matrix. Typically, an outer polystyrene shell isolates the iron and provides a surface for adsorption of antibodies or other reactive groups.

The magnetic properties of the 2.8 μm diameter M280 beads from Dynal Inc. were analyzed [87] and found to have an average momentum per bead of 1.42×10^{-13} Am^2 at a saturation field of 1.4×10^4 A/m with no detection of hysteresis. The standard deviation in the magnetization of a sample of particles was 40%, because of a variation in the number of Fe_2O_3 grains within the polystyrene matrix of individual beads. The average grain diameter was estimated to be 15 nm, half of the single-domain size limit, with a bead containing ~200,000 grains with a mean separation of 50 nm. Without an applied force, the density of the beads is 1.3 g/cm^3, and the mass is 15 pg, resulting in a downward force of 34 fN in solution due to gravity. The magnetic susceptibility of the beads quoted by the manufacturer was 0.1, but was found to be closer to 1 at low field strengths. Thus a saturated M280 bead in a 1 T/cm field gradient would experience a force of approximately 14 pN, the threshold for pulling apart the two strands of dsDNA at room temperature. Using a larger 4.5 μm diameter bead, forces of up to 200 pN are possible with a single NIB magnet.

3.4.2 Ferrofluids

Ferrofluids, discovered and popularized by NASA in the 1960s, are typically composed of ~10 nm magnetite ferromagnetic particles in a nonpolar medium containing a surfactant. The liquid is superparamagnetic with low hysteresis, while the small size of the particles and the surfactant helps prevent agglomeration. Ferrofluids have found many diverse applications in damping loudspeaker cones and in forming liquid seals in hard discs, and are also starting to find applications in biotechnology, in particular as a drug carrier and hyperthermia target. Related to ferrofluids are magnetorheological fluids, which contain micron-sized superparamagnetic particles able to overcome Brownian fluctuations and form chains in the presence of a magnetic field, causing an increase in viscosity. This has led to their use in brakes, actuators, and valves.

Although both types of liquids and magnetostrictive solids have been used for generating force in the macroscopic [105] and microscopic [106] world, they have not been used in force spectroscopy at the nanoscale; in part because the particles are difficult to functionalize, to image with white light, and to keep dispersed in physiological buffers.

While magnetic tweezers are attractive because the force transducer is decoupled from the imaging path, and optical tweezers are attractive because it is significantly easier to dynamically trap and manipulate an object in three dimensions, both techniques are excellent for non-contact force spectroscopy in the range 0.1–200 pN, with a high degree of flexibility and adaptability. The similarity in these techniques often permits the same imaging and tracking techniques to be used for resolving displacements and forces. These common construction and characterization techniques will be discussed in the next section.

4 Construction and Characterization of an Optical/Magnetic Trap

4.1 Building a Non-Contact Micromanipulator

A basic optical or magnetic tweezers system can be constructed from scratch or added to a good research microscope for less than $10,000. A number of practical primers have been published on how optical [107, 108, 109, 110] and magnetic [78, 79] systems can be constructed, aligned, and characterized. The major limitations of basic systems are in the strength and stability of the trap produced, which places constraints on force calibration and analysis. With increased funds, a more powerful laser with improved intensity and pointing stability, an objective with a higher numerical aperture and chromatic correction in the infrared, an anti-vibration table, a piezoelectric stage, and better image capture and processing software could provide single-molecule tracking and manipulation with nanometer and millisecond accuracy.

In addition to the generic components of an optical or magnetic trap considered in this section, the properties of a trap are also determined by the size, shape, and composition of the particles to be trapped and the surrounding medium. For most single-molecule experiments the medium is generally a physiological buffer, such as phosphate buffer saline (PBS) or TRIS-EDTA buffer, containing 100 mM Na^+ and possibly other ions, with the optical properties of a dilute salt solution. The particles to be manipulated are mostly if not all polystyrene, around a micron in size and best imaged with a broadband light source.

4.1.1 Trap Configuration

The spatial arrangement of the imaging path and force transducer is determined by the requirements of the experiments to be carried out and the limitations of the equipment available. Inverted microscope designs are the most prevalent for measuring biological forces because gravity ensures surface wetness, and that particles will congregate on the surface, liquids can be easily exchanged, and the use of oil or water immersion lenses will be straightforward. In the case of optical traps, the imaging light path is most commonly co-linear with the laser light used to form the trap, maintaining constant illumination if the trap, stage, or objective is translated. For attractive magnetic tweezers, the magnetic field gradient is typically perpendicular to the optical axis.

Arrangements in other orientations offer different trade-offs. For example, side-imaging into a flow cell reduces background at the expense of increased difficulty in finding particles to trap and in maintaining objective immersion. An upright design can permit direct liquid immersion, reducing the impact of the coverslip, but can be limited by fast evaporation of the small liquid volume. While commercial microscopes can readily and quickly be modified to include optical or magnetic tweezers, home-built systems typically offer the highest degree of stability and flexibility.

4.1.2 Objective

For imaging there is a trade-off to be considered between the magnification of the objective, and the working distance. An increase in magnification, generally associated with an increase in numerical aperture, decreases the working distance and field of view, but also increases the axial trapping force for optical tweezers. The primary constraint is that the working distance must exceed the thickness of the coverslip used as a substrate to hold the liquid sample. A coverslip is typically 50–200 μm thick, so in order to have a sufficient gradient to prevent axial escape, the trap is typically limited to within 100 μm of a surface for conventional high magnification (> 60x) objectives (NA > 1.2) and a diffraction limited beam in an inverted design. Higher numerical apertures are possible, though they do not offer improvement for regular white light imaging, instead being used for total internal reflection studies. Long working distance, high magnification objectives are also available, though this involves sacrificing numerical aperture, which may not be critical if photon counts are not limited.

4.1.3 Light Source

The choice of laser for an optical trap determines and is determined by a number of considerations including the required force and trap stiffness, sample, imaging, and trap geometry, incorporation of dynamic elements, and the wavelengths of least perturbation to the objects to be trapped. Cost plays a significant role, with additional money often correlating with improved pointing and amplitude stability, mode quality, reduced thermal drift, and increased wavelength tunability. As a rule of thumb, the maximum trapping force increases 0.1 pN/mW of light in a diffraction limited beam delivered to the specimen plane with a high NA objective, and scales to a maximum incident optical power of ~2 W before damage becomes an issue [21].

Expanding the input laser beam to overfill the back aperture of the objective can help increase the ratio of axial to radial force at the expense of overall intensity [20]. For high power lasers, this aperturing can result in heating and thermal fluctuations of the objective mount.

The majority of modern optical traps employ a near-infrared laser with a wavelength of 700–1100 nm. This range generally corresponds to the lowest risk of optical damage for biological specimens, though even within this range there is a large variability; for example, there is a small absorption peak around 980 nm. The growth in optical data storage and communications has provided a wealth of inexpensive and high power visible and infrared lasers, though they often do not have a single transverse mode in both orthogonal directions and have a high degree of astigmatism. Diode-pumped solid state lasers, such as Nd:YAG operating at 1064 nm and its siblings, are the workhorse of many single-molecule experiments because of their high output powers and higher quality optical outputs. At the most expensive end of the laser spectrum, Ti:sapphire lasers provide the most versatile light source, with wavelength tunability and a range of modes of operation, including mode-locking, dual wavelength operation, and synchronization with other light sources, useful for spectroscopic techniques such as Raman scattering.

For particle imaging in both optical and magnetic traps, a high brightness source is desirable, whether in bright-field, fluorescence, or image contrast mode. Halogen and mercury lamps are staples of microscope design, though as discussed later, light emitting diodes and lasers operating below and above threshold offer advantages for high resolution position detection. For complementary confocal and two-photon imaging, a separate imaging path is required, complicating, though not necessarily limiting, trap design.

4.1.4 Piezoelectric Stage

One of the most significant advances in optical trap design in the past decade has been the development of three-dimensional piezoelectric stages with capacitive feedback. These stages provide subnanometer position control over distances up to two hundred microns and submicro-radian straightness during translation, for driving frequencies up to 1 kHz. The introduction of capacitive sensing and feedback has dramatically improved the positioning and reproducibility of piezoelectric stages, eliminating the traditional problems of hysteresis and drift. The accuracy and speed of these stages can be incorporated as part of a force-clamp loop for optical tweezers, to maintain centered quadrant photodiode tracking or adding dithering to improve resolution.

4.1.5 Imaging Systems

High resolution imaging systems are required to track submicron displacements and quantify piconewton forces. Modern, digital, silicon-based charge-coupled devices (CCD cameras), which can provide megapixel images at tens of hertz and are sensitive to light from

400 nm to 1000 nm, are the workhorses of video microscopy. Cooled, slow-scan CCDs push the quantum detection efficiency to more than 80%, allowing the spatial motion of single fluorphors to be observed [111]. Alternatively, by binning and defining regions of interest, the capture rate can increase to more than 100 Hz. Using CMOS chips, the partial frame rates can be pushed up to and beyond 100 kHz, though it is worth remembering that there is an increased requirement for light intensity and image storage working at these speeds. At the fastest end, streak cameras can be used to measure picosecond events. Flash photography, using the light source rather than the camera frame rate, can also be used to study dynamics, for example, fluorescence recovery after photobleaching (FRAP).

Temporal statistics of low brightness sources can be collected at up to gigahertz frequencies using sensitive photomultiplier tubes and photodiodes, though these systems do not provide spatial information. Linear photodiode arrays and quadrant photodetectors offer the compromise of limited spatial information while maintaining high sensitivity with fast readout rates.

Simple spectral information can be collected using RGB filters like those found in color CCD cameras. For more detailed spectral information, light from the sample can be diffracted using a grating and imaged onto a 1D array of photodiodes, or the grating can be tuned and a single photodiode used. Single shot spectra can be obtained from Fourier transform spectrometers.

Moving away from silicon and detection of visible light, and still at the exploratory stage, high-resolution and fast imaging can be pushed into the mid-infrared using InGaAs and InSb detector arrays and photodiodes. The development of optics in these regions still has to catch up with the quality available in the visible range; however, there are a wide range of phenomena which can be detected and manipulated with micron to millimeter wavelengths, which may drive development.

4.2 Particle Tracking

Force cannot be measured directly in either an optical or magnetic tweezer setup; instead it is inferred either from displacement measurements from the center of a three-dimensional trap, or from the extension of a calibrated elastic molecule tethered to a surface.

Measuring these displacements and extensions for accurate calibration requires nanometer resolution. Bright-field imaging of the tagged polystyrene particles is the most common method, tracking the center of mass using digital image processing of a CCD image. Sufficient light needs to be refracted or scattered from these particles to form an image, limiting their size to >100 nm. In this size range a functionalized bead can tag one or more of the desired, smaller biomolecules and can potentially constrain the biomolecule's activity through steric effects.

Tagging the biomolecules of interest with fluorescence markers, such as dyes, quantum dots, or fluorescent proteins can remove size limitations and permit fluorescence microscopy. However, each of these techniques has a disadvantage in terms of duration, consistency, or intensity of the fluorescence it produces; and none of them contribute towards exerting a force. Their dipole nature does however make them extremely versatile, providing sensitivity to their environment—for example, in ion concentration sensing or distance to other fluorphors through resonant energy transfer. The small size of these flurophors is also advantageous in intracellular and diffusion studies or in advanced microscopy techniques such as two-photon, confocal, or stimulated emission/depletion microscopy. The development of fluorescent proteins and their incorporation as gene reporters has provided a powerful, quantifiable technique for in vivo gene expression and will aid in understanding the effects of force on cells. To date there are no magnetic equivalents to the diversity of optical reporters and probes, though undoubtedly they will be developed.

For any optical tracking system, the exposure time should be sufficiently short compared to the correlated motion time $\left(\tau_0 = \dfrac{\gamma}{\kappa}\right)$ of the particle to prevent blurring, while the time between acquisitions should be longer than the correlation time to prevent oversampling and correlated measurements. As the spring constant decreases, corresponding to a weaker optical trap or longer tether, the desirable sampling rate can fall to a few hertz, limiting detection of fast phenomena. Both conditions can be fulfilled by video and photodiode imaging systems, and the merits of both are now considered.

4.2.1 Video Tracking Systems

Video-based tracking systems can make use of an existing CCD camera attached to a microscope port to give ~5 nm accuracy, at the video acquisition rate (< 120 Hz). For the highest accuracy, the pixel scaling needs to be calibrated against a length standard, the image needs to be digitally processed, and a centroid-finding algorithm needs to be used [112]. Typically these algorithms are implemented in offline image processing packages such as IDL and ImageJ and require large quantities of data storage or in real time using LabView or custom-written applications.

Calibration of both the camera and the stage can be carried out by imagining first a length standard and then translating a fixed bead with the stage in each of the three orthogonal directions. Axial calibration can be obtained by measuring the diffraction rings visible around a fixed particle on a surface as a function of distance above the surface, again leading to a resolution of approximately 10 nm.

Real time tracking of multiple objects in three dimensions using a single camera has been demonstrated with ~10 nm resolution [79] and is indispensable for long-term studies of complicated dynamics. While stiff traps cannot be optimally imaged by CCD cameras because their correlated motion time is < 10 ms, video tracking is suitable for tracking multiple particles, tracking events involving discontinuous or large displacements, offline analysis, absolute displacements, and as a reference for other detection schemes.

4.2.2 Position-Sensitive Detectors

Quadrant detectors have proved to be a convenient, sensitive, high speed method for tracking single particles. The basic idea is to use the magnification of the microscope to image a micron-sized symmetric particle onto a millimeter-sized four-quadrant photodetector. The symmetric image of the particle will create equal voltages in each of the four quadrants. If the particle moves down, then more light will fall on the upper quadrants, which can be observed by summing and subtracting the relevant quadrants. Similarly, movement can be detected in the other three directions of the focal plane. If the particle moves out of focus, this can also be detected by a change in the overall light level falling on all four quadrants. In this way, nanometer movements in the position can be observed.

Although the response of a quadrant photodiode is linear for small displacements, it quickly becomes nonlinear within the range of typical usage; and therefore for accurate measurements the voltage outputs of the quadrants need to be calibrated with known displacements using a piezoelectric stage. Typically this is done by raster scanning a fixed bead with the stage and averaging a number of runs or beads to create a calibration map for the detector, taking care to calibrate axial positions, in particular the offset of an optical trap from the focus.

Axial displacements are more challenging to measure with the same resolution, because microscopes are designed to image in two dimensions and calibration of focal shifts is non-trivial. A number of methods have been described for determining focal shifts, including the intensity of scattered light [113], two-photon fluorescence [114], and evanescent wave excitation of fluorescence [115]. Perhaps the simplest approaches, however, are to sum the intensities of the four quadrants [116], or use a linear photodiode array or CCD to analyze the diffraction rings [79].

The laser light used to form the trap can also be used to find the axial position of a trapped particle by one of two methods. Using polarization interferometry [117], displacement of the particle results in a phase delay in one of the arms of a Wollaston-based interferometer, though in practice this technique is limited to one dimension and small displacements. Alternatively, by looking at the interference between the forward and unscattered trapping laser light at a phase conjugate of the back focal plane, particle displacements can be tracked in both lateral directions [118]. Often it is more practical to implement a second laser for detection, separated in wavelength from the trapping laser, particularly in characterizing multiple or moving traps [119].

The major distinctions between video-based tracking and the use of quadrant detectors are that video tracking provides a measure of absolute displacement at low frequencies for multiple particles, while quadrant detectors are more sensitive and have a higher bandwidth for measuring relative displacements of a single particle in a stiff trap.

4.2.3 Limitations to Tracking Resolution

The sensitivity of all detection systems is ultimately limited by noise. Noise sources include light source fluctuations, mechanical vibrations, electronic noise, stray light, and thermal drift. Acoustic and mechanical noise can be minimized by working in a quiet laboratory with the microscope system mounted on a vibration-isolation table and enclosing the observation area to reduce air currents, acoustic noise, and stray light. Thermal drift is more problematic for long-term observation, and requires isolation of heat sources and stable air conditioning.

Light source fluctuations, both from power and pointing variations, can register as anomalous particle movements, but even with a continuous light level, shot noise will limit detection accuracy at short time scales and low light levels. For example, with arc-lamp illumination, photon shot noise can introduce an uncertainty of several nanometers for a one micron bead imaged on a quadrant detector with a bandwidth of 15 kHz [108].

Although significantly spatially and spectrally brighter, laser illumination of particles is hampered by coherent artifacts, such as speckle, and anomalous correlated Brownian motion. Therefore, for passive position tracking, a laser diode operating just below threshold is used to provide a high brightness, incoherent illumination source [120], with a high light level which can be more precisely filtered than an arc lamp, while avoiding the coherent artifacts of a laser operating above threshold.

Low light levels increase the susceptibility of the detection system to electronic dark noise and Johnson noise. With careful circuit design, suitable choice of bandwidth, and isolation, these noise sources can be minimized. Random noise can be filtered digitally using arbitrary cut-off frequency filtering, local averaging, wavelet-based filtering, and nonlinear filters to remove noise outside the frequency range of interest and excessive noise within the range of interest. What remains then as the major source of noise is Brownian motion in determining a particle's position in a well configured tweezers, which is both an advantage and a disadvantage.

4.2.4 Photo-Induced Damage

Often the desired light level and high signal-to-noise is sacrificed in order to minimize photo-induced damage. Biological specimens absorb the most weakly in the wavelength region 800–1000 nm, and many optical tweezers operate in this range to minimize photodamage and optocution [121]. However, even in this range, photodamage of cells [122, 123] and proteins [124] occurs within a few minutes and has been linked to modification of ion channels [125]. As a rule of thumb, cells directly illuminated with 1064 nm light will experience a temperature rise of ~1 °C /100 mW [126], which can be calibrated more accurately using a low melting temperature wax [108]. This limits the power of direct illumination to a few hundred milliwatts in the infrared [127] and less than 100 mW in the visible [5] before cell damage or death occurs.

Photodamage and heating can be minimized by working at a wavelength with the lowest absorption and scattering and by minimizing impurities and imperfections. An alternative approach is to use hollow beam traps and low field seeking particles; in this case the trapping region is dark, minimizing heating because inelastic scattering is minimized. As with almost any disadvantage, heating can also be employed productively, for example in locally melting DNA base-pairs, if required.

Particle tracking in itself is useful for understanding intracellular transport, such as the movement of vesicles and membrane diffusion, in terms of velocity and correlation; and group behavior, such as the swarming of E. coli and cellular taxis. For single-molecule force spectroscopy, the primary goal is to understand biological interactions in terms of their energy landscape, which can be measured by position and force. In optical and magnetic tweezers there is no direct readout of force and it must be inferred from motion. Thus particle tracking is central to both position and force measurement and is a critical part of any optical or magnetic tweezer setup.

4.3 Force Calibration

Force information can be extracted from a system using two methods. Either the force can be kept constant ("force clamp") and the extension measured, or a constant rate of extension applied ("position clamp") and the displacement measured. The first approach describes the basic operation of a simple attractive gradient magnetic tweezers, while the latter describes optical tweezers. Using feedback in either setup can reverse which variable is clamped.

In a position clamp experiment, the rate at which the trap is displaced provides control of the loading rate. As might be expected, loading rate can have a profound effect on the viscoelastic behavior of biomolecules. This has caused some concern over the range of loading rates which should be studied and has opened up a distinction between dynamic and static force spectroscopy [128], in particular for studies using atomic force microscopy. Static force spectroscopy experiments, using a constant force, tend to operate closer to equilibrium. The energy landscape of a complex can start to be established by measuring dissociation under different load conditions and extrapolating to zero load, and though this plot may not be linear, often the thermal dissociation rates for a bound system can be determined [129].

A special case of passive particle tracking exists when the force is always adjusted to be zero by centering the trap continuously to the particle's position. In contrast, negative feedback can be used to enhance trap stiffness and change the time constants involved in maintaining a trapped particle's position ("position" clamp), to the point of underdamping. The peak-to-peak thermally driven movements of a trapped bead are approximately 20–60 nm, but with negative PID feedback the position can be clamped to approximately 1 nm [130], with the required trap displacements characteristic of the forces acting on the bead.

A number of in-depth reviews for calibrating the force in optical [21, 118, 131] and magnetic [79] tweezers have been published. Currently theoretical analysis is limited to calibrating only spherical particles, though of course irregular particles can be trapped and tracked and the force estimated. There are three commonly used methods for measuring the force in both optical and magnetic tweezers which will now be briefly considered.

4.3.1 Equipartition Theorem Analysis

Analysis of the motion of particles underlies the field of microrheology. Laser tracking microrheology is an extension of this concept, using low intensity light to illuminate and measure particle motion without exerting a significant force.

Unbound particles in a medium and in thermal equilibrium with their surroundings will undergo Brownian motion, executing a random walk as they are bombarded by the molecules making up the medium, if they are unhindered. In an optical or magnetic trap, the bead will also undergo Brownian motion, but there is a restoring force created by the potential of the field. Thus, because we know the energy associated with Brownian motion and the motion of the particles can be tracked, the gradient of the potential can be estimated. The restoring force experienced by a trapped bead can be modeled in one dimension by Hooke's law:

$$F_x = -\kappa(x - x_0) \tag{32}$$

where κ is the spring constant, or trap stiffness. In a well-aligned optical tweezers system, the trap is radially symmetric, with the axial spring constant typically an order of magnitude lower than the radial constant. The stiffness of a three-dimensional magnetic trap is dependent on the PID of the active feedback loop, but is approximately the same order of magnitude.

Decoupling the three orthogonal directions so they can be considered independently and using the equipartition theorem, the time-averaged potential energy in the x-dimension associated with the trap is:

$$\langle U(x) \rangle = \tfrac{1}{2}\kappa \left\langle (x - x_0)^2 \right\rangle = \tfrac{1}{2} k_B T \tag{33}$$

where x_0 is the equilibrium position, k_B is Boltzmann's constant, and T is the temperature. Thus the random distribution of the particle's position within the parabolic trap is a Gaussian:

$$p(x) \; e^{-\frac{U(x)}{k_B T}} \tag{34}$$

and the trap stiffness can be found by dividing the thermal energy by the mean square of the displacement. Furthermore, the validity of the Hookean spring approximation can be tested by plotting the probability density function and comparing it to a Boltzmann distribution.

For optical traps the radial displacements are easier to calibrate than axial excursions; though with more complicated image analysis and comparison to immobilized particles, the same resolution can be achieved axially. By observing the bead position at video rate and using image analysis software, the trap stiffness can be calibrated in several seconds.

In the case of magnetic particles attached to a tether, the force can be found by considering the assembly as a pendulum. Using the equipartition theorem, the transverse fluctuations of the bead relative to a tether of length L are related to the force on the bead:

$$F_x = \frac{k_B T L}{\left\langle (x - x_0)^2 \right\rangle} \tag{35}$$

This calibration technique assumes there is no directed motion, for example due to gravity, underlying the Brownian fluctuations, or any inhibition due to a tether. Any issue will become apparent by deviations in the Gaussian distribution, and this can be used as a helpful diagnostic. Measurements can be made relatively quickly using this method, with typical errors in the range of 1–5%, although calibration in all three dimensions requires more effort.

4.3.2 Viscosity Calibration

A number of methods for calibrating the force on a particle use knowledge of the viscosity of the surrounding medium. The viscous drag force on small spheres in the laminar flow regime (Reynolds number for the beads is ~10–5) can be found from the Stokes formula:

$$F_{vd} = 6\pi\eta r_p v = \gamma v \tag{36}$$

where η is the fluid viscosity, γ is the drag coefficient, r_p is the bead radius, and v is the relative velocity of the fluid to the bead. Thus by measuring the terminal velocity of spheres of known size in a liquid with a known viscosity, the force on the free bead can be measured.

Beads can be dispersed in a range of liquids with known viscosities, from water (1 cps) to silicone fluids (> 12500 cps) and tracked using video microscopy to determine both their size and velocities. Increasing the liquid viscosity decreases the terminal velocity, keeping speeds within the range which can be measured in the field of view of the camera, typically < 0.1 mm/s, and allows higher forces to be measured. Near a surface or other solid object, the drag is modified, but can be estimated by Fraxen's law [21].

Equating the drag and magnetic forces for a magnetic particle far from a surface gives the velocity of the particle relative to the gradient of the magnetic field [132]:

$$v = \frac{\Delta \chi r_p^2}{9\mu_0 \eta} \nabla \left(B^2 \right) = \frac{\xi}{\mu_0} \nabla \left(B^2 \right) \qquad (37)$$

where ξ is the magnetophoretic mobility of the particle, and a measure of how easy it is to manipulate. It is worth noting that as the force on a particle increases as the volume, the viscous drag force increases only linearly with the radius (inertial drag is dependent on the cross-section), making magnetic separation with large beads extremely fast and efficient in low viscosity liquids like water.

Many variations of the viscous drag method exist: particles can be dragged through the liquid at a known rate by displacing a trap repetitively at a known velocity [20]; the liquid can be displaced past a stationary trap by moving the stage or flowing the surrounding medium; or for magnetic particles, their terminal velocity at varying distances from the magnet can be measured. When the measurement is carried out with low particle density, in an isodense liquid in a region far from a surface, the estimation of the force can be better than 5%.

One advantage of these methods is that a trap can be characterized in all three dimensions both by a static force and by rapid displacement right to the edges of the trapping volume, where the response is no longer linear. In addition, measurement of the escape force provides a crude measurement of the trap depth to complement knowledge of its shape. The disadvantages include the need for a calibrated translation stage, imaging system, and surrounding liquid, long calibration and analysis times, as well as uncertainty in the drag coefficient because of irregularly shaped particles or the presence of nearby surfaces. This last issue can however also be used to provide height calibration to within ± 50 nm [133].

4.3.3 Power Spectrum Analysis

The previous methods have required a calibrated position detector to determine trap stiffness; however, it is possible to also calculate the spring constant of a trap using relative position measurements and knowledge of the particle's viscous drag coefficient. The motion of a trapped particle of mass m, undergoing Brownian motion by a random, fluctuating force F_L, can be described by Langevin's stochastic differential equation:

$$\frac{d^2 x}{dt^2} + \frac{\gamma}{m} \frac{dx}{dt} + \frac{\kappa}{m} x = \frac{F_L}{m} \qquad (38)$$

where x is the displacement of the particle from its equilibrium position, $\gamma = 6\pi\eta r_p$ is the Stoke's viscous drag coefficient, and κ is the trap stiffness in the x-direction. Assuming the system is overdamped and the inertial term can be omitted (low Reynold's number approximation), the two-sided power spectrum of the particle is of the Lorentzian form:

$$|S(v)| = \frac{|F_L|}{\pi^2 \gamma \left(f_c^2 + v^2 \right)} \tag{39}$$

where f_c is the 3dB corner frequency:

$$f_c = \frac{\kappa}{2\pi\gamma} \tag{40}$$

If no distinction is made between positive and negative fluctuations, then the one-sided power spectrum or power spectral density (PSD) is:

$$PSD = 2|S(v)|^2 \tag{41}$$

A typical power spectrum of a trapped bead is illustrated in Figure 12. Physically, this is the power spectrum corresponding to a bead attached to a spring immersed in a viscous liquid and excited by a thermal white noise source. At frequencies below the corner frequency, the power spectrum is that of a Hookean spring of stiffness κ, whereas at higher frequencies the inverse square frequency dependence is characteristic of the viscosity.

Experimentally the spring constant is found from a Lorentzian fit of the displacement power spectrum measured by a quadrant photodiode. The bandwidth of the detector system should be tens of kilohertz, substantially above the corner frequency (~1 kHz for a 0.8 μm-diameter sphere in water) in order to resolve the rollover in the power spectrum and to avoid problems with phase lag.

The accuracy of the stiffness measurement, $\sqrt{1/f_c T_m}$, increases inversely with the number of samples. For stiff traps, the corner frequency and hence the sampling rate are in

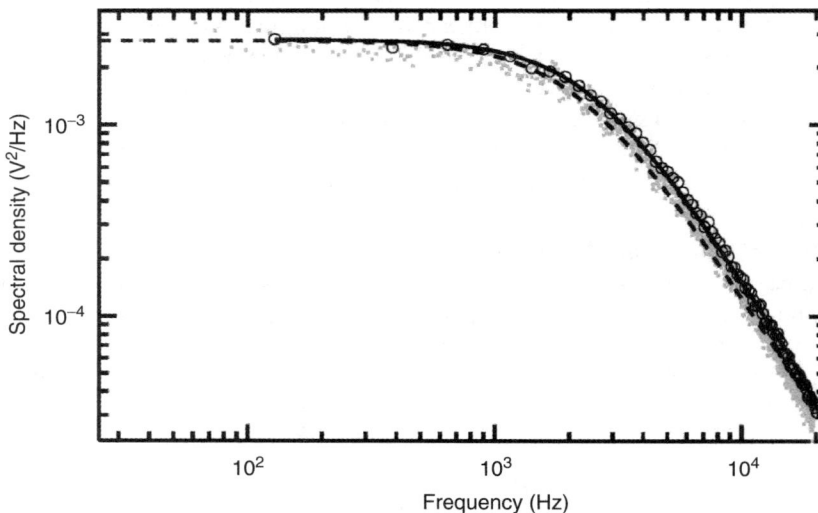

Figure 12. Power spectrum of an optically trapped 0.5 μm diameter polystyrene bead trapped 1.2 μm above a surface. The result is fitted with a corrected Lorentzian with a roll-off frequency of 2.43 kHz and a stiffness of 0.08 pN/nm [reprinted, with permission, from Ref. 110. Copyright 2004 American Institute of Physics].

the kilohertz range. To obtain good statistics and an accuracy of $<5\%$ only a few seconds of sampling is required. For weak traps ($<1\,\mathrm{pN/\mu m}$) the acquisition rate is much slower and requires many minutes before the error is reduced below 10%. The power spectrum is also a powerful diagnostic, with a non-Lorentzian spectrum or peaks in the spectrum characteristic of noise and misalignment.

Also, an analysis of the autocorrelation function, the inverse Fourier transform of the power spectral density, can be used to calculate the trap stiffness. The correlation function can be rapidly calculated as:

$$\langle x(t)x(t+\tau)\rangle = \frac{k_b T}{\kappa} e^{-\frac{\tau}{\tau_0}} \tag{42}$$

where τ_0 is the trap correlation period defined in equation 15. Again, analysis of the correlation function can be used to diagnose problems and to determine sampling and exposure periods in order to avoid oversampling and blurring.

These high sampling rate methods are well suited to characterizing the force on strongly trapped particles, which can be carried out within a few seconds to an arbitrary degree of accuracy. The disadvantages are that the viscosity needs to be well calibrated and only stiff three-dimensional traps can be calibrated in this way.

The construction and characterization of optical and magnetic tweezers is well documented and straightforward, forming part of many undergraduate laboratory classes. The complexity of the setup can be scaled to meet resources and requirements, but even with the most basic systems there is a degree of wonder in the ability to manipulate micron-sized particles with light alone and in measuring distances and forces many orders of magnitude below those of our normal day-to-day experiences. The application of these techniques to measure and study the influence of force in biological systems is perhaps the most fascinating aspect and has acted as a draw for people from all backgrounds.

5 Applications of Optical and Magnetic Tweezers in Force Spectroscopy

5.1 Introduction

Force spectroscopy is a powerful, dynamic technique for measuring the mechanical properties and interactions of individual molecules. Many cell functions have a positional dependence and mechanical component, whether through bond rearrangement or movement. Requirements for access in biological systems make probing and manipulating without a sub-micron, non-contact technique difficult, especially for intracellular work, the topic of the second half of this section and an ideal regime for force spectroscopy with optical and magnetic tweezers.

At the beginning of this chapter the forces and energies involved in determining the structure and interactions of biomolecules were discussed, with an emphasis on low energy processes. In the first half of this section, focusing on in vitro single-molecule experiments, the diversity of molecules which have been studied will help illustrate that most biological interactions and structures are determined by these relatively low energies.

For a physics audience it should be noted that the force spectroscopy of biomolecules is rarely simple to predict, analyze, or carry out. Biological adhesion is not typically mediated by a single, high-affinity bond, but by many weak ones, some of which may be nonspecific, adding diversity and also frustrating analysis, often requiring prescreening of results in order to make sure the system being observed is the one expected. Contamination and environmental changes are a constant battle in working with a single molecule for long periods of time, as well as coping with transients and drift. That said, nanometer and piconewton scale phenomena such as the stepwise motion of single kinesin motors [117] and the discrete rotation of ATP synthase [134] have been observed, highlighting the power and excitement associated with force spectroscopy carried out by optical and magnetic tweezers.

5.2   Inter- and Intra-Molecular Force Spectroscopy

Not surprisingly, experiments have focused on commonly available, simple systems, in particular biopolymers and proteins which can be purified easily and have defined properties which have been characterized by other techniques. Over the past decade, optical and magnetic tweezers have most often been applied to measuring the tensile strength of ligand/receptor interactions, measuring the elastic properties of biopolymers, following and manipulating the activity of enzymatic proteins and investigating molecular motors. Each of these areas will now be considered briefly.

5.2.1   Ligand/Receptor Interactions

Interactions between biomolecules are mediated by a complex array of intermolecular and intersurface forces, which form a number of specific, noncovalent bonds. For example, antibodies binding to cell surface antigens form a complex with specific lock-and-key interactions which can include steric, electrostatic, and van der Waals forces. These forces are governed by the chemical and physical structure of the molecules, both of which are influenced by their environment. The net interaction profile of all the specific and nonspecific forces provides a complex energy potential landscape. To infer the strength and nature of these interactions, equilibrium and kinetic measurements of bulk samples can be carried out by flow shearing, circular dichroism spectroscopy, surface plasmon resonance, calorimetry, titration, and a wealth of binding assays.

Ligands, referring to extracellular substances, bind to membrane-bound receptors, providing recognition of intracellular signaling molecules and metabolites. Ligands are small molecules, such as vitamins and hormones, which can be isolated relatively easily from the environment. Complete transmembrane receptor proteins can be difficult to isolate, though complete peripheral and anchored receptors or functional subunits can be isolated, providing source material for in vitro analysis. In the second half of this section, experiments to measure receptor binding in vivo will also be discussed.

The primary goals of single-molecule studies of ligand/receptor systems are to confirm the binding affinity or dissociation constant measured by ensemble methods, to determine whether loading rate significantly alters dissociation rate, and to determine the number of binding sites and presence of intermediate states.

The dissociation rate of a simple bond under an applied force F is related to the zero force rate by

$$k_{off}(F) = k_{off}(0)e^{\frac{Fx}{k_B T}}$$

(43)

with x interpreted to be the width of the potential bound state and $k_{off}(0)$ the thermal dissociation rate. Thus the dissociation rate is expected to increase exponentially with force, under the Bell model. Single-molecule measurements can be related to ensemble measurements by considering the unbinding rate as a function of applied force [135]

$$F = \frac{k_B T}{x} \ln \left(\frac{r_l}{k_{off}(0)\dfrac{k_B T}{x}} \right)$$

(44)

where r_l is the loading rate. From a plot of F vs. ln (r_l), the zero force rate and width of the potential can be found from the sloped and intercept, respectively. Multiple transition states can be modeled in a similar way by evaluating the association and dissociation rates between each state.

The majority of ligand/receptor measurements have been carried out with atomic force microscopes because of the large range in loading rate. For example, the interacting force between two adhesion proteoglucan carbohydrates, used as an adhesion system in aggregating cells, was found to be 40 ± 15 pN using an AFM [136]. Optical tweezer systems, with their higher force sensitivity, could potentially provide a more accurate description, though they do not have the same loading rate range.

5.2.1.1 Biotin/(Strept)avidin: There has been a lot of experimental effort, particularly with AFMs, in studying the interaction of the small ligand biotin, with a pair of closely related receptor proteins, avidin and streptavidin. Avidin, derived from egg white, and biotin, or vitamin H, are widely used as a model system because of their unusually high affinity (the thermal dissociation rate of biotin/strepavidin is 2.4×10^{-6} s^{-1} [137]) and ready availability, as well as extensive structural and thermodynamic data.

The binding free energy of biotin is 22 kcal/mol, corresponding to pulling with a constant force of ~160 pN over the binding pocket size of 1.0 nm. Experimental measurements of biotin/avidin using an AFM has suggested intermediate transition states because of an increase in rupture force from 5 pN to 170 pN for loading rates 0.05–60000 pN/s [138], while a later study measured forces from 120 to 300 pN for loading rates 100–5000 pN/s [139].

The dissociation of biotin/strepavidin has recently been explored using magnetic tweezers in a parallel system, measuring dissociation at a constant force (Figure 13) [140]. The dissociation rate of $k_{off} = 0.9 - 1.4 \times 10^{-4}$ s^{-1} at zero force was calculated by observing the rate at which beads functionalized with streptavidin unbound from a surface coated with BSA-biotin. This dissociation rate is nearly two orders of magnitude larger than that from an ensemble experiment and is attributed to unfavorable steric interactions.

5.2.1.2 Antibody/Antigen: Antibodies, proteins which are produced as part of the immune system, are designed to specifically recognize target molecules, antigens, and mediate the immune system response. It is therefore useful to look at this lock-and-key interaction using force spectroscopy to determine the operating range and constraints of this process.

Optical tweezers have been used as an immunoassay strategy for detecting bovine serum albumin (BSA) [141]. Latex beads 4.5 µm in diameter were covalently labeled with different concentrations of BSA and immobilized on a surface with anti-BSA antibodies. The force required for dissociation was found to be concentration dependent, with a minimum detectable BSA concentration of 1.5×10^{-15} mol/L. Optical tweezers have been used to passively observe the dissociation rate of a microsphere coated with IgE from a sphere coated with N-ε-2,4-dinitrophenyl-L-lysine, for different coating concentrations and with and without a tether [142]. The presence of the tether reduced the dissociation rate and caused the dissociation rate to decrease with concentration, while the absence of the tether increased the dissociation rate and caused it to increase with increasing concentration. These results show how important experimental design and the effect of surfaces are in determining what exactly is being measured.

There are still many ligand/receptor combinations to be probed at the single-molecule level, with optical tweezers providing an excellent way of measuring loading rate dependent dissociation for weakly bound complexes, while magnetic tweezers provide an excellent way of collecting statistics in parallel. Inhibition of binding needs to be more clearly understood to provide a truly complementary analysis to ensemble experiments.

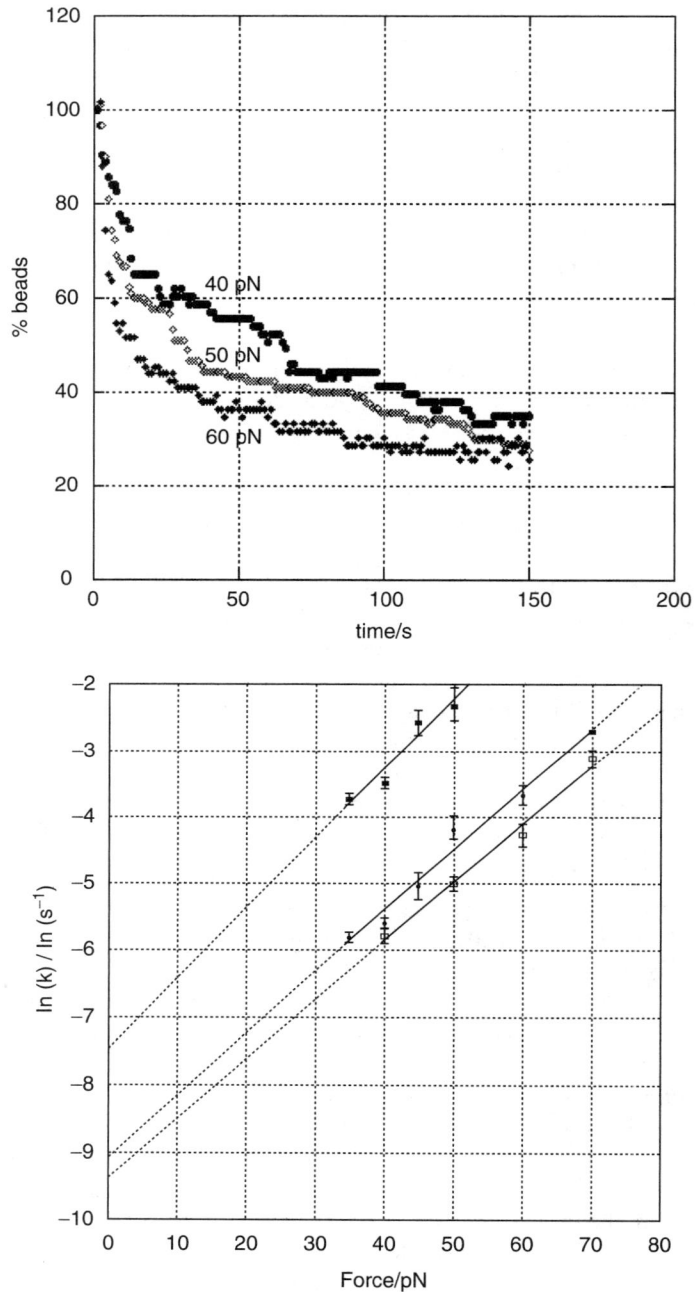

Figure 13. Dissociation of (strept)avidin-coated, 4.5 μm diameter magnetic beads from BSA-biotin immobilized on a surface. (A) With increasing constant force the dissociation rate of avidin coated beads decreases to the non-specifically bound baseline. (B) The rate constants for avidin (open white diamond) and streptavidin (solid black diamond) can be found by extrapolating the force dependent dissociation rate to zero force [reproduced with permission from Ref. 140. Copyright 2005 American Chemical Society].

5.2.2 Biopolymers

Long-chain biopolymers provide many central services in cellular life. From the information coding DNA and RNA, to functional proteins and structural microtubules, the chemical and physical properties of polymers are exploited in many ways. Structurally, biomolecules do not fold randomly but adopt configurations based on specific intramolecular interactions, and an understanding of the folding process will help in understanding the structure and function of unknown proteins and in fine tuning models of molecular dynamics.

The mechanical properties of polymers are typically characterized by the contour length (overall end-to-end length) and the persistence length, a measure of the correlation length. For stiff biopolymers, the persistence length is much longer than the monomer length, and they are modeled with some form of the worm-like chain model; while flexible biopolymers, where the persistence length is close to the monomer length, are best described by the freely-jointed chain model. There is good agreement for a quadratic relationship between the linear density of a polymer and the persistence length, as well as between the fourth power of the radius and the persistence length, though there are exceptions to both relations.

Many of the common biopolymers have already been examined using optical and magnetic tweezers as described below. There are a large number of less common biopolymers including mucin [143], amyloid fibrils [144] which have been characterized by other techniques but not yet by optical or magnetic tweezers.

5.2.2.1 DNA: The right-handed helical structure of deoxyribonucleic acid (DNA), discovered by Watson and Crick more than fifty years ago, is a classic example of intramolecular interactions determining structure. Hydrogen bonding between complementary base pairs in the two strands of the DNA and minimization of hydrophobicity provides a stable structure in a wide range of environmental conditions, while the electrostatic repulsion of the two negatively charged phosphate-sugar backbones determines width and twist. Steric hindrance between the sugar residues attached to the backbone favor a twisted stacking, and particular sequences and conditions can result in either an A or Z conformation rather than the usual B form. It is no coincidence that these unusual conformations have biological relevance, for example, in protecting against radiation damage and activating interferons. Despite being composed of a pseudo-random sequence of four different base monomers, DNA is remarkably uniform and stable, acting as a superior genetic storage carrier. Yet it is capable of being processed by protein enzymes powered by the hydrolysis of phosphate bonds, suggesting that many of its properties and interactions can be probed by optical and magnetic tweezers.

To complement ensemble measurements and passive imaging, single-molecule manipulation of DNA started in 1992 [145] using magnetic beads to measure the force-extension relation for double stranded DNA (dsDNA). A persistence length of 50–400 nm for B-DNA, dependent on salt concentration, was observed, with a value of around 50 nm now generally accepted for physiological buffers at room temperature. Four years later, this time using a dual beam optical tweezer to access higher force, the Bustamante group complemented their earlier work by reporting a structural transition in dsDNA for forces greater than 60 pN [34]. If the dsDNA is not free to rotate, a much less cooperative transition at a force of 110 pN is observed [146]. Despite significant work on this S-DNA phase, the exact nature of its structure remains a subject for debate [147], perhaps to be resolved using other techniques such as electron microscopy. In addition to S-DNA, twisting the dsDNA while stretching using magnetic beads has revealed three more forms of DNA: a Pauling-like form P-DNA, a Z-DNA form with a left-handed helical structure, and an over-twisted, supercoiled sc-P form [146], resulting in a phase diagram of DNA for stretching and twisting (Figure 14a).

Single-stranded DNA (ssDNA) has also been characterized by a force-extension curve, producing a response at high force similar to that of the overstretched S-DNA, a stretching modulus of 800 pN, and a persistence length of 0.75 nm [34] to 3 nm [148]. However, deviations from the normally well-fitting FJC model in the stretching of ssDNA at high and low salt concentrations have been observed [149]. More in-depth reviews on the theoretical and experimental work that has been carried out on the mechanical stretching of DNA can be found in a number of recently published reviews [150, 151, 152].

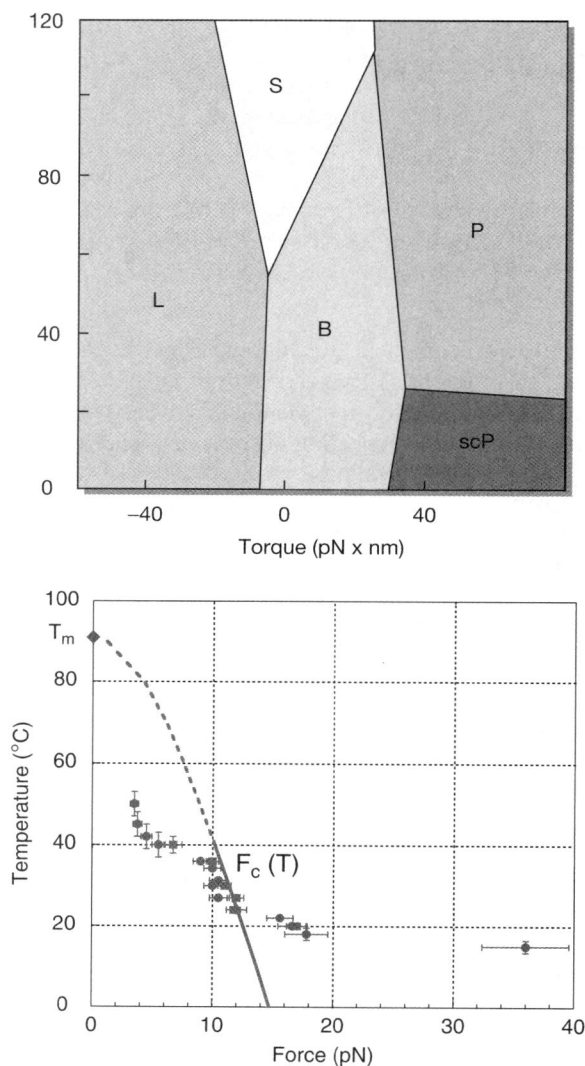

Figure 14. (A) Force-supercoiling phase diagram for dsDNA identifying five phases of double-stranded DNA and (B) Force-temperature phase diagram for unzipping dsDNA, where the blue line represents a simple thermodynamic model [(A) reprinted from Ref. 152 by permission of Macmillan Publishers Ltd; (B) reprinted Figure, with permission, from Ref. 160 by the American Physical Society]. (*See Color Plates*)

As mentioned earlier, the winding and unwinding of DNA has also been extensively explored. A typical extension versus supercoiling curve is shown in Figure 15. Supercoiling of DNA and the formation of plectonemes are important in DNA packing and processing because of the change in conformation which can be induced by or is a result of enzyme activity. Positive and negative supercoiling have been observed to have similar effects at low force and to agree with the expectations of twisting an elastic rod in a thermal bath [153], but the limits of elastic behavior are evident at higher torques (> +34 pN nm, > −9.6 pN nm) with different behavior and possibly different structure [76]. The torsional modulus of dsDNA has been measured to be C = 344 pN nm^2 with magnetic tweezers [154] and C = 410 pN nm^2 with optical tweezers [155], somewhat larger than the generally accepted value of 300 pN nm^2 from ensemble measurements. As well as studying the effect of supercoiling DNA, rotation can also be used as a diagnostic for determining the number of tethered molecules and also to braid multiple strands of DNA [156]. A good primer on the protocols involved in these single-molecule DNA experiments using magnetic beads has been published recently [157].

In addition to stretching and twisting DNA, the two strands of the double helix can be pulled apart by attaching one strand to a fixed object and the other to a bead which can be manipulated. Sequence dependent force induced unzipping, which is strongly temperature dependent [160], has been measured using both optical [158] and magnetic [159] tweezers. At room temperature partial unzipping starts with an applied force around 12 pN, in agreement with a consideration of the free energy difference between dsDNA and ssDNA. The phase diagram for dsDNA melting by force and temperature, illustrated in Figure 14b cannot, however, be modeled by simple thermodynamic considerations.

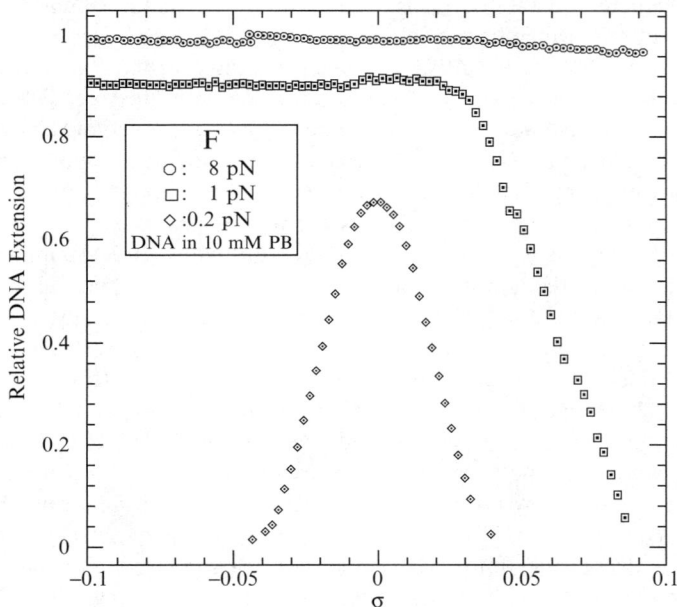

Figure 15. Relative extension of lambda phage DNA versus supercoiling for three different stretching forces, illustrating three different regimes. For forces < 0.4 pN (open diamonds) the response is symmetric, while at intermediate forces (> 0.5 pN, < 3 pN) the extension is insensitive to negative supercoiling, and finally at high force (> 3 pN) the extension is independent of both positive and negative supercoiling [reprinted from Ref. 76 by permission of the Biophysical Society].

The trapping and manipulation of DNA in chromosomes has been investigated [161], with the unfolding of individual nucelosomes observed [162]. Using an optical trap to pull on dsDNA wound round nucleosomes, dissociation of the nucleosomes was observed to increase linearly with the logarithm of the loading rate for forces above 15 pN, releasing 76–80 bp/nucleosome [163], in agreement with the earlier study [162]. There is still much to be discovered about how DNA is packed into a variety of structures, and whether it is purely a stochastic process or whether there is an underlying structure to the process, for which single-molecule approaches may provide useful insight.

One of the unresolved challenges is to tie this single-molecule work on the mechanical properties of DNA to physiologically relevant scenarios. For example, can single-molecule studies provide useful insight into the packing of DNA and the impact of particular sequences, such as those involved in creating unusual conformations, promoter regions, and repetitive tracts?

5.2.2.2 RNA: Over the last two decades, the diversity of the roles ribonucleic acid (RNA) plays within the life cycle has become apparent. In addition to the three major RNA types (mRNA, tRNA, rRNA), small, non-coding RNAs are now recognized as playing a major role in the cell cycle [164]; and a number of viruses transfer their genetic material in the form of double stranded (dsRNA) or single stranded (ssRNA) ribonucleic acid. RNA also plays an active role through RNA interference (RNAi) and micro-RNA (miRNA) in eukaryotes.

dsRNA is structurally similar to A-DNA, forming a right-handed helical structure with a hydrodynamic radius of 2.6 nm and a rise per base pair of 0.27 nm. However, it has a slightly increased rigidity and is less unstable than DNA, primarily due to the 2′ OH on RNA which intermediates under basic conditions and the higher energy of the backbone structure of RNA. The secondary structure of ssRNA is also rich, with many intramolecular bonds forming a range of hairpin motifs, which have a role in function.

The persistence length of dsRNA has been measured using a number of techniques, though without the same accuracy as dsDNA [165]. Recently the force extension curve and persistence length of dsRNA was measured (Figure 16) using magnetic tweezers and AFM to be 63 nm [166], less than the 70–75 nm from ensemble measurements and larger than dsDNA. The torsional properties of RNA have not yet been explored, though they are perhaps of less interest because RNA is not processed in the same way by enzymes.

Instead, work has focused on the folding and unfolding of RNA molecules which is critical to understanding the activation and incorporation of RNA but which is complicated by free energy landscapes. A RNA molecule can fold along many pathways and into many intermediate states before arriving at a stable, low energy state, which may or may not be unique. Single-molecule studies, using the techniques of FRET (fluorescence resonance energy transfer) [167] and FCS (fluorescence correlation spectroscopy) in particular, have helped provide significant thermodynamic and kinetic information [168]. Reversible unfolding of RNA, dependent on the concentration of multivalent ions, was first observed using a microneedle [169] with follow-up research using optical tweezers identifying kinetic barriers in RNA, which require forces of 10–30 pN to cross [170]. Work from the same group showing the reversible and irreversible unfolding of small RNA hairpins using optical tweezers [169] required a force of about 15 pN to pull apart the structures, similar to the force required to unzip dsDNA. The pulling and relaxing force curves for some of the hairpins were found to be indistinguishable, suggesting reversibility, while in other RNA sequences the metal binding pockets could be identified.

The diversity and importance of RNA structure will lead to more in-depth studies of RNA under different conditions, to understand phenomena like self-splicing. The interactions of RNA and proteins, RNA and DNA, and DNA and proteins are also important, and optical and magnetic tweezers are the ideal tools to map out their energy landscapes.

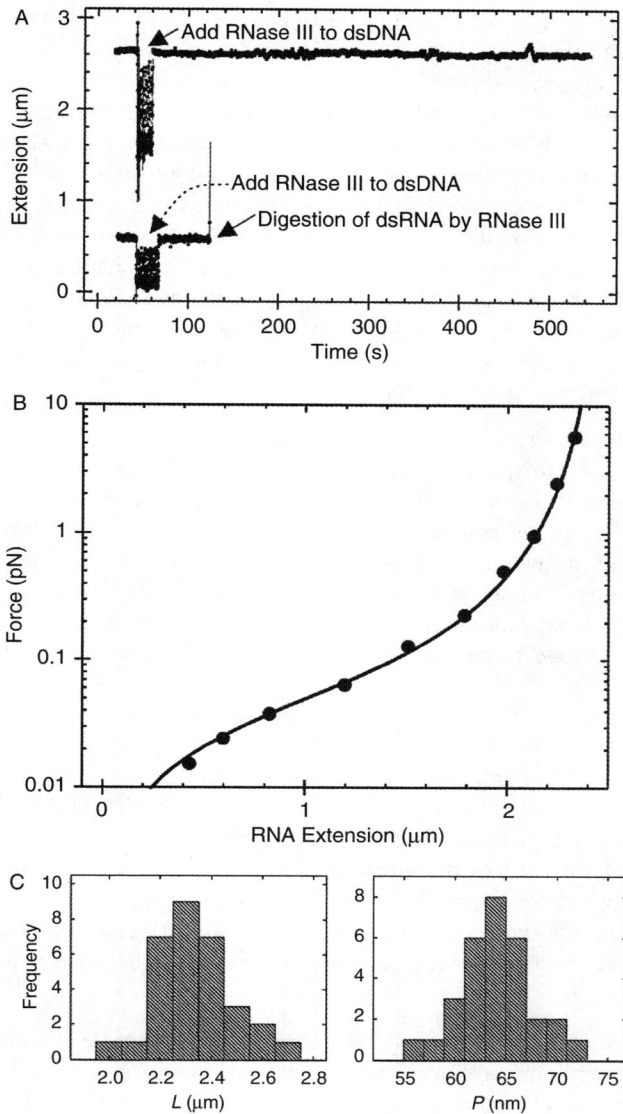

Figure 16. (A) Stretching of a 8.3 kb dsRNA molecule using magnetic tweezers. (B) The force extension curve for a single dsRNA polymer is well characterized by the worm-like chain model, while (C) for a sample of 31 strands the mean contour and persistence length were 2.33 μm and 64 nm, respectively [Reprinted from Ref. 166 by permission of the Biophysical Society].

5.2.2.3 Actin: Filamentous actin (F-actin), a polarized biopolymer of globular actin (G-actin) monomers, is one of the main components of a cell's cytoskeleton, and is involved in the construction of dynamic structures such as filopodia, microvilli, and lamellipodia. The filaments have a right-handed double helix structure with a diameter of 5–7 nm and are up to a few microns in length with a persistence length of ~15 μm [171], making them semiflexible. The stretching modulus of actin is ~400 pN, less than DNA, while the torsional modulus is 6000–7000 pN nm², an order of magnitude larger than DNA [172]. Interestingly, the breaking force of the actin-actin bonds decreased from 600 pN to 320 pN under twist, but was independent of twist direction.

The dynamics of the construction, destruction, and cross-linking of actin filaments has been explored extensively by the microrheology community. Microrheometry at the molecular level was pioneered by Ziemann and co-workers in 1994 [88], who measured the viscoelastic moduli of an F-actin network along a single axis using magnetic particles. The magnetically driven beads which were tracked optically provided a quantitative measurement of the frequency-dependent storage and loss moduli at low frequencies (0.1–5 Hz). These measurements were later extended to two dimensions using a magnetic micromanipulator which also controlled rotation [173]. For more active manipulation, the flexibility of a dual beam optical tweezer has been used to tie a knot with a diameter of less than 400 nm in an actin filament in a sucrose solution using 1 pN of force [174]. The knots often broke within 10 s, and this raises the question of how often actin filaments spontaneously break or are actively broken in cells, because only relatively small perpendicular forces are required (1.6 pN), within the range of myosin motors.

5.2.2.4 Microtubules: Protofilaments of polymerized tubulin subunits are complex structural components used for many purposes within eukaryotic cells. Thirteen protofilaments assemble into microtubule assemblies with a 25 nm outside diameter and a 15 nm hollow interior, used in the mitotic spindle during cell division, intracellular transport, and ciliary and flagella motion. In the case of cilia and flagella, the microtubules are arranged in nine pairs of microtubules forming the outer wall and a pair of microtubules in the center (9+2). Microtubules are polarized and have the interesting property of being dynamically unstable, associated with a change in conformation and a negative feedback role in translation and transcription.

The mechanical properties of microtubules have been investigated extensively using optical tweezers [175, 176], with the rigidity found to be dependent on contour length, with a persistence length of the order of 1–10 mm [177], making them stiff on the size scale of a cell. The rigidity of microtubules is 4–34 pN μm^2, and it has been suggested that this variation is due to growth rate and potential bending modes [178].

The manipulation of microtubules and their dynamic instability are of great interest for intracellular work, for example, to explore nuclear positioning [179]. Further work to include microtubule associated proteins and their influence on both the static and dynamic characteristics of microtubule networks may provide further insight into cellular functions and structure.

5.2.2.5 Intermediate Filaments: In addition there are a range of flexible, nonpolar filaments which provide support in metazoan creatures and a range of eukaryotic cell types. This superfamily of self-assembled filaments are generally cytoplasmic, typically 10–12 nm in diameter, and composed of a range of proteins (40–280 kDa) which include keratins, neurofilaments, and desmin. The intricate structure of these filaments is still debated, though the persistence length of vimentin has been measured with an AFM and is surprisingly short at 0.3–1 μm [180]. Magnetic microrheology has suggested that intermediate filaments play a role in cell stiffening, both naturally and under applied loads [181], though there is still much to discover about this large family of structural biopolymers.

5.2.2.6 Proteins: Proteins are single-stranded biopolymers composed of up to several thousand amino acid residues which fold into a three-dimensional structure which is the key to its function. As a rule of thumb, the persistence length of a protein chain is approximately 5 amino acids, though the presence of charged residues, secondary structures, and background buffer expands the range to 1–10 amino acids [182]. An understanding of protein folding has been a key driving force in single-molecule imaging and manipulation, in part because of the difficulty in determining their structure from sequence alone and in crystallizing many proteins. A number of in-depth reviews of single-molecule protein studies have been published [183], and only a brief selection of results on particular proteins are considered here.

The first studies of the mechanical unfolding of proteins were carried out on titin, the protein responsible for the passive elasticity of muscle, using AFM [184] and optical tweezers [185, 186], resulting in a force extension (Figure 17) corresponding to a persistence length of 1 nm in the folded state. In the later paper, forces above 400 pN were used to completely unfold the titin molecules. Follow-up work in the Bustamante group observed mechanical failure in titin under repeated stretching [187], another noteworthy result.

Spectrin, a structural protein found in human erythrocytes, consists of two pairs of chains with an end-to-end length of approximately 200 nm. The persistence length has been measured by holding an erythrocyte in a flow chamber with optical tweezers and observing the shape of the cytoskeleton [188]. The persistence length was found to be 5–20 nm depending on salt concentration, and the shear modulus approximately 8.3 μN/m [189].

Collagen, the primary protein in connective tissue, forms a variety of mat and rope structures. Type I collagen is a 1.5 nm diameter helix of three amino acids per turn, and three such strands form a larger molecule called tropocollagen. This has been stretched using optical tweezers [190], and the spring constant measured to be approximately 0.3 pN/nm, though the short length of the collagen (290 nm) and small sample size (n = 5) limited accuracy. Follow-up work, using a three-bead design and interferometric imaging, measured the force extension curve of collagen (Figure 18), and fitted it to a worm-like chain model with a contour length of 300 nm and a persistence length of 14 nm [191]. Type II collagen filaments have also been stretched by the same group and found to have a similar persistence length (11.2 nm) and contour length (296 nm), while the cartilage polysaccharide hyaluronan has a smaller persistence length of 4.5 nm [192]. The higher organization of tropocollagen into collagen fibrils, with a diameter of 10–300 nm has yet to be probed at the single-molecule level.

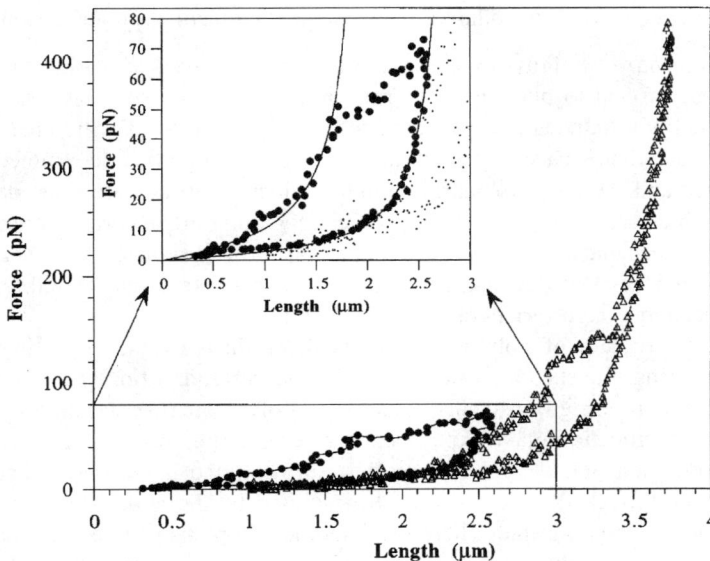

Figure 17. Stretching of a multi-domain titin protein using optical tweezers at a constant rate (65 nm/s). Both partial (closed circles) and complete (open triangles) denaturation were observed. Fitting the partial denaturing results with a worm-like chain model (solid lines) gave persistence lengths of 0.4 nm and 0.9 nm during stretch and release, respectively [reprinted from Ref. 186 by permission of Elsevier].

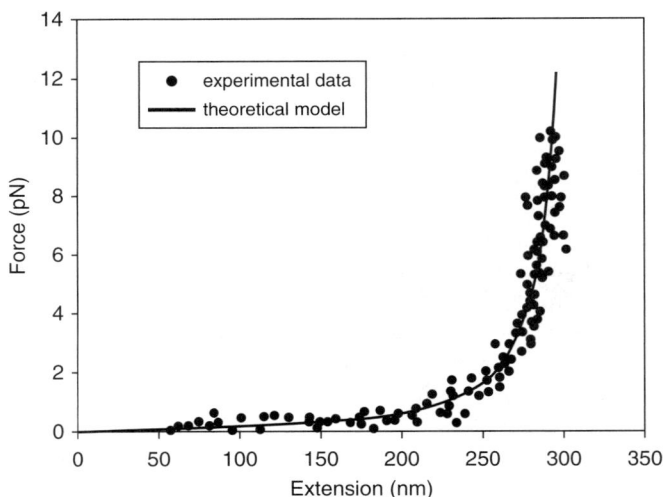

Figure 18. Force extension curve of collagen I stretched using optical tweezers. Fitting with a worm-like chain model gives a contour length of 320 nm and a persistence length of 14 nm [reprinted from Ref. 191 by permission of Elsevier].

As mentioned earlier, membrane proteins can be difficult to isolate and study, but the pigment protein bacteriorhodopsin, composed of seven transmembrane helices, has been unfolded by forces of 100–200 pN [193]. In addition, a number of extracellular proteins, including fibrin with a persistence length of 0.5 μm [194], myelin with a contour length of 53 nm and a persistence length of 0.5 nm [195], and the related bovine A1 protein with a persistence length of 1.6 nm [196] and fibronectin with three subdomains and a persistence length of 0.5 nm [197] have all been characterized. The diversity of proteins and their interactions provide a wealth of systems which can be studied in vitro, in addition to the complex problem of protein folding.

5.2.2.7 Polysaccharides: Polysaccharides, originally conceived as biologically inert molecules, have been found to play an active role in a number of important cellular functions, from storing energy to intracellular communication and adhesion. As with other biopolymers, these carbohydrate chains have structural and mechanical properties determined by the glycosidic linkages and the monosaccharide building blocks, but they are often decorated with other groups, which are likely to provide specific recognition in the correct conformation, and also modify the mechanical properties. Unlike proteins or nucleic acids, however, polysaccharides are not genetically encoded, and unlike microtubules or actin, the variability in monomer sequence can lead to a large variety of structures [198].

The basic structure of polysaccharides is determined by the conformation of each individual sugar ring, the stacking orientation of these rings, and finally the coiling induced by entropic or decoration groups. This variation in polysaccharide structure and challenges in isolation and purification make experimental determination more technically demanding. Structural information of bulk polysaccharides can be found from mass spectrometry and X-ray diffraction, though this final approach is limited by the ability to grow crystals and purify a sufficiently large sample. Therefore, there is a hope that force spectroscopy can be used to provide a level of characterization for unknown polysaccharide chains based on their force-extension curves.

The persistence lengths of a number of polysaccharide chains, including cellulose (5 nm), xyloglucan (6–8 nm), xanthan (120 nm), and schizophyllan polymers (180 nm) have been measured using ensemble methods [199], but not yet tackled at the single-molecule level.

The majority of bacterial polysaccharides have repeating motifs of 2–8 residues in helical structures, and can be decomposed into oligomers by mild acid hydrolysis, simplifying sequence analysis. At a more basic level, researchers have focused on monomers which have been purified and reconstituted as homopolymers and naturally occurring oligomers. The persistence length of homopolymers, such as dextran, sclerox, and alginate, has been observed to be 1–20 nm [200], similar to nucleic acids, which are also stacked sugar rings. Since 1997, when dextran chains were stretched by AFM and significant conformational changes observed for forces above 250 pN [201], a number of other polysaccharides have been examined. These experiments have focused on the high force (>1 nN) structural transitions and have not revealed significant information on changes at the low forces around equilibrium, likely to be relevant to cellular interactions. Complementary work with optical and magnetic tweezers would provide a more complete understanding of polysaccharide dynamics in this low force regime.

5.2.3 Force Spectroscopy of Enzymes

The catalytic behavior of some proteins was recognized one hundred and fifty years ago by Louis Pasteur, but not characterized until the work of J.B. Sumner sixty years later. Enzymes can accelerate the rate of a reaction by several orders of magnitude, though the complexity of their structure leads to strong dependence on temperature, pH, pressure, and ion concentrations. The field of proteomics is accelerating our knowledge of proteins, their structure and their function, revealing a dynamic and complicated environment which includes separate but necessary co-factors, activators, prosthetic groups, and inhibitors.

Enzymes perform six basic operations, but of greatest interest for single-molecule studies are the processive enzymes, which are generally associated with creating, repairing, or destroying biopolymers, in particular DNA, RNA, and proteins. These enzymes carry out many rounds of catalysis often by translocation along the substrate before dissociation, making their observation and manipulation more observable than trying to detect single events on a single molecule.

Single events, such as protein binding and dissociation, can be measured in a single-molecule experiment, for example by measuring the dissocation rate k_{off} of a labeled bead from a substrate on a surface as a function of force [140], or by examining the effect of force on the footprint of enzymes which bind specific DNA sequences. Force microscopy has started to provide an enhanced understanding of the dynamics involved in each step of an enzyme's pathway [202]. The majority of enzymes perform only a single catalytic cycle between association and dissociation, generally creating or breaking a covalent bond, making single-molecule observation less beneficial, so research has focused on some of the processive enzymes now described.

5.2.3.1 Exonuclease: Exonucleases are part of the DNA repair process, individually operating to delete mismatched regions, as well as being built into other complexes such as DNA polymerase to allow editing during DNA and RNA synthesis. They are enzymes which hydrolyze nucleotides one at a time from the end of a polymer chain, in contrast to endonucleases which cleave the backbone once at a particular site in the chain. Digestion of DNA or RNA can take place from either the 5′ end or the 3′ end with a blunt end or an overhang, depending on the flavor on enzyme used.

Of greatest interest have been the enzymes which only convert dsDNA to ssDNA, including lambda, T7, and Exo III exonucleases. The digestion of single dsDNA strand by lambda exonuclease was first studied using magnetic tweezers [203] and more recently by flow [204] and optical tweezers [205], observing a sequence and motif dependent rate. The rate of digestion observed for lambda exonuclease was approximately 6–12 nt/s punctuated by pauses, with a processivity of more than 15000 bp. In contrast, a broad distribution of rates centered around 100 nt/s has been reported for exonuclease I, also a highly processive DNA exonuclease [206]. At low forces (< 3 pN) the digestion rate was found to be insensitive to loading force, though it is not clear what happens at higher loading forces, due to failure of the tethering complex. The T7 gene 6 exonuclease which is non-processive and the ExoIII nuclease which is processive and conformationally sensitive have yet to be extensively examined at the single-molecule level.

5.2.3.2   RNA Polymerase: Central to the current dogma of biology, the enzyme RNA polymerase transcribes DNA into a mRNA template which is subsequently translated into proteins by the ribosome. The one bacterial and three eukaryotic polymerases are relatively large protein complexes which bind to specific DNA promoter regions and processively transcribe the DNA template using NTP hydrolysis, until termination is initiated by a number of mechanisms.

The force generated by RNA polymerase during transcription was first measured ten years ago using optical tweezers [207]. As with many of these enzymatic experiments, the polymerase was immobilized to a surface, while the biopolymer substrate was attached to a bead held in a trap. The stall force has been measured for a variety of RNAP to be 14 pN to 34 pN [208], with an average velocity of 7–16 bp/s with no applied force. Interestingly, the enzyme utilizes only a small fraction of the energy liberated from ATP hydrolysis for mechanical work [209].

Translocation of the RNAP is not a rate-limiting step because an opposing force up to 10–20 pN does not slow the rate; but ≥ 30 pN of opposing force triggers a conformational change in RNAP that allows backtracking over 5–10 bp [210]. More recently, the binding, elongation, and escape of RNAP has been observed, with a $k_{off} = 2.9\,s^{-1}$ and a processivity of $k_{pol} = 43$ nt/s [211].

There is much concerning the stability and processivity of RNAP that is poorly understood. Beyond working towards a more complete understanding of the molecular mechanisms underlying transcription [212], a more complete analysis of conditions for the initial ~8 bp hybrid formation before formation of a stable elongation complex in bacteria would distinguish between a number of competing models, and the prediction of pauses and backtracking [213] would assist in understanding transcription fidelity. The more complex eukaryotic polymerases pose as many challenging questions, among which are the interactions between the transcription complex and nucleosomes, and whether propagation causes torsional tension and the activation of DNA promoter sites.

5.2.3.3   Helicases: The stable double-stranded nature of DNA requires that for many processes to occur, unwinding needs to be amongst the first steps. There are tens of enzymes in the helicase family that carry out unwinding of DNA or RNA as part of recombination, repair, replication, and splicing. They are flexible in polarity and substrate, and processive, consuming ATP as they translocate along the strands. Recent structural data on these enzymes has helped elucidate the nature of both the dimeric and hexameric subfamilies, though many questions remain on how hydrolysis couples to unwinding and translocation and whether a rolling model or inchworm model is appropriate [214]. Mutations in the RecQ helicase have been associated with Bloom, Werner, and Rothmund-Thompson's syndromes, though how these affect the different roles of RecQ is not clear.

A number of helicase enzymes have been examined using single-molecule force spectroscopy, observing a variety of results. RecBCD, a processive, DNA motor enzyme with both helicase and nuclease activities, can unwind DNA at rates of 500 bp/s, has a processivity of more than 40000 bp [215], and has been characterized using optical tweezers (Figure 19) [216]. Stop/start motion and backtracking were observed, as well as a substantial variation in unwinding rates, with the enzyme migrating at a rate of 21 bp/s for extended periods, substantially different from previous results.

c

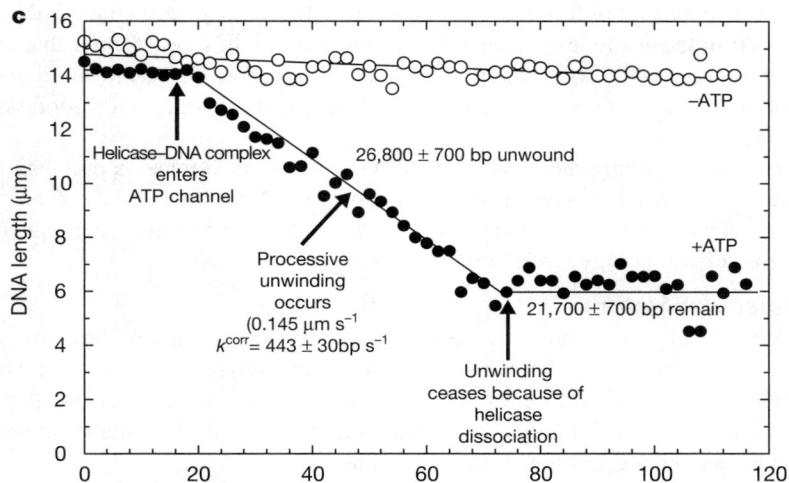

Figure 19. Unwinding of dsDNA by a RecBCD enzyme tethered to a bead held in an optical trap in the flow cell. The dsDNA, stretched by the flow, was fluorescently tagged to determine its length [reprinted from Ref. 215 by permission of Macmillan Publishers Ltd].

Another member of the family, the RuvAB helicase complex, which mediates the Holliday junction migration by denaturing and reannealing, has been studied recently with magnetic tweezers [217, 218]. The branch migration rate was observed to be 43–98 bp/s with pauses of inactivity and a processivity of 7000 bp. Positive and negative loading forces had minimal impact, up to the maximum applied force of 23 pN.

UvrD, a DNA repair enzyme with helicase activity, has also been studied using magnetic tweezers [219]. The number of bases unwound per catalytic cycle was estimated to be 6 bp, with a slow unzipping and rezipping rate of approximately 250 bp/s, and fast rehybridization events of 2300 bp/s with complicated burst patterns. Results also suggest that the enzyme unwinds as it moves 3′ to 5′ on one strand and allows hybridization as it switches across to the other strand, as opposed to dissociating. The purpose of strand switching is not clear, or whether stretching the DNA substrate has a significant effect on the rate and efficiency of the enzyme, an important consideration in any of these enzymatic studies.

5.2.3.4 Topoisomerase and Gyrase: In the course of unwinding DNA for processing, additional coiling can cause the formation of secondary structures which could potentially interfere with processing. Topoisomerases guard against excessive positive and negative supercoiling of the DNA in vivo. The subfamilies are distinguished by whether the cleavage used to relax the supercoiling is single-stranded or doubled-stranded [220]. The importance of these enzymes is exemplified by the role reverse gyrase plays in stabilizing the DNA structure of extremophiles and by the fact that many antibiotic and chemotherapy drugs interfere with topoisomerase function. An in-depth article describing the single-molecule research carried out on topoisomerase using optical and magnetic tweezers has recently been published [221] and highlights are only briefly mentioned here.

Using magnetic tweezers, the supercoiling of DNA [76], and the operation of IA [222] IB [223], II [224] and IV [225] topoisomerases involved in the formation or relaxing of supercoils have been examined. Each enzyme has been found to have its own operating characteristics, preference for relaxing left- or right-handed supercoiling, and rates. For example, the rate of supercoil relaxation of eurkaryotic type II topoisomerase decays exponentially from the maximum of ~3 cycles/s with applied force, whereas prokaryotic type IV has a rate

~2.3 cycles/s independent of force up to 2 pN, but rapidly drops to zero above 2 pN [221]. The rates measured in these single-molecule experiments are eight times greater than equivalent ensemble measurements [226], though it should be noted that single-molecule experiments focus on the movement on the enzyme once it is bound and do not provide good statistics on the unbound activity.

As with many single-molecule experiments, work on topoisomerase has started to answer questions about the physiological functions and limitations of these enzymes, but in the process have raised as many questions, such as those concerning their specificity and sensitivity to global topology while acting at a local level.

5.2.4 Molecular Motors

A wide range of motor proteins exist in nature, with a diversity of structures and functions. From the linear cytoskeleton motors, kinesin and myosin, to the rotary membrane motors, ATP synthase and bacterial flagella, motors provide active transport, defeating the limitations of diffusion, and allowing the construction of macromolecular structures such as muscle and axons many times longer than a normal cell.

The majority of motors bind, hydrolyse, and unbind ATP or GTP to induce conformational changes, providing motion through a process of detachment, movement, and rebinding. Some motors, such as ATP synthase, can run the process in reverse, generating conformation changes to store chemical energy. In their operation, motors trade off speed for processivity, step size for load size, and overall size with all other factors. Each ATP hydrolysis supplies approximately $\sim 20 k_B T$, or 80–120 pN nm, which places an upper limit on the step size and torques possible in a single hydrolysis cycle. The maximum catalytic rate is determined by availability and diffusion ($\sim 10 \mu m^2/s$) of ATP and by the hydrolysis cycle time (~ 1–100 ms).

Beyond the motors considered here, a number of other motors have been examined at the single-molecule level, most notably the portal motor of the phi-29 virus which has been investigated using dual optical tweezers [227]. The motor packs the viral DNA into the capsid with such efficiency that the pressure inside is about 60 atmospheres, working against forces of up to 57 pN, slightly higher than the estimated 50 pN repulsive force from the fully packed 6.6 μm long viral genome. The step size per ATP cycle was found to be less than 5 bp with a rate of packing fluctuating up to a maximum of 20 bp/s and pauses of variable duration as well as occasional slippages.

Our knowledge of molecular motors, as with most proteins, is inversely proportional to their molecular weight, with kinesin being the most closely studied. We will now consider some of the more common motors.

5.2.4.1 ATPase: The ATP synthase enzyme is one of the most conserved enzymes in eukaryotes, acting as the source for the ubiquitous energy currency adenosine triphosphate (ATP). Generally, ATP synthase is a membrane rotational motor which catalyses the synthesis of ATP from ADP, driven by a flux of protons due to the electrochemical potential across the membrane, and operates with an efficiency close to 100%. The motor is fully reversible and stoichiometric measurements had suggested, assuming integer values, that three protons were required to synthesize one ATP in mitochondria, while more surprisingly four protons were required in chloroplasts.

One of the simplest forms of ATP synthase is that of E. coli, the soluble portion of which, F_1 ATP-ase, can be dissociated from the hydrophobic F_0 proton channel by relatively mild salt treatment. Dramatic support of the rotational model came in 1997 from Masasuke Yosida's group in Japan [134]. They tethered the β subunit of F_1 ATP-ase to a glass surface and attached a fluorescently-labelled actin filament to the central γ subunit, and observed counterclockwise rotation only under conditions of ATP hydrolysis. In the same year, an alternative approach using a small chromophore on the γ subunit instead of the long actin filament, observed the polarization anisotropy on activation, again consistent with the three-step rotary model [228]. This work was followed by Yasuda et al. [229], who observed discrete 120° rotations under low ATP concentrations, as illustrated in Figure 20, and judged the distribution of the dwell times to be concentration dependent, with a torque of up to 80 pN nm generated on ATP hydrolysis. The reversibility of the motor was demonstrated more recently [230] using a magnetic field to rotate a magnetic particle attached to the stalk to generate ATP.

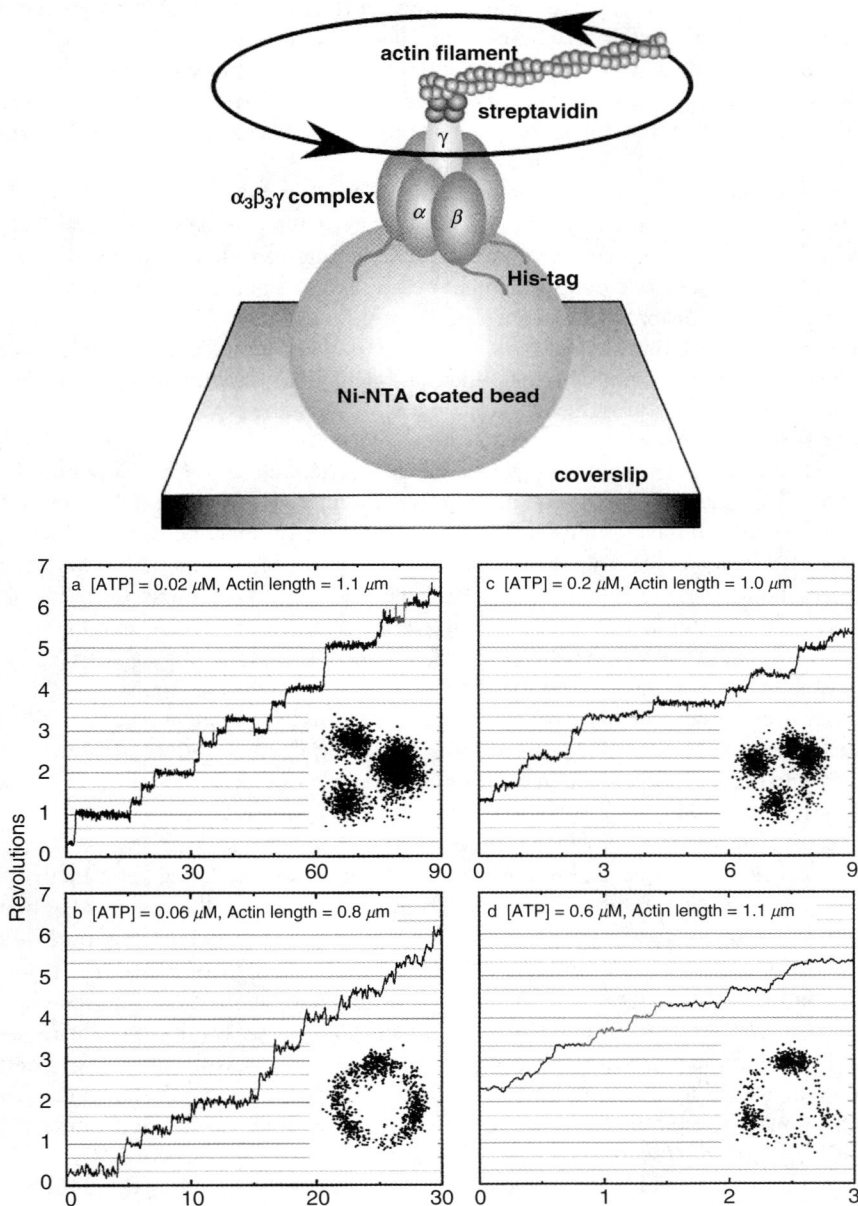

Figure 20. (A) Experimental setup used to measure the rotation of the F1 ATPase subunit. (B) Discrete rotations of 120° can be seen in the time course of the rotating filament [reprinted from Ref. 229 by permission of Elsevier]. (*See Color Plates*)

Isolation of the full F_0F_1 complex is now possible, and substeps of $40°$ and $80°$ have been observed along with the concentration dependent rotation rate of up to 350 rps under saturating ATP concentrations [231].

Although there is now good understanding of the structure and kinetics of the F_1 subunit, there are still many interesting questions to be addressed concerning the mechanism of the F_0 subunit and whether the ATP synthase assembly in eukaryotes and chloroplasts is significantly different from those in bacteria.

5.2.4.2 Myosin: The myosins are a huge family of linear motors which move non-reversibly toward the positive end of actin microfilaments. Actin is the principle component of thin filaments in muscle fiber, which slides past the thick filaments consisting mainly of myosin, driven by ATP and the presence of Ca^{2+}. Polymerized F-actin forms a right-handed helix with a 72 nm pitch, with each individual myosin in the thick filament binding only briefly and each head contributing an average force of 3–6 pN.

A representative member of the family is myosin II, a two-headed motor involved in cell movement and also found in skeletal muscle, which shares many of the same structural features as kinesin [232], but is nonprocessive. This makes measurement more problematic because a single molecule will not track along a filament over a long distance; instead each catalytic cycle is punctuated by a diffusional pause. Therefore a myosin molecule must be artificially held in close proximity to an actin filament by a weak trap to observe multiple steps in a reasonable time period. It is worth remembering that thermal fluctuations on a trapped particle can result in displacements of tens of nanometers, further complicating the process.

The first single-molecule measurements of the myosin II step size used an actin filament stretched out between two polystyrene beads and held close to myosin bound on a surface-immobilized bead [233]. Although there were numerous experimental issues with this setup, many of which have continued to plague these measurements, a mean displacement of 11 nm of the beads holding the actin filament was observed. A complementary variability in step size, in the range of 3–17 nm, has also been observed by a number of optical tweezer studies [130, 234]. Complementary work using a microneedle [235] measured the peak force to be 6 pN, with an average of 3 pN and a cycle rate of 10–20 Hz with step sizes of 20 nm. A more careful analysis, using aligned mixtures of actin and myosin as opposed to the random orientations used earlier, found that the step size was strongly dependent on orientation [236]. The effect of load has been difficult to quantify; however, there is evidence that the normally low duty cycle myosin II may become a high duty-ratio motor under force [237], though the effect of force on the kinetic and molecular mechanics is not well understood.

The diversity of the myosin family is illustrated by the contrast between myosin II and V. Myosin V is a two-headed motor which transports vesicles along actin filaments over long distances. The ADP release step is rate limiting, resulting in the bound state being the predominant intermediate. Several groups have tried to identify what determines the step size of myosin-V with a wide range of results [238]. The length of the arm is determined by the IQ motifs and this in turn is thought to determine the overall transport velocity [239]. Experiments have observed 36 nm [240] and 74 nm [241] steps leading to consideration of both inchworm and hand-over-hand models, while experiments with myosin-Vs containing reduced or increased numbers of IQ repeats (and thus lever arm length) concluded that the longer this arm, the faster the transport velocity in vivo [239]. More recent results [242, 243] have confirmed the 72–74 nm step, hand-over-hand model.

In addition to myosin II and V, the mechanics of myosin VI [244] and I [245] have also been the subject of single-molecule experiments, as well as the biochemistry of the power-stroke cycle [246, 247]. The adaptation of the power stroke by each member of the myosin family to meet its biological function has provided a rich diversity in similar experiments. Many challenges remain in characterizing these linear motors, but further quantitative measurements will provide insight into a wide range of active transport processes, from muscle contraction to intracellular transport. These studies help underline a major quest of the single-molecule biophysics community, that of assigning potential cellular functions based on mechanical characterization.

5.2.4.3 Kinesin: Kinesin is a processive, linear cytoskeleton motor which moves non-reversibly toward the positive end of microtubules. Microtubules are 30 nm diameter polar helices composed of thirteen tubulin subunits per turn and critical to structures such as flagella, cilia, and the mitotic spindle, as well as maintaining cell shape and a transport network. Kinesin is a tetramer, composed of two light chains connecting to the cargo and two heavy chains containing the motor domain heads, one of which was believed to be bound to the microtubule at all times [248] to give the ability to transverse many microns at a rate of ~800 nm/s before dissociation [249].

Both tubulin and kinesin can be purified and reconstituted in vitro, making single-molecule experiments possible. The first direct measurement of the proposed 8 nm step size, corresponding to the tubulin repeat distance, came from the Block lab, with a silica bead attached to a kinesin and tracked as it advanced along microtubules immobilized on a coverslip [117]. The rate of movement was found to slow linearly with increasing force, up to a stall force of 7 pN, with the occasional step backwards [250] explained by the increasing load decreasing the probability of the catalytic cycle producing a mechanical step. Analysis of the dwell time [251] and of the fluctuations [252] concluded that an ATP binding, rate limiting process occurred before the 8 nm advance, as illustrated in Figure 21.

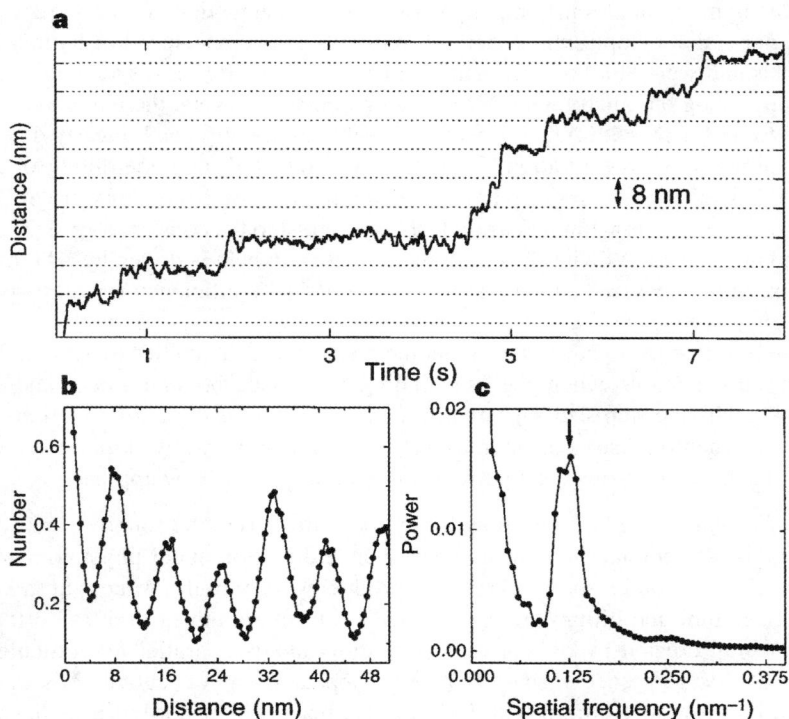

Figure 21. (A) Position of an optically trapped bead being pulled by a single kinesin protein. 8 nm center of mass steps are observable with variable dwell times. (B) Histogram of a pairwise comparison of distances from (A), clearly illustrating a 8 nm periodicity. (C) Power spectrum of the data from (B) showing a peak at the reciprocal of 8 nm. [reprinted from Ref. 252 by permission of Macmillan Publishers Ltd].

As with myosin there has been controversy over whether kinesin moves by a hand-over-hand motion or an inchworm motion where one head always leads. Recent measurements [253] strongly support the hand-over-hand model, with tracking of fluorescent-labeled heads supporting alternate 17 nm and 0 nm step sizes for each head, and a corresponding 8 nm movement of the center-of-mass for each ATP cycle. Further, both heads are bound or interacting with the filament during the dwell time between ATP binding events, and the hand-over-hand motion is likely asymmetric to avoid twisting the cargo, though the expected limping due to twisting of the stalk in this model has not been observed in wild-type kinesin [254]. Direct observation of the motion of the heads and stalk during the step process requires faster time resolution than currently available, but would help solve some of the unresolved uncertainties.

Work is ongoing [255], studying a range of other kinesins, including mitotic kinesin and the four-headed BimC, as well as the effect of mutations, some of which are associated with neurological diseases. Beyond these single-molecule experiments, the question of how a number of kinesins can carry cargo cooperatively over much longer distances still has to be addressed.

5.2.4.4 Dynein: Complementing kinesin, dynein is a linear cytoskeleton motor which move non-reversibly toward the negative end of microtubules. It is a large motor protein derived from the AAA family and well conserved. Dynein comes in two flavors: axonemal dynein, which provides the connective bridge and motor in cilia and sperm tails; and cytoplasmic dynein, which complements kinesin in subcellular vesicle transport.

An early experiment showed that latex beads coated with cytoplasmic dynein moved smoothly along microtubules in comparison to the step-wise motion of kinesin [256]. Follow-up reports noted that cytoplasmic dynein moved with an average speed of 5 μm/s along the microtubules and were able to backtrack distances of up to 100 nm [257], with some diffusing randomly along the microtubule [258]. Using optical tweezers, the maximum stall force was found to be 1.1 pN with ATP concentrations above saturation and linearly dependent on ATP concentration below saturation [259], similar to the stall force measured for axonemal dynein [260]. The step size has also been observed to decrease from a mixture of 24 nm and 32 nm steps down to 8 nm with increasing load, in contrast to the constant step sizes of kinesin and myosin under load. As with the other linear motors, dynein is driven by ATP hydrolysis, though from these measurements it operates with only 10% efficiency in comparison to the 50% efficiency of kinesin.

There is still much to be learned about the mechanics and kinetics of dynein [261] in its various roles: for example, where the ATP binding takes place, both physically and within the power stroke; why the step size is geared to load; and what the effect of mutations linked to inherited motor neuron disease is on transport. These are all unresolved questions which can potentially be answered using a single-molecule, force spectroscopy approach.

5.2.4.5 Cilia and Flagella: The flagella motor, with variations found in many bacteria and protists, is a rotational motor which can change direction under physiological control. It is driven by a proton or sodium gradient, producing very high torque with a small step size. The flagellum and cilium are constructed from microtubules in a 9+2 arrangement, with force being created by dynein motors pushing against parallel microtubules to create motion. Cilia, shortened forms of flagella, appear in all vertebrate cells in the form of a primary, non-motile cilium with 9+0 arrangement, while motile forms have the 9+2 arrangement. Cilia play an important role in our sensory perception as well as enabling protists to move.

Motile bacteria like E. coli are propelled by a number of flagella forming a bundle several microns long. Block et al. [262], making one of the first quantitative measurements with optical tweezers, measured the nonlinear torsional compliance of tethered Escherichia coli and a motile Streptococcus. Follow-up work in the Berg lab, using a piezoelectric stage to rotate the tethered E.coli, found that the motor generated ~4500 pN nm of torque at all angles, regardless of whether or not it was stalled [263].

In addition to E.coli, the swimming force of a wide range of other motile cell types have been characterized, including human sperm [264], Chlamydomonas [265], and spiroplasma [266]. In the case of sperm, near-infrared light at 760 nm was found to lead to paralysis and cell death within a minute [267]; but this has not prevented optical tweezers operating at 1064 nm being used to insert sperm into a trapped oocyte [268], and to study the influence of pentoxifyl-line, to decrease the viscosity of blood, on the swimming behavior of sperm [269]. Potentially there are many other force characterizations which have yet to be carried out on the many varie-ties of algae and ciliates, perhaps addressing unresolved questions on the beating of cilia and the influence of drugs on motile opportunistic bacteria.

5.3 Cellular and System-Level Force Spectroscopy

One of the greatest challenges working within a cell or organism is to interact with and only with the targets of interest in a controlled way. Optical and magnetic tweezers require an object of at least 100 nm in diameter with favorable properties in order to exert a useful force. The presence of such a particle would displace at least 10,000 biomolecules within a cell and be constrained by the cytoskeleton. Imaging in and around live cells is also hampered by the refractive index changes and increased scattering, limiting resolution and signal to noise. In addition, the presence of an uncontrolled environment can lead to multi-ple specific and nonspecific interactions, complicating analysis. Therefore the majority of in vivo studies with optical and magnetic tweezers have focused on the plasma membrane and on the manipulation of whole cells.

5.3.1 Membrane and Cytoplasm Force Spectroscopy

The ability to apply a mechanical force to a cell or one of its components is of fundamen-tal interest in understanding the influence of environment on cell development. Biomechanics has made great strides in describing the effects of force at the macroscopic ensemble level; and contact/fluid flow techniques have been used to observe the effects of force at the cel-lular level, though these approaches are limited to applying force to the outside of the plasma membrane over a relatively large area. The non-contact manipulation of submicron particles with optical and magnetic tweezers has refined these studies, creating as many new questions as answering existing ones, thanks to the complexity of the cellular environment.

Following Crick and Hughes's twisting, dragging, and prodding of the cytoplasm of an explant using magnetic particles in the 1950s [1], Yagi examined the properties of the protoplasm of an amoeba [270], again exploiting the then unique ability to apply a force to a particle within a cell. In the 1970s, Cohen [271] and Valberg et al. [272] started study-ing the retention of magnetic aerosols of ferromagnetic particles in the lung macrophages, which led to the development of magnetic twisting cytometry by Valberg and Butler [273] in 1987. Butler went on to work with Wang and Ingber [274] to study the effect of mechan-otransduction using magnetic particles and to propose a tensegrity model for cell structure. Askhin began the optical manipulation of cells, studying the viscoelastic properties of plant cells in 1989 [121], leading over the next fifteen years to a wide range of studies of motile and immotile cells, focused on characterizing cell morphology and manipulating cell func-tions. Recent reviews of cell mechanics and mechanotransduction [275] and of mechanical manipulation of cell membrane [276] have been published, and only some notable results are considered here.

Measurement of the mechanical forces involved in focal adhesions created by integrin proteins, the bridging focal adhesion proteins, and their associated complexes has been carried out with both optical [277] and magnetic tweezers [91]. Force applied to transmembrane proteins involved in these processes is met with increased traction forces and less displacement [278], while other membrane proteins not involved can be dragged around the membrane to probe structures which cannot freely diffuse [279]. A permanent magnetic microneedle has been used recently to probe a single beta1 integrin adhesion site [81]. Small ($< 0.1\,\mu m$) displacements for forces as high as 130 pN were observed and were dependent on the degree of assembly of the focal adhesion site. The formation and disassembly of cell adhesion sites involves complicated dynamics and assemblies of proteins, but good progress has been made toward pulling together single-molecule force spectroscopy and the results from more traditional molecular biology.

Other membrane proteins have been investigated using optical or magnetic tweezers, including the cell adhesion proteins LFA-1 [280], NrCAM [281], and E-cadherin [282]; the lambda receptor in E.coli [283]; the transferin receptor [284]; the transmembrane band 3 in erythrocytes [285]; and the adhesion of the von Willebrand factor (found in the subendothelium) to wild-type and mutant platelet membrane receptors [286]. The tracking of membrane proteins is particularly relevant in understanding how cells react to and survive mechanical stress and how movement, adhesion, and repair processes are governed [287].

The influence of drugs on the plasma membrane has begun to be investigated and will be of growing interest as the properties of cells and proteins become better known. The effect of the amphipathic drug salicylate on the mechanical properties of the plasma membrane was found to be minimal, discounting one of the competing theories for a side effect of the drug [288]. Using similar approaches, optical tweezers may help to play a role in pharmacology, providing a way of screening drugs [289].

Observation and manipulation of intracellular transport using optical and magnetic tweezers has been of interest to a small but growing number of microbiologists. Optical tweezers have the advantage here of being able to apply force directly on an organelle. Ashkin used this advantage in 1990 to manipulate organelles and estimate the force used to translate them along microtubules at 2.6 pN [290]. Optical tweezers have also been used to measure the forces involved in lipid vesicle trafficking in Drosophilia embryogenesis, observing discrete steps in the applied force up to the stall force of 1.1 pN, associated with decreasing numbers of motors [291].

In addition to tracking and manipulating organelles, intense light can also be used for optoporation [292], and to cut [12], manipulate [293], and isolate [294] chromosomes and human gametes [295]. Optical scalpels have also been used on trapped plant cells to perforate the cell wall, the mitochondria, and chloroplasts, and to move them around [296]. A more comprehensive review of the destructive use of lasers in cell biology has recently been published [297].

There are many unresolved questions concerning the dynamics of membranes and associated proteins, lipids, and polysaccharides which optical and magnetic tweezers can help address. For example, questions on the nature of lipid rafts and the function of polysaccharides in adhesion, defense, and signaling could be addressed by force spectroscopy in the range of 1–100 pN. Although mechano-transduction and light-directed cell growth have already demonstrated the ability to manipulate cell functions, there is significant interest in manipulating tagged transmembrane proteins and organelles to characterize and control more defined cell functions, for example, cell growth, division, signaling, and movement.

5.3.3 Cell Manipulation and Characterization

A wide range of cells have been trapped and manipulated, in particular with optical tweezers, to explore motility and other cell functions. The ability to trap a cell without damage and either manipulate or observe it using another technique is a very powerful asset in understanding cellular properties.

As mentioned early in the chapter, motile cells have been examined at various levels, from the force generated by swimming sperm to the coordination of flagella in E.coli. Motile cells sensitive to their environment, for example, the gravisensing Chara rhizoids [298] and magnetotactic bacterium Magnetospirillum magnetotacticum [299], have been explored using optical and magnetic tweezers, respectively. The vast diversity of archae, protists, and multi-cellular organisms that can sense and respond to fields means there are many more examples to explore which will further our understanding of many processes, including how organisms can navigate according to the Earth's magnetic field.

The interaction of immotile cells with their environment and each other can also be probed using optical tweezers. The binding of cells to a surface, such as the binding of fibroblasts to a fibronectin coated surface [300] and E.coli to a self-assembled monolayer of mannose [301], has been explored. Contact between effector cells of the immune system and their target cells held in optical traps can be used to establish the attack kinetics of a natural killer on an erythroleukemia cell [296]. There are many interesting interactions to explore, from the gruesome predator skills of Didinium to the more benign questions of plant cell adhesion under the environmental extremes to which they are exposed.

The morphology of cells can be characterized passively by their light refractive and scattering properties, and the results routinely used in automated cell sorting machines. It is unclear whether the information gained by magnetic or optical tweezer assays can outweigh the assays' complexity and limitations in throughput, when compared to competing cell characterization techniques such as fluorescent labeling, scattering, and dielectric spectros-copy, all of which offer single-molecule analysis and fast, straightforward screening. But in large cell populations where traditional cell sorting is not preferable, then optical tweezers can be used to isolate single cells [302], though throughput is substantially lower. Optical tweezers have also been used to isolate archaea according to 16S RNA labeling and to carry out PCR to identify the population [303].

The mechanical deformation of red blood cells, typically 7–9 µm in diameter, has been of particular interest because the cells can deform and flow in capillaries less than 3 µm in diameter. Intense light has been used to actively deform cells and to observe their recovery, in particular red blood cells [304] and chloroplasts in leaf tissue [305]. The spread in elasticity of sickle-shaped red blood cells has been found to be marginally different from that of normal red blood cells [306]. However, the story is not straightforward as red blood cells are themselves subject to a range of factors affecting their elasticity [307]. Although a powerful demonstration of the ability to manipulate and quantify the mechanical properties of individual cells, the complexity of and poor distinction between normal and sickle cells are unlikely to lead to a competitive assay.

The dual role of particles for both passively measuring displacements and actively applying force is highlighted in a recent study of single cardiac myocytes [308]. Heart cells can beat spontaneously, or be induced to beat, causing micron deformations with an average contractile force of 13 µN, which can be observed by a labeled bead. The authors then pro-ceeded to apply a reported force of 5 µN to the 20 µm bead, two orders of magnitude higher than any other magnetic force reported on a single bead, using a 5 mm diameter permanent magnet at a distance of 5 mm, and observed a shortening in the contraction length of the myocyte.

These are only a few of the many examples where whole cells have been manipulated using optical and magnetic tweezers. Interactions cannot be quantified as clearly at the cel-lular level as in the single-molecule studies described earlier, though qualitative observations and estimations can be made.

(A)

(B)

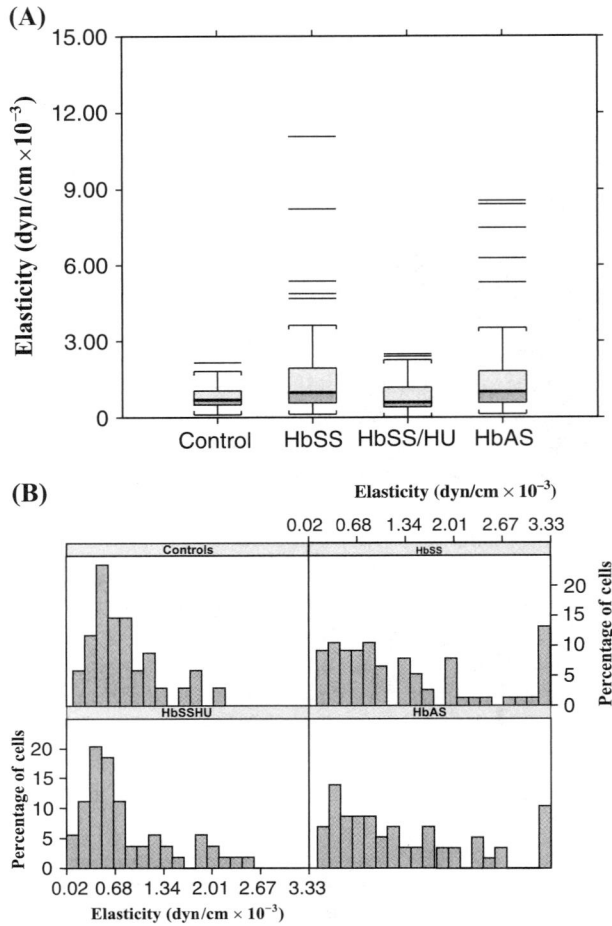

Figure 22. (A) Box plot and (B) histograms of the elasticity of red blood cells from subjects measured using optical tweezers. HbAS are homozygous sickle cell mutants, with HbSS showing characteristic traits, and HbSS/HU are subjects treated with hydroxyurea [reprinted from Ref. 306 by permission of Macmillan Publishers Ltd].

6   Conclusions and Future Prospects

Optical and magnetic tweezers are unique tools for force microscopy at the cellular level. They are unmatched in the sensitivity and flexibility of the force they can apply. The non-contact nature of both techniques removes complicating surface interactions or limiting environmental conditions and leads to the ability to rapidly and completely release trapped objects. As described in the first half of this chapter, the ability to develop techniques which can image particles with nanometer resolution and resolve piconewton forces at kilohertz frequencies has pushed back the limits in studying the force spectroscopy of single molecules and their interactions. Although their use in non-specialized labs has come about only in the last fifteen years, the success and commonplace usage of both techniques are perhaps best indicated by the decrease in their use as keywords in published papers.

The low force regime available with optical or magnetic tweezers means there are a range of forces, nanonewtons and higher, that can only be probed by contact probe methods.

In addition, the complexity and rate of dynamic processes can exceed the resolution and feedback possible in current optical detection systems. Furthermore, both of these non-contact techniques have a limited depth of operation, favoring single-molecule in vitro studies. However, from a biological perspective, the largest handicap to both approaches is the need to artificially attach a micron-sized handle with which to apply a significant force without causing modification of behavior. Steric constraints from the presence of these particles again favor minimally reconstituted biological systems and are always a consideration in interpreting results. Nonetheless, these disadvantages are not show-stoppers, as illustrated by the depth and breadth of the studies described in the second half of this chapter. Table 2 summarizes the physical properties of a number of biomolecules examined with optical or magnetic tweezers and discussed in this chapter.

There are many challenges and opportunities to meet in the coming decade. There is still a wide diversity of biomolecules and interactions to be characterized in vitro at the single-molecule level. With the rapidly growing commercial availability of biomolecules and new techniques to isolate and purify them, reproducibility and range of measurement will improve our understanding of molecular structure and dynamics. Controlled manipulation of intracellular organelles, membrane proteins, and cells themselves will open avenues for exploring systems biology at the single-cell level, linking with the ensemble methods currently used.

Table 2. Dimensions and persistence lengths of biopolymers.

	Length (nm)	Diameter (nm)	Persistence Length (nm)
Nucleic Acids			
dsDNA	$1 - 10^7$	2	50
ssDNA	$1 - 10^7$	1	$0.75 - 3$
dsRNA	$1 - 10^4$	2.6	$60 - 75$
ssRNA	$1 - 10^4$	1	3.5
Actin	$1 - 10^5$	$5 - 7$	15000
Microtubules	$1 - 10^4$	25	$1 - 8 \times 10^6$
Intermediate Filaments	$1 - 10^4$	$10 - 12$	$300 - 1000$
Proteins	$1 - 10^3$	0.5	
Titin	1200	2.5	$1 - 9$
Spectrin	200	6	$5 - 20$
Collagen	290	1.5	$4.5 - 14$
Fibrin	45	10	500
Myelin	53		0.5
Fibronectin	$120 - 160$		0.5
Polysaccharides			
Cellulose			5
Xyloglucan			$6 - 8$
Xanthan			120
Schizophyllan Polymers			180
Dextran, Sclerox, Alginate			$1 - 20$

Symbols used in this chapter:

α is the relative complex polarizability

κ is the viscosity of the surrounding medium (water is 10^{-3} s/m^3)

κ is the trap stiffness

λ is the wavelength of incident light

K is the wave vector of the light $\left(k = \dfrac{2\pi}{\lambda} \right)$

c is the speed of light in a vacuum ($\sim 3 \times 10^8$ m/s)

σ is the optical scattering cross section (cm^2)

h is Planck's constant (6.6×10^{-34} Js)

n_p is the particle refractive index

n_m is the surrounding medium refractive index

n_c is the refractive index contrast $\left(n_c = \dfrac{n_p}{n_m} \right)$

r_p is the particle radius

R is the reflectivity coefficient

TF is the transmission coefficient

T is the Maxwell stress tensor

p is the electric dipole moment

m is the magnetic dipole moment

α is the complex polarizability

B is the magnetic induction (Webers/m^2)

H is the external magnetic field strength (A/m)

M is the magnetization (A/m)

E is the electric field

D is the electric field displacement

ε_0 is the permittivity of free space (8.85×10^{-12} F/m)

μ_0 is the permeability of free space ($4\pi \times 10^{-7}$ Wb A^{-1} m^{-1})

χ_m is the magnetic susceptibility

$\Delta\chi$ is the susceptibility contrast ($\Delta\chi = \chi_{particle} - \chi_{liquid}$)

Acknowledgements

RC acknowledges the support of a National Research Council Research Associateship Award, held at NINDS.

References

1. Crick, F. H. C.; Hughes, A. F. W., The physical properties of cytoplasm. *Experimental Cell Research* **1950**, (1), 37–80.
2. Ashkin, A., Acceleration and Trapping of Particles by Radiation Pressure. *Phys. Rev. Lett.* 24, 156–159 (1970).
3. Lebedev, P. N., Experimental Examination of Light Pressure. *Annalen der Physik* **1901**, 6, 433–458.
4. Ashkin, A.; Dziedzic, J. M.; Bjorkholm, J. E.; Chu, S., Observation of a Single-Beam Gradient Force Optical Trap for Dielectric Particles. *Optics Letters* **1986**, 11, (5), 288–290.
5. Ashkin, A.; Dziedzic, J. M., Optical Trapping and Manipulation of Viruses and Bacteria. Science 1987, 235, (4795), 1517–1520.
6. Ashkin, A.; Dziedzic, J. M.; Yamane, T., Optical Trapping and Manipulation of Single Cells Using Infrared-Laser Beams. Nature 1987, 330, (6150), 769–771.
7. Burke, P. J., Encyclopedia of Nanoscience and Nanotechnology. American Scientific: Stevenson Ranch, CA, 2004; Vol. 6, p 623–641.
8. Markelz, A. G.; Roitberg, A.; Heilweil, E. J., Pulsed terahertz spectroscopy of DNA, bovine serum albumin and collagen between 0.1 and 2.0 THz. Chemical Physics Letters 2000, 320, (1–2), 42–48.

9. Naumann, D., *Encyclopedia of Analytical Chemistry*. John Wiley & Sons Ltd: Chichester, **2000**; p 102–131.

10. Buican, T. N.; Smyth, M. J.; Crissman, H. A.; Salzman, G. C.; Stewart, C. C.; Martin, J. C., Automated Single-Cell Manipulation and Sorting by Light Trapping. *Applied Optics* **1987**, 26, (24), 5311–5316.

11. Ashkin, A.; Dziedzic, J. M., Internal Cell Manipulation Using Infrared-Laser Traps. *Proceedings of the National Academy of Sciences of the United States of America* **1989**, 86, (20), 7914–7918.

12. Liang, H.; Wright, W. H.; Cheng, S.; He, W.; Berns, M. W., Micromanipulation of Chromosomes in Ptk2 Cells Using Laser Microsurgery (Optical Scalpel) in Combination with Laser-Induced Optical Force (Optical Tweezers). *Experimental Cell Research* **1993**, 204, (1), 110–120.

13. Yodh, A. G.; Lin, K. H.; Crocker, J. C.; Dinsmore, A. D.; Verma, R.; Kaplan, P. D., Entropically driven self-assembly and interaction in suspension. *Philosophical Transactions of the Royal Society of London Series a-Mathematical Physical and Engineering Sciences* **2001**, 359, (1782), 921–937.

14. Galajda, P.; Ormos, P., Complex micromachines produced and driven by light. *Applied Physics Letters* **2001**, 78, (2), 249–251.

15. Wang, G. M.; Sevick, E. M.; Mittag, E.; Searles, D. J.; Evans, D. J., Experimental demonstration of violations of the second law of thermodynamics for small systems and short time scales. *Physical Review Letters* 89, (5): Art No. 050601 JUL 29 **2002**.

16. Crocker, J. C.; Grier, D. G., Microscopic measurement of the pair interaction potential of charge-stabilized colloid. Physical Review Letters **1994**, 73, (2), 352–355.

17. Molloy, J. E.; Padgett, M. J., Lights, action: optical tweezers. *Contemporary Physics* **2002**, 43, (4), 241–258.

18. Lang, M. J.; Block, S. M., Resource letter: LBOT-1: Laser-based optical tweezers. *American Journal of Physics* **2003**, 71, (3), 201–215.

19. Williams, M. C., Optical Tweezers: Measuring Piconewton Forces. In http://www.biophysics.org/education/williams.pdf: 1992.

20. Ashkin, A., Forces of a Single-Beam Gradient Laser Trap on a Dielectric Sphere in the Ray Optics Regime. *Biophysical Journal* **1992**, 61, (2), 569–582.

21. Svoboda, K.; Block, S. M., Biological Applications of Optical Forces. Annual Review of *Biophysics and Biomolecular Structure* **1994**, 23, 247–285.

22. Tlusty, T.; Meller, A.; Bar-Ziv, R., Optical gradient forces of strongly localized fields. *Physical Review Letters* **1998**, 81, (8), 1738–1741.

23. Cox, A. J.; DeWeerd, A. J.; Linden, J., An experiment to measure Mie and Rayleigh total scattering cross sections. *American Journal of Physics* **2002**, 70, (6), 620–625.

24. Smith, S. B.; Cui, Y.; Bustamante, C., Optical-trap force transducer that operates by direct measurement of light momentum. *Methods Enzymol* **2003**, 361, 134–62.

25. Conroy, R. S.; Mayers, B. T.; Vezenov, D. V.; Wolfe, D. B.; Prentiss, M. G.; Whitesides, G. M., Optical waveguiding in suspensions of dielectric particles. *Applied Optics* **2005**, 44, (36), 7853–7857.

26. Svoboda, K.; Block, S. M., Optical Trapping of Metallic Rayleigh Particles. *Optics Letters* **1994**, 19, (13), 930–932.

27. Stamper-Kurn, D. M.; Andrews, M. R.; Chikkatur, A. P.; Inouye, S.; Miesner, H. J.; Stenger, J.; Ketterle, W., Optical confinement of a Bose-Einstein condensate. *Physical Review Letters* **1998**, 80, (10), 2027–2030.

28. Nieminen, T. A.; Rubinsztein-Dunlop, H.; Heckenberg, N. R., Calculation and optical measurement of laser trapping forces on non-spherical particles. *Journal of Quantitative Spectroscopy & Radiative Transfer* **2001**, 70, (4–6), 627–637.

29. Harada, Y.; Asakura, T., Radiation forces on a dielectric sphere in the Rayleigh scattering regime. *Optics Communications* **1996**, 124, (5–6), 529–541.

30. Rohrbach, A.; Stelzer, E. H., Trapping forces, force constants, and potential depths for dielectric spheres in the presence of spherical aberrations. *Appl Opt* **2002**, 41, (13), 2494–507.

31. Pralle, A.; Florin, E. L.; Stelzer, E. H. K.; Horber, J. K. H., Localized diffusion measurements by 3D-SPT provide support for membrane microdomains. *Biophysical Journal* **1999**, 76, (1), A390-a390.

32. Guck, J.; Ananthakrishnan, R.; Mahmood, H.; Moon, T. J.; Cunningham, C. C.; Kas, J., The optical stretcher: A novel laser tool to micromanipulate cells. Biophysical Journal **2001**, 81, (2), 767–784.

33. Guck, J.; Schinkinger, S.; Lincoln, B.; Wottawah, F.; Ebert, S.; Romeyke, M.; Lenz, D.; Erickson, H. M.; Ananthakrishnan, R.; Mitchell, D.; Kas, J.; Ulvick, S.; Bilby, C., Optical deformability as an inherent cell marker for testing malignant transformation and metastatic competence. *Biophysical Journal* **2005**, 88, (5), 3689–3698.

34. Smith, S. B.; Cui, Y. J.; Bustamante, C., Overstretching B-DNA: The elastic response of individual double-stranded and single-stranded DNA molecules. *Science* **1996**, 271, (5250), 795–799.

35. Williams, M. C.; Wenner, J. R.; Rouzina, I.; Bloomfield, V. A., Entropy and heat capacity of DNA melting from temperature dependence of single molecule stretching. *Biophysical Journal* **2001**, 80, (4), 1932–1939.

36. Visscher, K.; Brakenhoff, G. J.; Krol, J. J., Micromanipulation by Multiple Optical Traps Created by a Single Fast Scanning Trap Integrated with the Bilateral Confocal Scanning Laser Microscope. *Cytometry* **1993**, 14, (2), 105–114.

37. Visscher, K.; Gross, S. P.; Block, S. M., Construction of multiple-beam optical traps with nanometer-resolution position sensing. *Ieee Journal of Selected Topics in Quantum Electronics* **1996**, 2, (4), 1066–1076.

38. Sasaki, K.; Koshioka, M.; Misawa, H.; Kitamura, N.; Masuhara, H., Pattern-Formation and Flow-Control of Fine Particles by Laser-Scanning Micromanipulation. *Optics Letters* **1991**, 16, (19), 1463–1465.

39. Dufresne, E. R.; Spalding, G. C.; Dearing, M. T.; Sheets, S. A.; Grier, D. G., Computer-generated holographic optical tweezer arrays. *Review of Scientific Instruments* **2001**, 72, (3), 1810–1816.

40. Liesener, J.; Reicherter, M.; Haist, T.; Tiziani, H. J., Multi-functional optical tweezers using computer-generated holograms. *Optics Communications* **2000**, 185, (1–3), 77–82.

41. Curtis, J. E.; Koss, B. A.; Grier, D. G., Dynamic holographic optical tweezers. *Optics Communications* **2002**, 207, (1–6), 169–175.

42. Sanford, J. L.; Greier, P. F.; Yang, K. H.; Lu, M.; Olyha, R. S.; Narayan, C.; Hoffnagle, J. A.; Alt, P. M.; Melcher, R. L., A one-megapixel reflective spatial light modulator system for holographic storage. *Ibm Journal of Research and Development* **1998**, 42, (3–4), 411–426.

43. Kawata, S.; Sugiura, T., Movement of Micrometer-Sized Particles in the Evanescent Field of a Laser-Beam. *Optics Letters* **1992**, 17, (11), 772–774.

44. Taylor, R. S.; Hnatovsky, C., Particle trapping in 3-D using a single fiber probe with an annular light distribution. *Optics Express* **2003**, 11, (21), 2775–2782.

45. Garces-Chavez, V.; Dholakia, K.; Spalding, G. C., Extended-area optically induced organization of microparticies on a surface. *Applied Physics Letters* **2005**, 86, (3), -.

46. Moothoo, D. N.; Arlt, J.; Conroy, R. S.; Akerboom, F.; Voit, A.; Dholakia, K., Beth's experiment using optical tweezers. *American Journal of Physics* 69, (3): 271–276 MAR **2001**.

47. Konig, K.; Liang, H.; Berns, M. W.; Tromberg, B. J., Cell damage in near-infrared multimode optical traps as a result of multiphoton absorption. *Optics Letters* **1996**, 21, (14), 1090–1092.

48. O'Neil, A. T.; Padgett, M. J., Three-dimensional optical confinement of micron-sized metal particles and the decoupling of the spin and orbital angular momentum within an optical spanner. *Optics Communications* **2000**, 185, (1–3), 139–143.

49. Rubinsztein-Dunlop, H.; Nieminen, T. A.; Friese, M. E. J.; Heckenberg, N. R., Optical trapping of absorbing particles. *Advances in Quantum Chemistry, Vol 30* **1998**, 30, 469–492.

50. Gahagan, K. T.; Swartzlander, G. A., Trapping of low-index microparticles in an optical vortex. *Journal of the Optical Society of America B-Optical Physics* **1998**, 15, (2), 524–534.

51. O'Neill, A. T.; Padgett, M. J., Axial and lateral trapping efficiency of Laguerre-Gaussian modes in inverted optical tweezers. *Optics Communications* **2001**, 193, (1–6), 45–50.

52. Allen, L.; Beijersbergen, M. W.; Spreeuw, R. J. C.; Woerdman, J. P., Orbital Angular-Momentum of Light and the Transformation of Laguerre-Gaussian Laser Modes. *Physical Review A* **1992**, 45, (11), 8185–8189.

53. Gahagan, K. T.; Swartzlander, G. A., Optical vortex trapping of particles. *Optics Letters* 1996, 21, (11), 827–829.

54. Simpson, N. B.; Allen, L.; Padgett, M. J., Optical tweezers and optical spanners with Laguerre-Gaussian modes. *Journal of Modern Optics* **1996**, 43, (12), 2485–2491.

55. Arlt, J.; Padgett, M. J., Generation of a beam with a dark focus surrounded by regions of higher intensity: the optical bottle beam. *Optics Letters* **2000**, 25, (4), 191–193.

56. Arlt, J.; Garces-Chavez, V.; Sibbett, W.; Dholakia, K., Optical micromanipulation using a Bessel light beam. *Optics Communications* **2001**, 197, (4–6), 239–245.

57. McGloin, D.; Dholakia, K., Bessel beams: diffraction in a new light. *Contemporary Physics* **2005**, 46, (1), 15–28.

58. Garces-Chavez, V.; McGloin, D.; Melville, H.; Sibbett, W.; Dholakia, K., Simultaneous micromanipulation in multiple planes using a self-reconstructing light beam. *Nature* **2002**, 419, (6903), 145–147.

59. Ke, P. C.; Gu, M., Characterization of trapping force on metallic Mie particles. *Applied Optics* **1999**, 38, (1), 160–167.

60. Plewa, J.; Tanner, E.; Mueth, D. M.; Grier, D. G., Processing carbon nanotubes with holographic optical tweezers. *Optics Express* **2004**, 12, (9), 1978–1981.

61. Gardel, M. L.; Valentine, M. T.; Weitz, D. A., *Microscale Diagnostic Techniques*. Springer-Verlag: Berlin, 2005.

62. *Advances in Biomagnetic Separation*. Eaton Publishing Company: Natick, 1994.

63. Shapiro, E. M.; Skrtic, S.; Sharer, K.; Hill, J. M.; Dunbar, C. E.; Koretsky, A. P., MRI detection of single particles for cellular imaging. *Proceedings of the National Academy of Sciences of the United States of America* **2004**, 101, (30), 10901–10906.

64. Bulte, J. W. M.; Kraitchman, D. L., Iron oxide MR contrast agents for molecular and cellular imaging. *Nmr in Biomedicine* **2004**, 17, (7), 484–499.

65. Berry, C. C.; Curtis, A. S. G., Functionalisation of magnetic nanoparticles for applications in biomedicine. *Journal of Physics D-Applied Physics* **2003**, 36, (13), R198-R206.

66. Takahashi, T.; Dimitrov, A. S.; Nagayama, K., Two-dimensional patterns of magnetic particles at air-water or glass-water interfaces induced by an external magnetic field: Theory and simulation of the formation process. *Journal of Physical Chemistry* **1996**, 100, (8), 3157–3162.

67. Assi, F.; Jenks, R.; Yang, J.; Love, C.; Prentiss, M., Massively parallel adhesion and reactivity measurements using simple and inexpensive magnetic tweezers. *Journal of Applied Physics* **2002**, 92, (9), 5584–5586.

68. Rife, J. C.; Miller, M. M.; Sheehan, P. E.; Tamanaha, C. R.; Tondra, M.; Whitman, L. J., Design and performance of GMR sensors for the detection of magnetic microbeads in biosensors. *Sensors and Actuators a-Physical* **2003**, 107, (3), 209–218.

69. Besse, P. A.; Boero, G.; Demierre, M.; Pott, V.; Popovic, R., Detection of a single magnetic microbead using a miniaturized silicon Hall sensor. *Applied Physics Letters* **2002**, 80, (22), 4199–4201.

70. Barbic, M., Magnetic wires in MEMS and bio-medical applications. *Journal of Magnetism and Magnetic Materials* **2002**, 249, (1–2), 357–367.

71. Kim, K. S.; Park, J. K., Magnetic force-based multiplexed immunoassay using superparamagnetic nanoparticles in microfluidic channel. *Lab on a Chip* **2005**, 5, (6), 657–664.

72. Romano, G.; Sacconi, L.; Capitanio, M.; Pavone, F. S., Force and torque measurements using magnetic micro beads for single molecule biophysics. *Optics Communications* **2003**, 215, (4–6), 323–331.

73. van der Heijden, T.; van Noort, J.; van Leest, H.; Kanaar, R.; Wyman, C.; Dekker, N.; Dekker, C., Torque-limited RecA polymerization on dsDNA. *Nucleic Acids Research* **2005**, 33, (7), 2099–2105.

74. Strick, T. R.; Charvin, G.; Dekker, N. H.; Allemand, J. F.; Bensimon, D.; Croquette, V., Tracking enzymatic steps of DNA topoisomerases using single-molecule micromanipulation. *Comptes Rendus Physique* **2002**, 3, (5), 595–618.

75. Charvin, G.; Allemand, J. F.; Strick, T. R.; Bensimon, D.; Croquette, V., Twisting DNA: single molecule studies. *Contemporary Physics* **2004**, 45, (5), 383–403.

76. Strick, T. R.; Allemand, J. F.; Bensimon, D.; Croquette, V., Behavior of supercoiled DNA. *Biophysical Journal* **1998**, 74, (4), 2016–2028.

77. Simson, D. A.; Ziemann, F.; Strigl, M.; Merkel, R., Micropipet-based pico force transducer: In depth analysis and experimental verification. *Biophysical Journal* **1998**, 74, (4), 2080–2088.

78. Haber, C.; Wirtz, D., Magnetic tweezers for DNA micromanipulation. *Review of Scientific Instruments* **2000**, 71, (12), 4561–4570.

79. Gosse, C.; Croquette, V., Magnetic tweezers: Micromanipulation and force measurement at the molecular level. *Biophysical Journal* **2002**, 82, (6), 3314–3329.

80. Strick, T.; Allemand, J. F. O.; Croquette, V.; Bensimon, D., The manipulation of single biomolecules. *Physics Today* **2001**, 54, (10), 46–51.

81. Matthews, B. D.; Overby, D. R.; Alenghat, F. J.; Karavitis, J.; Numaguchi, Y.; Allen, P. G.; Ingber, D. E., Mechanical properties of individual focal adhesions probed with a magnetic microneedle. *Biochemical and Biophysical Research Communications* **2004**, 313, (3), 758–764.

82. Meyer, A.; Hansen, D. B.; Gomes, C. S. G.; Hobley, T. J.; Thomas, O. R. T.; Franzreb, M., Demonstration of a strategy for product purification by high-gradient magnetic fishing: Recovery of superoxide dismutase from unconditioned whey. *Biotechnology Progress* **2005**, 21, (1), 244–254.

83. Matthews, B. D.; LaVan, D. A.; Overby, D. R.; Karavitis, J.; Ingber, D. E., Electromagnetic needles with submicron pole tip radii for nanomanipulation of biomolecules and living cells. *Applied Physics Letters* **2004**, 85, (14), 2968–2970.

84. Mirowski, E.; Moreland, J.; Russek, S. E.; Donahue, M. J., Integrated microfluidic isolation platform for magnetic particle manipulation in biological systems. *Applied Physics Letters* **2004**, 84, (10), 1786–1788.

85. Vengalattore, M.; Conroy, R. S.; Rooijakkers, W.; Prentiss, M., Ferromagnets for integrated atom optics. *Journal of Applied Physics* **2004**, 95, (8), 4404–4407.

86. de Vries, A. H. B.; Krenn, B. E.; van Driel, R.; Kanger, J. S., Micro magnetic tweezers for nanomanipulation inside live cells. *Biophysical Journal* **2005**, 88, (3), 2137–2144.

87. Amblard, F.; Yurke, B.; Pargellis, A.; Leibler, S., A magnetic manipulator for studying local rheology and micromechanical properties of biological systems. *Review of Scientific Instruments* **1996**, 67, (3), 818–827.

88. Ziemann, F.; Radler, J.; Sackmann, E., Local Measurements of Viscoelastic Moduli of Entangled Actin Networks Using an Oscillating Magnetic Bead Micro-Rheometer. *Biophysical Journal* **1994**, 66, (6), 2210–2216.

89. Bausch, A. R.; Ziemann, F.; Boulbitch, A. A.; Jacobson, K.; Sackmann, E., Local measurements of viscoelastic parameters of adherent cell surfaces by magnetic bead microrheometry. *Biophysical Journal* **1998**, 75, (4), 2038–2049.

90. Bausch, A. R.; Moller, W.; Sackmann, E., Measurement of local viscoelasticity and forces in living cells by magnetic tweezers. *Biophysical Journal* **1999**, 76, (1), 573–579.

91. Bausch, A. R.; Hellerer, U.; Essler, M.; Aepfelbacher, M.; Sackmann, E., Rapid stiffening of integrin receptor-actin linkages in endothelial cells stimulated with thrombin: A magnetic bead microrheology study. Biophysical Journal **2001**, 80, (6), 2649–2657.

92. Moore, L. R.; Zborowski, M.; Nakamura, M.; McCloskey, K.; Gura, S.; Zuberi, M.; Margel, S.; Chalmers, J. J., The use of magnetite-doped polymeric microspheres in calibrating cell tracking velocimetry. *Journal of Biochemical and Biophysical Methods* **2000**, 44, (1–2), 115–130.

93. Todd, P.; Cooper, R. P.; Doyle, J. F.; Dunn, S.; Vellinger, J.; Deuser, M. S., Multistage magnetic particle separator. *Journal of Magnetism and Magnetic Materials* **2001**, 225, (1–2), 294–300.

94. Ghiringhelli, F.; Schmitt, E., Cellular and molecular purification processes based on the use of magnetic micro- and nanobeads. *Annales De Biologie Clinique* **2004**, 62, (1), 73–78.

95. Tibbe, A. G. J.; de Grooth, B. G.; Greve, J.; Liberti, P. A.; Dolan, G. J.; Terstappen, L. W. M. M., Optical tracking and detection of immunomagnetically selected and aligned cells. *Nature Biotechnology* **1999**, 17, (12), 1210–1213.

96. Zigeuner, R. E.; Riesenberg, R.; Pohla, H.; Hofstetter, A.; Oberneder, R., Isolation of circulating cancer cells from whole blood by immunomagnetic cell enrichment and unenriched immunocytochemistry in vitro. *Journal of Urology* **2003**, 169, (2), 701–705.

97. Morisada, S.; Miyata, N.; Iwahori, K., Immunomagnetic separation of scum-forming bacteria using polyclonal antibody that recognizes mycolic acids. *Journal of Microbiological Methods* **2002**, 51, (2), 141–148.

98. Mura, C. V.; Becker, M. L.; Orellana, A.; Wolff, D., Immunopurification of Golgi vesicles by magnetic sorting. *Journal of Immunological Methods* **2002**, 260, (1–2), 263–271.

99. Winkleman, A.; Gudiksen, K. L.; Ryan, D.; Whitesides, G. M.; Greenfield, D.; Prentiss, M., A magnetic trap for living cells suspended in a paramagnetic buffer. *Applied Physics Letters* **2004**, 85, (12), 2411–2413.

100. Simon, M. D.; Geim, A. K., Diamagnetic levitation: Flying frogs and floating magnets (invited). *Journal of Applied Physics* **2000**, 87, (9), 6200–6204.

101. Toussaint, R.; Akselvoll, J.; Helgesen, G.; Skjeltorp, A. T.; Flekkoy, E. G., Interaction model for magnetic holes in a ferrofluid layer. *Physical Review* 69, (1): Art No. 011407 Part 1 JAN **2004**.

102. Lyuksyutov, I. F.; Lyuksyutova, A.; Naugle, D. G.; Rathnayaka, K. D. D., Trapping microparticles with strongly inhomogeneous magnetic fields. *Modern Physics Letters* B **2003**, 17, (17), 935–940.

103. Lecommandoux, S. B.; Sandre, O.; Checot, F.; Rodriguez-Hernandez, J.; Perzynski, R., Magnetic nanocomposite micelles and vesicles. *Advanced Materials* **2005**, 17, (6), 712-+.

104. Gatteschi, D.; Caneschi, A.; Pardi, L.; Sessoli, R., Large Clusters of Metal-Ions - the Transition from Molecular to Bulk Magnets. Science **1994**, 265, (5175), 1054–1058.

105. Genç, S. Synthesis and Properties of Magnetorheological (MR) Fluids. University of Pittsburgh, Pittsburgh, **2002**.

106. Hatch, A.; Kamholz, A. E.; Holman, G.; Yager, P.; Bohringer, K. F., A ferrofluidic magnetic micropump. *Journal of Microelectromechanical Systems* **2001**, 10, (2), 215–221.

107. Smith, S. P.; Bhalotra, S. R.; Brody, A. L.; Brown, B. L.; Boyda, E. K.; Prentiss, M., Inexpensive optical tweezers for undergraduate laboratories. *American Journal of Physics* **1999**, 67, (1), 26–35.

108. Laser Tweezers in Cell Biology. Academic Press: San Diego, 1998.

109. Bechhoefer, J.; Wilson, S., Faster, cheaper, safer optical tweezers for the undergraduate laboratory. *American Journal of Physics* **2002**, 70, (4), 393–400.

110. Neuman, K. C.; Block, S. M., Optical trapping. *Review of Scientific Instruments* **2004**, 75, (9), 2787–2809.

111. Schmidt, T.; Schutz, G. J.; Baumgartner, W.; Gruber, H. J.; Schindler, H., Imaging of single molecule diffusion. *Proceedings of the National Academy of Sciences of the United States of America* **1996**, 93, (7), 2926–2929.

112. Thompson, R. E.; Larson, D. R.; Webb, W. W., Precise nanometer localization analysis for individual fluorescent probes. *Biophysical Journal* **2002**, 82, (5), 2775–2783.

113. Ghislain, L. P.; Switz, N. A.; Webb, W. W., Measurement of Small Forces Using an Optical Trap. *Review of Scientific Instruments* **1994**, 65, (9), 2762–2768.

114. Jonas, A.; Zemanek, P.; Florin, E. L., Single-beam trapping in front of reflective surfaces. *Optics Letters* **2001**, 26, (19), 1466–1468.

115. Clapp, A. R.; Ruta, A. G.; Dickinson, R. B., Three-dimensional optical trapping and evanescent wave light scattering for direct measurement of long range forces between a colloidal particle and a surface. *Review of Scientific Instruments* **1999**, 70, (6), 2627–2636.

116. Rohrbach, A.; Kress, H.; Stelzer, E. H. K., Three-dimensional tracking of small spheres in focused laser beams: influence of the detection angular aperture. *Optics Letters* **2003**, 28, (6), 411–413.

117. Svoboda, K.; Schmidt, C. F.; Schnapp, B. J.; Block, S. M., Direct Observation of Kinesin Stepping by Optical Trapping Interferometry. *Nature* **1993**, 365, (6448), 721–727.

118. Gittes, F.; Schmidt, C. F., Interference model for back-focal-plane displacement detection in optical tweezers. *Optics Letters* **1998**, 23, (1), 7–9.

119. Visscher, K.; Block, S. M., Versatile optical traps with feedback control. *Methods Enzymol* **1998**, 298, 460–89.

120. Rice, S. E.; Purcell, T. J.; Spudich, J. A., Building and using optical traps to study properties of molecular motors. *Biophotonics, Pt B* **2003**, 361, 112–133.

121. Ashkin, A.; Dziedzic, J. M., Optical Trapping and Manipulation of Single Living Cells Using Infrared-Laser Beams. *Berichte Der Bunsen-Gesellschaft-Physical Chemistry Chemical Physics* **1989**, 93, (3), 254–260.

122. Liu, Y.; Sonek, G. J.; Berns, M. W.; Tromberg, B. J., Physiological monitoring of optically trapped cells: Assessing the effects of confinement by 1064-nm laser tweezers using microfluorometry. *Biophysical Journal* **1996**, 71, (4), 2158–2167.

123. Neuman, K. C.; Chadd, E. H.; Liou, G. F.; Bergman, K.; Block, S. M., Characterization of photodamage to Escherichia coli in optical traps. *Biophysical Journal 1999*, 77, (5), 2856–2863.

124. Wuite, G. J. L.; Davenport, R. J.; Rappaport, A.; Bustamante, C., An integrated laser trap/flow control video microscope for the study of single biomolecules. *Biophysical Journal* **2000**, 79, (2), 1155–1167.

125. Schmitt, K. E. Optical neuronal guiding on the hypothalamic GnRH cell line GT1. The University of Texas at Austin, Austin, **2003**.

126. Liu, Y.; Cheng, D. K.; Sonek, G. J.; Berns, M. W.; Chapman, C. F.; Tromberg, B. J., Evidence for Localized Cell Heating Induced by Infrared Optical Tweezers. *Biophysical Journal* **1995**, 68, (5), 2137–2144.

127. Berns, M. W.; Aist, J. R.; Wright, W. H.; Liang, H., Optical Trapping in Animal and Fungal Cells Using a Tunable, near-Infrared Titanium-Sapphire Laser. *Experimental Cell Research* **1992**, 198, (2), 375–378.

128. Underhill, P. T.; Doyle, P. S., Development of bead-spring polymer models using the constant extension ensemble. *Journal of Rheology* **2005**, 49, (5), 963–987.

129. Evans, E.; Ritchie, K., Dynamic strength of molecular adhesion bonds. *Biophysical Journal* **1997**, 72, (4), 1541–1555.

130. Molloy, J. E.; Burns, J. E.; Kendrickjones, J.; Tregear, R. T.; White, D. C. S., Movement and Force Produced by a Single Myosin Head. *Nature* **1995**, 378, (6553), 209–212.

131. Wright, W. H.; Sonek, G. J.; Berns, M. W., Parametric Study of the Forces on Microspheres Held by Optical Tweezers. *Applied Optics* **1994**, 33, (9), 1735–1748.

132. Pankhurst, Q. A.; Connolly, J.; Jones, S. K.; Dobson, J., Applications of magnetic nanoparticles in biomedicine. *Journal of Physics D-Applied Physics* **2003**, 36, (13), R167-R181.

133. Wang, M. D.; Yin, H.; Landick, R.; Gelles, J.; Block, S. M., Stretching DNA with optical tweezers. *Biophysical Journal* **1997**, 72, (3), 1335–1346.

134. Noji, H.; Yasuda, R.; Yoshida, M.; Kinosita, K., Direct observation of the rotation of F-1-ATPase. *Nature* **1997**, 386, (6622), 299–302.

135. Guthold, M.; Mullin, J.; Lord, S.; Superfine, R.; Taylor, R.; Erie, D., Investigating the mechanical properties of individual fibrin fibers with the nanomanipulator AFM. *Biophysical Journal* **2001**, 80, (1), 307A-307A.

136. Dammer, U.; Popescu, O.; Wagner, P.; Anselmetti, D.; Guntherodt, H. J.; Misevic, G. N., Binding Strength between Cell-Adhesion Proteoglycans Measured by Atomic-Force Microscopy. *Science* **1995**, 267, (5201), 1173–1175.

137. Piran, U.; Riordan, W. J., Dissociation Rate-Constant of the Biotin-Streptavidin Complex. *Journal of Immunological Methods* **1990**, 133, (1), 141–143.

138. Merkel, R.; Nassoy, P.; Leung, A.; Ritchie, K.; Evans, E., Energy landscapes of receptor-ligand bonds explored with dynamic force spectroscopy. *Nature* **1999**, 397, (6714), 50–53.

139. Yuan, C. B.; Chen, A.; Kolb, P.; Moy, V. T., Energy landscape of streptavidin-biotin complexes measured by atomic force microscopy. *Biochemistry* **2000**, 39, (33), 10219–10223.

140. Danilowicz, C.; Greenfield, D.; Prentiss, M., Dissociation of ligand-receptor complexes using magnetic tweezers. *Analytical Chemistry* **2005**, 77, (10), 3023–3028.

141. Helmerson, K.; Kishore, R.; Phillips, W. D.; Weetall, H. H., Optical tweezers-based immunosensor detects femtomolar concentrations of antigens. *Clinical Chemistry* **1997**, 43, (2), 379–383.

142. Kulin, S.; Kishore, R.; Hubbard, J. B.; Helmerson, K., Real-time measurement of spontaneous antigen-antibody dissociation. *Biophysical Journal* **2002**, 83, (4), 1965–1973.

143. Round, A. N.; Berry, M.; McMaster, T. J.; Stoll, S.; Gowers, D.; Corfield, A. P.;Miles, M. J., Heterogeneity and persistence length in human ocular mucins. *Biophysical Journal* **2002**, 83, (3), 1661–1670.

144. Sagis, L. M. C.; Veerman, C.; van der Linden, E., Mesoscopic properties of semiflexible amyloid fibrils. *Langmuir* **2004**, 20, (3), 924–927.

145. Smith, S. B.; Finzi, L.; Bustamante, C., Direct Mechanical Measurements of the Elasticity of Single DNA-Molecules by Using Magnetic Beads. *Science* **1992**, 258, (5085), 1122–1126.

146. Leger, J. F.; Romano, G.; Sarkar, A.; Robert, J.; Bourdieu, L.; Chatenay, D.; Marko, J. F., Structural transitions of a twisted and stretched DNA molecule. *Physical Review Letters* **1999**, 83, (5), 1066–1069.

147. Allemand, J. F.; Bensimon, D.; Croquette, V., Stretching DNA and RNA to probe their interactions with proteins. *Current Opinion in Structural Biology* **2003**, 13, (3), 266–274.

148. Tinland, B.; Pluen, A.; Sturm, J.; Weill, G., Persistence length of single-stranded DNA. *Macromolecules* **1997**, 30, (19), 5763–5765.

149. Bustamante, C.; Smith, S. B.; Liphardt, J.; Smith, D., Single-molecule studies of DNA mechanics. *Current Opinion in Structural Biology* **2000**, 10, (3), 279–285.

150. Conroy, R. S.; Danilowicz, C., Unravelling DNA. *Contemporary Physics* **2004**, 45, (4), 277–302.

151. Bockelmann, U., Single-molecule manipulation of nucleic acids. *Current Opinion in Structural Biology* **2004**, 14, (3), 368–373.

152. Bustamante, C.; Bryant, Z.; Smith, S. B., Ten years of tension: single-molecule DNA mechanics. *Nature* **2003**, 421, (6921), 423–427.

153. Marko, J.; Propperova, A., Environmental Monitoring within the Slovak Republic. *Environmental Monitoring and Assessment* **1995**, 34, (2), 131–136.

154. Strick, T. R.; Bensimon, D.; Croquette, V., Micro-mechanical measurement of the torsional modulus of DNA. *Genetica* **1999**, 106, (1–2), 57–62.

155. Bryant, Z.; Stone, M. D.; Gore, J.; Smith, S. B.; Cozzarelli, N. R.; Bustamante, C., Structural transitions and elasticity from torque measurements on DNA. *Nature* **2003**, 424, (6946), 338–341.

156. Charvin, G.; Vologodskii, A.; Bensimon, D.; Croquette, V., Braiding DNA: Experiments, simulations, and models. *Biophysical Journal* **2005**, 88, (6), 4124–4136.

157. Revyakin, A.; Ebright, R. H.; Strick, T. R., Single-molecule DNA nanomanipulation: Improved resolution through use of shorter DNA fragments. *Nature Methods* **2005**, 2, (2), 127–138.

158. Bockelmann, U.; Thomen, P.; Essevaz-Roulet, B.; Viasnoff, V.; Heslot, F., Unzipping DNA with optical tweezers: high sequence sensitivity and force flips. *Biophysical Journal* **2002**, 82, (3), 1537–1553.

159. Danilowicz, C.; Coljee, V. W.; Bouzigues, C.; Lubensky, D. K.; Nelson, D. R.; Prentiss, M., DNA unzipped under a constant force exhibits multiple metastable intermediates. *Proceedings of the National Academy of Sciences of the United States of America* **2003**, 100, (4), 1694–1699.

160. Danilowicz, C.; Conroy, R.; Kafri, Y.; Coljee, V.; Prentiss, M., Measurement of the phase diagram of DNA unzipping in the temperature-force plane. *Physical Review Letters* **2004**, 93, (7), 078101.

161. Vorobjev, I. A.; Hong, L.; Wright, W. H.; Berns, M. W., Optical Trapping for Chromosome Manipulation - a Wavelength Dependence of Induced Chromosome Bridges. *Biophysical Journal* **1993**, 64, (2), 533–538.

162. Bennink, M. L.; Leuba, S. H.; Leno, G. H.; Zlatanova, J.; de Grooth, B. G.; Greve, J., Unfolding individual nucleosomes by stretching single chromatin fibers with optical tweezers. *Nature Structural Biology* **2001**, 8, (7), 606–610.

163. Brower-Toland, B. D.; Smith, C. L.; Yeh, R. C.; Lis, J. T.; Peterson, C. L.; Wang, M. D., Mechanical disruption of individual nucleosomes reveals a reversible multistage release of DNA. *Proceedings of the National Academy of Sciences of the United States of America* **2002**, 99, (4), 1960–1965.

164. Storz, G., An expanding universe of noncoding RNAs. *Science* **2002**, 296, (5571), 1260–1263.

165. Hagerman, P. J., Flexibility of RNA. *Annual Review of Biophysics and Biomolecular Structure* **1997**, 26, 139–156.

166. Abels, J. A.; Moreno-Herrero, F.; van der Heijden, T.; Dekker, C.; Dekker, N. H., Single-molecule measurements of the persistence length of double- stranded RNA. *Biophysical Journal* **2005**, 88, (4), 2737–2744.

167. Ha, T.; Zhuang, X. W.; Kim, H. D.; Orr, J. W.; Williamson, J. R.; Chu, S., Ligand-induced conformational changes observed in single RNA molecules. *Proceedings of the National Academy of Sciences of the United States of America* **1999**, 96, (16), 9077–9082.

168. Zhuang, X. W., Single-molecule RNA science. *Annual Review of Biophysics and Biomolecular Structure* **2005**, 34, 399–414.

169. Liphardt, J.; Onoa, B.; Smith, S. B.; Tinoco, I.; Bustamante, C., Reversible unfolding of single RNA molecules by mechanical force. *Science* **2001**, 292, (5517), 733–737.

170. Onoa, B.; Dumont, S.; Liphardt, J.; Smith, S. B.; Tinoco, I.; Bustamante, C., Identifying kinetic barriers to mechanical unfolding of the T-thermophila ribozyme. *Science* **2003**, 299, (5614), 1892–1895.

171. Ott, A.; Magnasco, M.; Simon, A.; Libchaber, A., Measurement of the Persistence Length of Polymerized Actin Using Fluorescence Microscopy. *Physical Review* E **1993**, 48, (3), R1642-R1645.

172. Tsuda, Y.; Yasutake, H.; Ishijima, A.; Yanagida, T., Torsional rigidity of single actin filaments and actin-actin bond breaking force under torsion measured directly by in vitro micromanipulation. *Proceedings of the National Academy of Sciences of the United States of America* **1996**, 93, (23), 12937–12942.

173. Amblard, F.; Maggs, A. C.; Yurke, B.; Pargellis, A. N.; Leibler, S., Subdiffusion and anomalous local viscoelasticity in actin networks. Physical Review Letters **1996**, 77, (21), 4470–4473.

174. Arai, Y.; Yasuda, R.; Akashi, K.; Harada, Y.; Miyata, H.; Kinosita, K.; Itoh, H., Tying a molecular knot with optical tweezers. *Nature* **1999**, 399, (6735), 446–448.

175. Kurachi, M.; Hoshi, M.; Tashiro, H., Buckling of a Single Microtubule by Optical Trapping Forces - Direct Measurement of Microtubule Rigidity. *Cell Motility and the Cytoskeleton* **1995**, 30, (3), 221–228.

176. Felgner, H.; Frank, R.; Schliwa, M., Flexural rigidity of microtubules measured with the use of optical tweezers. *Journal of Cell Science* **1996**, 109, 509–516.

177. Gittes, F.; Mickey, B.; Nettleton, J.; Howard, J., Flexural Rigidity of Microtubules and Actin-Filaments Measured from Thermal Fluctuations in Shape. *Journal of Cell Biology* **1993**, 120, (4), 923–934.

178. Janson, M. E.; Dogterom, M., A bending mode analysis for growing microtubules: Evidence for a velocity-dependent rigidity. *Biophysical Journal* **2004**, 87, (4), 2723–2736.

179. Tolic-Norrelykke, I. M.; Sacconi, L.; Stringari, C.; Raabe, I.; Pavone, F. S., Nuclear and division-plane positioning revealed by optical micromanipulation. *Current Biology* **2005**, 15, (13), 1212–1216.

180. Mucke, N.; Kreplak, L.; Kirmse, R.; Wedig, T.; Herrmann, H.; Aebi, U.; Langowski, J., Assessing the flexibility of intermediate filaments by atomic force microscopy. *Journal of Molecular Biology* **2004**, 335, (5), 1241–1250.

181. Wang, N.; Stamenovic, D., Contribution of intermediate filaments to cell stiffness, stiffening, and growth. *American Journal of Physiology-Cell Physiology* **2000**, 279, (1), C188-C194.

182. Bright, J. N.; Hoh, J. H.; Woolf, T. B., Computational investigation of confined unstructured proteins. *Biophysical Journal* **2001**, 80, (1), 407A-408A.

183. Kellermayer, M. S. Z., Visualizing and manipulating individual protein molecules. *Physiological Measurement* **2005**, 26, (4), R119-R153.

184. Rief, M.; Gautel, M.; Oesterhelt, F.; Fernandez, J. M.; Gaub, H. E., Reversible unfolding of individual titin immunoglobulin domains by AFM. *Science* **1997**, 276, (5315), 1109–1112.

185. Tskhovrebova, L.; Trinick, J.; Sleep, J. A.; Simmons, R. M., Elasticity and unfolding of single molecules of the giant muscle protein titin. *Nature* **1997**, 387, (6630), 308–312.

186. Kellermayer, M. S. Z.; Smith, S. B.; Bustamante, C.; Granzier, H. L., Complete unfolding of the titin molecule under external force. *Journal of Structural Biology* **1998**, 122, (1–2), 197–205.

187. Kellermayer, M. S. Z.; Smith, S. B.; Bustamante, C.; Granzier, H. L., Mechanical fatigue in repetitively stretched single molecules of titin. *Biophysical Journal* **2001**, 80, (2), 852–863.

188. Svoboda, K.; Schmidt, C. F.; Branton, D.; Block, S. M., Conformation and Elasticity of the Isolated Red-Blood-Cell Membrane Skeleton. *Biophysical Journal* **1992**, 63, (3), 784–793.

189. Li, J.; Dao, M.; Lim, C. T.; Suresh, S., Spectrin-level modeling of the cytoskeleton and optical tweezers stretching of the erythrocyte. *Biophysical Journal* **2005**, 88, (5), 3707–3719.

190. Luo, Z. P.; Bolander, M. E.; An, K. N., A method for determination of stiffness of collagen molecules. *Biochemical and Biophysical Research Communications* **1997**, 232, (1), 251–254.

191. Sun, Y. L.; Luo, Z. P.; An, K. N., Stretching short biopolymers using optical tweezers. *Biochemical and Biophysical Research Communications* **2001**, 286, (4), 826–830.

192. Luo, Z. P.; Sun, Y. L.; Fujii, T.; An, K. N., Single molecule mechanical properties of type II collagen and hyaluronan measured by optical tweezers. *Biorheology* **2004**, 41, (3–4), 247–254.

193. Oesterhelt, F.; Oesterhelt, D.; Pfeiffer, M.; Engel, A.; Gaub, H. E.; Muller, D. J., Unfolding pathways of individual bacteriorhodopsins. *Science* **2000**, 288, (5463), 143–146.

194. Storm, C.; Pastore, J. J.; MacKintosh, F. C.; Lubensky, T. C.; Janmey, P. A., Nonlinear elasticity in biological gels. *Nature* **2005**, 435, (7039), 191–194.

195. Mueller, H.; Butt, H. J.; Bamberg, E., Force measurements on myelin basic protein adsorbed to mica and lipid bilayer surfaces done with the atomic force microscope. *Biophysical Journal* **1999**, 76, (2), 1072–1079.

196. Krigbaum, W. R.; Hsu, T. S., Molecular-Conformation of Bovine a-1 Basic-Protein, a Coiling Macromolecule in Aqueous-Solution. *Biochemistry* **1975**, 14, (11), 2542–2546.

197. Oberdorfer, Y.; Schrot, S.; Fuchs, H.; Galinski, E.; Janshoff, A., Impact of compatible solutes on the mechanical properties of fibronectin: a single molecule analysis. *Physical Chemistry Chemical Physics* 2003, 5, (9), 1876–1881.

198. Sharon, N.; Lis, H., Carbohydrates in Cell Recognition. *Scientific American* 1993, 268, (1), 82–89.

199. Picout, D. R.; Ross-Murphy, S. B.; Errington, N.; Harding, S. E., Pressure cell assisted solubilization of xyloglucans: Tamarind seed polysaccharide and detarium gum. *Biomacromolecules* **2003**, 4, (3), 799–807.

200. Sletmoen, M.; Maurstad, G.; Sikorski, P.; Paulsen, B. S.; Stokke, B. T., Characterisation of bacterial polysaccharides: steps towards single-molecular studies. *Carbohydrate Research* **2003**, 338, (23), 2459–2475.

201. Rief, M.; Oesterhelt, F.; Heymann, B.; Gaub, H. E., Single molecule force spectroscopy on polysaccharides by atomic force microscopy. *Science* **1997**, 275, (5304), 1295–1297.

202. Janicijevic, A.; Ristic, D.; Wyman, C., The molecular machines of DNA repair: scanning force microscopy analysis of their architecture. *Journal of Microscopy-Oxford* **2003**, 212, 264–272.

203. Dong, C.; So, P. T. C.; Mahadevan, L.; Kaizuka, Y.; Sutin, J. D.; Graton, E., Control of exonuclease digestion activities by microscopic mechanical forces. *Biophysical Journal* **1999**, 76, (1), A132-a132.

204. van Oijen, A. M.; Blainey, P. C.; Crampton, D. J.; Richardson, C. C.; Ellenberger, T.; Xie, X. S., Single-molecule kinetics of lambda exonuclease reveal base dependence and dynamic disorder. *Science* **2003**, 301, (5637), 1235–1238.

205. Perkins, T. T.; Dalal, R. V.; Mitsis, P. G.; Block, S. M., Sequence-dependent pausing of single lambda exonuclease molecules. *Science* **2003**, 301, (5641), 1914–1918.

206. Werner, J. H.; Cai, H.; Keller, R. A.; Goodwin, P. M., Exonuclease I hydrolyzes DNA with a distribution of rates. *Biophysical Journal* **2005**, 88, (2), 1403–1412.

207. Yin, H.; Wang, M. D.; Svoboda, K.; Landick, R.; Block, S. M.; Gelles, J., Transcription against an Applied Force. *Science* **1995**, 270, (5242), 1653–1657.

208. Wuite, G. J. L.; Smith, S. B.; Young, M.; Keller, D.; Bustamante, C., Single-molecule studies of the effect of template tension on T7 DNA polymerase activity. *Nature* **2000**, 404, (6773), 103–106.

209. Thomen, P.; Lopez, P. J.; Heslot, F., Unravelling the mechanism of RNA-polymerase forward motion by using mechanical force. *Physical Review Letters* **2005**, 94, (12), -.

210. Wang, M. D.; Schnitzer, M. J.; Yin, H.; Landick, R.; Gelles, J.; Block, S. M., Force and velocity measured for single molecules of RNA polymerase. *Science* **1998**, 282, (5390), 902–907.

211. Skinner, G. M.; Baumann, C. G.; Quinn, D. M.; Molloy, J. E.; Hoggett, J. G., Promoter binding, initiation, and elongation by bacteriophage T7 RNA polymerase - A single-molecule view of the transcription cycle. *Journal of Biological Chemistry* **2004**, 279, (5), 3239–3244.

212. Wang, H. Y.; Elston, T.; Mogilner, A.; Oster, G., Force generation in RNA polymerase. *Biophysical Journal* **1998**, 74, (3), 1186–1202.

213. Forde, N. R.; Izhaky, D.; Woodcock, G. R.; Wuite, G. J. L.; Bustamante, C., Using mechanical force to probe the mechanism of pausing and arrest during continuous elongation by Escherichia coli RNA polymerase. *Proceedings of the National Academy of Sciences of the United States of America* **2002**, 99, (18), 11682–11687.

214. Waksman, G.; Lanka, E.; Carazo, J. M., Helicases as nucleic acid unwinding machines. *Nature Structural Biology* **2000**, 7, (1), 20–22.

215. Bianco, P. R.; Brewer, L. R.; Corzett, M.; Balhorn, R.; Yeh, Y.; Kowalczykowski, S. C.; Baskin, R. J., Processive translocation and DNA unwinding by individual RecBCD enzyme molecules. *Nature* **2001**, 409, (6818), 374–378.

216. Perkins, T. T.; Li, H. W.; Dalal, R. V.; Gelles, J.; Block, S. M., Forward and reverse motion of single RecBCD molecules on DNA. *Biophysical Journal* **2004**, 86, (3), 1640–1648.

217. Dawid, A.; Croquette, V.; Grigoriev, M.; Heslot, F., Single-molecule study of RuvAB-mediated Holliday-junction migration. Proceedings of the National Academy of Sciences of the United States of America **2004**, 101, (32), 11611–11616.

218. Amit, R.; Gileadi, O.; Stavans, J., Direct observation of RuvAB-catalyzed branch migration of single Holliday junctions. *Proceedings of the National Academy of Sciences of the United States of America* **2004**, 101, (32), 11605–11610.

219. Dessinges, M. N.; Lionnet, T.; Xi, X. G.; Bensimon, D.; Croquette, V., Single-molecule assay reveals strand switching and enhanced processivity of UvrD. *Proceedings of the National Academy of Sciences of the United States of America* **2004**, 101, (17), 6439–6444.

220. Champoux, J. J., DNA topoisomerases: Structure, function, and mechanism. Annual Review of Biochemistry **2001**, 70, 369–413.

221. Charvin, G.; Strick, T. R.; Bensimon, D.; Croquette, V., Tracking topoisomerase activity at the single-molecule level. *Annual Review of Biophysics and Biomolecular Structure* **2005**, 34, 201–219.

222. Dekker, N. H.; Rybenkov, V. V.; Duguet, M.; Crisona, N. J.; Cozzarelli, N. R.; Bensimon, D.; Croquette, V., The mechanism of type IA topoisomerases. *Proceedings of the National Academy of Sciences of the United States of America* **2002**, 99, (19), 12126–12131.

223. Koster, D. A.; Croquette, V.; Dekker, C.; Shuman, S.; Dekker, N. H., Friction and torque govern the relaxation of DNA supercoils by eukaryotic topoisomerase IB. *Nature* **2005**, 434, (7033), 671–674.

224. Strick, T. R.; Croquette, V.; Bensimon, D., Single-molecule analysis of DNA uncoiling by a type II topoisomerase. *Nature* **2000**, 404, (6780), 901–904.

225. Charvin, G.; Bensimon, D.; Croquette, V., Single-molecule study of DNA unlinking by eukaryotic and prokaryotic type-II topoisomerases. *Proceedings of the National Academy of Sciences of the United States of America* **2003**, 100, (17), 9820–9825.

226. Crisona, N. J.; Strick, T. R.; Bensimon, D.; Croquette, V.; Cozzarelli, N. R., Preferential relaxation of positively supercoiled DNA by E-coli topoisomerase IV in single-molecule and ensemble measurements. *Genes & Development* **2000**, 14, (22), 2881–2892.

227. Smith, D. E.; Tans, S. J.; Smith, S. B.; Grimes, S.; Anderson, D. L.; Bustamante, C., The bacteriophage phi 29 portal motor can package DNA against a large internal force. *Nature* **2001**, 413, (6857), 748–752.

228. Sabbert, D.; Engelbrecht, S.; Junge, W., Functional and idling rotatory motion within F-1-ATPase. *Proceedings of the National Academy of Sciences of the United States of America* **1997**, 94, (9), 4401–4405.

229. Yasuda, R.; Noji, H.; Kinosita, K.; Yoshida, M., F-1-ATPase is a highly efficient molecular motor that rotates with discrete 120 degrees steps. *Cell* **1998**, 93, (7), 1117–1124.

230. Itoh, H.; Takahashi, A.; Adachi, K.; Noji, H.; Yasuda, R.; Yoshida, M.; Kinosita, K., Mechanically driven ATP synthesis by F-1-ATPase. *Nature* **2004**, 427, (6973), 465–468.

231. Ueno, H.; Suzuki, T.; Kinosita, K.; Yoshida, M., ATP-driven stepwise rotation of FOF1,-ATP synthase. *Proceedings of the National Academy of Sciences of the United States of America* **2005**, 102, (5), 1333–1338.

232. Kull, F. J.; Sablin, E. P.; Lau, R.; Fletterick, R. J.; Vale, R. D., Crystal structure of the kinesin motor domain reveals a structural similarity to myosin. *Nature* **1996**, 380, (6574), 550–555.

233. Finer, J. T.; Simmons, R. M.; Spudich, J. A., Single Myosin Molecule Mechanics - Piconewton Forces and Nanometer Steps. *Nature* **1994**, 368, (6467), 113–119.

234. Mehta, A. D.; Finer, J. T.; Spudich, J. A., Detection of single-molecule interactions using correlated thermal diffusion. *Proceedings of the National Academy of Sciences of the United States of America* **1997**, 94, (15), 7927–7931.

235. Ishijima, A.; Kojima, H.; Higuchi, H.; Harada, Y.; Funatsu, T.; Yanagida, T., Multiple- and single-molecule analysis of the actomyosin motor by nanometer piconewton manipulation with a microneedle: Unitary steps and forces. *Biophysical Journal* **1996**, 70, (1), 383–400.

236. Tanaka, H.; Ishijima, A.; Honda, M.; Saito, K.; Yanagida, T., Orientation dependence of displacements by a single one-headed myosin relative to the actin filament. *Biophysical Journal* **1998**, 75, (4), 1886–1894.

237. Geeves, M. A.; Holmes, K. C., Structural mechanism of muscle contraction. *Annual Review of Biochemistry* **1999**, 68, 687–728.

238. Vale, R. D., Myosin V motor proteins: marching stepwise towards a mechanism. *Journal of Cell Biology* **2003**, 163, (3), 445–450.

239. Schott, D. H.; Collins, R. N.; Bretscher, A., Secretory vesicle transport velocity in living cells depends on the myosin-V lever arm length. *Journal of Cell Biology* **2002**, 156, (1), 35–39.

240. Rief, M.; Rock, R. S.; Mehta, A. D.; Mooseker, M. S.; Cheney, R. E.; Spudich, J. A., Myosin-V stepping kinetics: A molecular model for processivity. *Proceedings of the National Academy of Sciences of the United States of America* **2000**, 97, (17), 9482–9486.

241. Yildiz, A.; Forkey, J. N.; McKinney, S. A.; Ha, T.; Goldman, Y. E.; Selvin, P. R., Myosin V walks hand-over-hand: Single fluorophore imaging with 1.5-nm localization. *Science* **2003**, 300, (5628), 2061–2065.

242. Snyder, G. E.; Sakamoto, T.; Hammer, J. A.; Sellers, J. R.; Selvin, P. R., Nanometer localization of single green fluorescent proteins: Evidence that myosin V walks hand-over-hand via telemark configuration. *Biophysical Journal* **2004**, 87, (3), 1776–1783.

243. Warshaw, D. M.; Kennedy, G. G.; Work, S. S.; Krementsova, E. B.; Beck, S.; Trybus, K. M., Differential Labeling of myosin V heads with quantum dots allows direct visualization of hand-over-hand processivity. *Biophysical Journal* **2005**, 88, (5), L30-L32.

244. Rock, R. S.; Ramamurthy, B.; Dunn, A. R.; Beccafico, S.; Rami, B. R.; Morris, C.; Spink, B. J.; Franzini-Armstrong, C.; Spudich, J. A.; Sweeney, H. L., A flexible domain is essential for the large step size and processivity of myosin VI. *Molecular Cell* **2005**, 17, (4), 603–609.

245. Veigel, C.; Coluccio, L. M.; Jontes, J. D.; Sparrow, J. C.; Milligan, R. A.; Molloy, J. E., The motor protein myosin-I produces its working stroke in two steps. *Nature* **1999**, 398, (6727), 530–533.

246. Tyska, M. J.; Warshaw, D. M., The myosin power stroke. *Cell Motility and the Cytoskeleton* **2002**, 51, (1), 1–15.

247. De La Cruz, E. M.; Ostap, E. M., Relating biochemistry and function in the myosin superfamily. *Current Opinion in Cell Biology* **2004**, 16, (1), 61–67.

248. Howard, J., Molecular motors: structural adaptations to cellular functions. Nature 1997, 389, (6651), 561–567.

249. Howard, J.; Hudspeth, A. J.; Vale, R. D., Movement of Microtubules by Single Kinesin Molecules. *Nature* **1989**, 342, (6246), 154–158.

250. Kojima, H.; Muto, E.; Higuchi, H.; Yanagida, T., Mechanics of single kinesin molecules measured by optical trapping nanometry. *Biophys J* **1997**, 73, (4), 2012–22.

251. Hua, W.; Young, E. C.; Fleming, M. L.; Gelles, J., Coupling of kinesin steps to ATP hydrolysis. *Nature* **1997**, 388, (6640), 390–393.

252. Schnitzer, M. J.; Block, S. M., Kinesin hydrolyses one ATP per 8-nm step. *Nature* **1997**, 388, (6640), 386–390.

253. Yildiz, A.; Tomishige, M.; Vale, R. D.; Selvin, P. R., Kinesin walks hand-over-hand. *Science* **2004**, 303, (5658), 676–678.

254. Higuchi, H.; Bronner, C. E.; Park, H. W.; Endow, S. A., Rapid double 8-nm steps by a kinesin mutant. *Embo Journal* **2004**, 23, (15), 2993–2999.

255. Yildiz, A.; Selvin, P. R., Kinesin: walking, crawling or sliding along? *Trends in Cell Biology* **2005**, 15, (2), 112–120.

256. Kuo, S. C.; Gelles, J.; Steuer, E.; Sheetz, M. P., A Model for Kinesin Movement from Nanometer-Level Movements of Kinesin and Cytoplasmic Dynein and Force Measurements. *Journal of Cell Science* **1991**, 135–138.

257. Wang, Z. H.; Khan, S.; Sheetz, M. P., Different Patterns of Kinesin and Cytoplasmic Dynein Movement - a Single Mechanism. *Biophysical Journal* **1995**, 68, (4), S328-S328.

258. Wang, Z. H.; Sheetz, M. P., One-dimensional diffusion on microtubules of particles coated with cytoplasmic dynein an immunoglobulins. *Cell Structure and Function* **1999**, 24, (5), 373–383.

259. Mallik, R.; Carter, B. C.; Lex, S. A.; King, S. J.; Gross, S. P., Cytoplasmic dynein functions as a gear in response to load. *Nature* **2004**, 427, (6975), 649–652.

260. Sakakibara, H.; Kojima, H.; Sakai, Y.; Katayama, E.; Oiwa, K., Inner-arm dynein c of Chlamydomonas flagella is a single-headed processive motor. *Nature* **1999**, 400, (6744), 586–590.

261. Gee, M.; Vallee, R., The role of the dynein stalk in cytoplasmic and flagellar motility. *European Biophysics Journal with Biophysics Letters* **1998**, 27, (5), 466–473.

262. Block, S. M.; Blair, D. F.; Berg, H. C., Compliance of Bacterial Flagella Measured with Optical Tweezers. *Nature* **1989**, 338, (6215), 514–518.

263. Berry, R. M.; Berg, H. C., Absence of a barrier to backwards rotation of the bacterial flagellar motor demonstrated with optical tweezers. *Proceedings of the National Academy of Sciences of the United States of America* **1997**, 94, (26), 14433–14437.

264. Tadir, Y.; Wright, W. H.; Vafa, O.; Ord, T.; Asch, R. H.; Berns, M. W., Force Generated by Human Sperm Correlated to Velocity and Determined Using a Laser Generated Optical Trap. *Fertility and Sterility* **1990**, 53, (5), 944–947.

265. McCord, R. P.; Yukich, J. N.; Bernd, K. K., Analysis of force generation during flagellar assembly through optical trapping of free-swimming Chlamydomonas reinhardtii. *Cell Motility and the Cytoskeleton* **2005**, 61, (3), 137–144.

266. Gilad, R.; Porat, A.; Trachtenberg, S., Motility modes of Spiroplasma melliferum BC3: a helical, wall-less bacterium driven by a linear motor. *Molecular Microbiology* **2003**, 47, (3), 657–669.

267. Konig, K.; Svaasand, L.; Liu, Y. G.; Sonek, G.; Patrizio, P.; Tadir, Y.; Berns, M. W.; Tromberg, B. J., Determination of motility forces of human spermatozoa using an 800 nm optical trap. *Cellular and Molecular Biology* **1996**, 42, (4), 501–509.

268. Schutze, K.; Clementsengewald, A.; Ashkin, A., Zona Drilling and Sperm Insertion with Combined Laser Microbeam and Optical Tweezers. *Fertility and Sterility* **1994**, 61, (4), 783–786.

269. Patrizio, P.; Liu, Y. G.; Sonek, G. J.; Berns, M. W.; Tadir, Y., Effect of pentoxifylline on the intrinsic swimming forces of human sperm assessed by optical tweezers. *Journal of Andrology* **2000**, 21, (5), 753–756.

270. Yagi, K., The mechanical and colloidal properties of Amoeba protoplasm and their relations to the mechanism of amoeboid movement. *Comp Biochem Physiol* **1961**, 3, 73–91.

271. Cohen, D., Ferromagnetic Contamination in Lungs and Other Organs of Human Body. *Science* **1973**, 180, (4087), 745–748.

272. Valberg, P. A.; Meyrick, B.; Brain, J. D.; Brigham, K. L., Phagocytic and Motile Properties of Endothelial-Cells Measured Magnetometrically - Effects of Endotoxin. *Tissue & Cell* **1988**, 20, (3), 345–354.

273. Valberg, P. A.; Butler, J. P., Magnetic Particle Motions within Living Cells - Physical Theory and Techniques. *Biophysical Journal* **1987**, 52, (4), 537–550.

274. Wang, N.; Butler, J. P.; Ingber, D. E., Mechanotransduction across the Cell-Surface and through the Cytoskeleton. *Science* **1993**, 260, (5111), 1124–1127.

275. Huang, H. D.; Kamm, R. D.; Lee, R. T., Cell mechanics and mechanotransduction: pathways, probes, and physiology. *American Journal of Physiology-Cell Physiology* **2004**, 287, (1), C1-C11.

276. Ikai, A.; Afrin, R.; Sekiguchi, H.; Okajima, T.; Alam, M. T.; Nishida, S., Nano-mechanical methods in bio-chemistry using atomic force microscopy. *Current Protein & Peptide Science* **2003**, 4, (3), 181–193.

277. Galbraith, C. G.; Yamada, K. M.; Sheetz, M. P., The relationship between force and focal complex development. *Journal of Cell Biology* **2002**, 159, (4), 695–705.

278. Choquet, D.; Felsenfeld, D. P.; Sheetz, M. P., Extracellular matrix rigidity causes strengthening of integrin-cytoskeleton linkages. *Cell* **1997**, 88, (1), 39–48.

279. Kusumi, A.; Sako, Y.; Fujiwara, T.; Tomishige, M., Application of laser tweezers to studies of the fences and tethers of the membrane skeleton that regulate the movements of plasma membrane proteins. *Methods in Cell Biology, Vol 55* **1998**, 55, 173–194.

280. Peters, I. M.; van Kooyk, Y.; van Vliet, S. J.; de Grooth, B. G.; Figdor, C. G.; Greve, J., 3D single-particle tracking and optical trap measurements on adhesion proteins. *Cytometry* **1999**, 36, (3), 189–194.

281. Falk, J.; Thoumine, O.; Dequidt, C.; Choquet, D.; Faivre-Sarrailh, C., NrCAM coupling to the cytoskeleton depends on multiple protein domains and partitioning into lipid rafts. *Molecular Biology of the Cell* **2004**, 15, (10), 4695–4709.

282. Sako, Y.; Nagafuchi, A.; Tsukita, S.; Takeichi, M.; Kusumi, A., Cytoplasmic regulation of the movement of E-cadherin on the free cell surface as studied by optical tweezers and single particle tracking: Corralling and tethering by the membrane skeleton. *Journal of Cell Biology* **1998**, 140, (5), 1227–1240.

283. Oddershede, L.; Flyvbjerg, H.; Berg-Sorensen, K., Single-molecule experiment with optical tweezers: improved analysis of the diffusion of the lambda-receptor in E-coli's outer membrane. *Journal of Physics-Condensed Matter* **2003**, 15, (18), S1737-S1746.

284. Sako, Y.; Kusumi, A., Barriers for Lateral Diffusion of Transferrin Receptor in the Plasma-Membrane as Characterized by Receptor Dragging by Laser Tweezers - Fence Versus Tether. *Journal of Cell Biology* **1995**, 129, (6), 1559–1574.

285. Tomishige, M.; Sako, Y.; Kusumi, A., Regulation mechanism of the lateral diffusion of band 3 in erythrocyte membranes by the membrane skeleton. *Journal of Cell Biology* **1998**, 142, (4), 989–1000.

286. Arya, M.; Lopez, J. A.; Romo, G. M.; Cruz, A. A.; Kasirer-Friede, A.; Shattil, S. J.; Anvari, B., Glycoprotein Ib-IX-mediated activation of integrin alpha(IIb)beta(3): effects of receptor clustering and von Willebrand factor adhesion. *Journal of Thrombosis and Haemostasis* **2003**, 1, (6), 1150–1157.

287. Stoltz, J. F.; Dumas, D.; Wang, X.; Payan, E.; Mainard, D.; Paulus, F.; Maurice, G.; Netter, P.; Muller, S., Influence of mechanical forces on cells and tissues. *Biorheology* **2000**, 37, (1–2), 3–14.

288. Ermilov, S. A.; Murdock, D. R.; El-Daye, D.; Brownell, W. E.; Anvari, B., Effects of salicylate on plasma membrane mechanics. *Journal of Neurophysiology* **2005**, 94, (3), 2105–2110.

289. Zahn, M.; Seeger, S., Optical tweezers in pharmacology. *Cellular and Molecular Biology* **1998**, 44, (5), 747–761.

290. Ashkin, A.; Schutze, K.; Dziedzic, J. M.; Euteneuer, U.; Schliwa, M., Force Generation of Organelle Transport Measured Invivo by an Infrared-Laser Trap. *Nature* **1990**, 348, (6299), 346–348.

291. Welte, M. A.; Gross, S. P.; Postner, M.; Block, S. M.; Wieschaus, E. F., Developmental regulation of vesicle transport in Drosophila embryos: Forces and kinetics. *Cell* **1998**, 92, (4), 547–557.

292. Schneckenburger, H.; Hendinger, A.; Sailer, R.; Strauss, W. S. L.; Schmitt, M., Laser-assisted optoporation of single cells. *Journal of Biomedical Optics* **2002**, 7, (3), 410–416.

293. Aufderheide, K. J.; Du, Q.; Fry, E. S., Directed Positioning of Micronuclei in Paramecium-Tetraurelia with Laser Tweezers - Absence of Detectable Damage after Manipulation. *Journal of Eukaryotic Microbiology* **1993**, 40, (6), 793–796.

294. Liu, X.; Wang, H.; Li, Y.; Tang, Y.; Liu, Y.; Hu, X.; Jia, P.; Ying, K.; Feng, Q.; Guan, J.; Jin, C.; Zhang, L.; Lou, L.; Zhou, Z.; Han, B., Preparation of single rice chromosome for construction of a DNA library using a laser microbeam trap. *J Biotechnol* **2004**, 109, (3), 217–26.

295. Conia, J.; Voelkel, S., Optical Manipulations of Human Gametes. *Biotechniques* **1994**, 17, (6), 1162–1165.

296. Leitz, G.; Weber, G.; Seeger, S.; Greulich, K. O., The Laser Microbeam Trap as an Optical Tool for Living Cells. *Physiological Chemistry and Physics and Medical Nmr* **1994**, 26, (1), 69–88.

297. Thalhammer, S.; Lahr, G.; Clement-Sengewald, A.; Heckl, W. M.; Burgemeister, R.; Schutze, K., Laser micro-tools in cell biology and molecular medicine. *Laser Physics* **2003**, 13, (5), 681–691.

298. Leitz, G.; Schnepf, E.; Greulich, K. O., Micromanipulation of Statoliths in Gravity-Sensing Chara Rhizoids by Optical Tweezers. *Planta* **1995**, 197, (2), 278–288.

299. Lee, H.; Purdon, A. M.; Westervelt, R. M., Micromanipulation of biological systems with microelectromagnets. *Ieee Transactions on Magnetics* **2004**, 40, (4), 2991–2993.

300. Thoumine, O.; Kocian, P.; Kottelat, A.; Meister, J. J., Short-term binding of fibroblasts to fibronectin: optical tweezers experiments and probabilistic analysis. *European Biophysics Journal with Biophysics Letters* **2000**, 29, (6), 398–408.

301. Liang, M. N.; Smith, S. P.; Metallo, S. J.; Choi, I. S.; Prentiss, M.; Whitesides, G. M., Measuring the forces involved in polyvalent adhesion of uropathogenic Escherichia coli to mannose-presenting surfaces. *Proceedings of the National Academy of Sciences of the United States of America* **2000**, 97, (24), 13092–13096.

302. Grimbergen, J. A.; Visscher, K.; Demesquita, D. S. G.; Brakenhoff, G. J., Isolation of Single Yeast-Cells by Optical Trapping. *Yeast* **1993**, 9, (7), 723–732.

303. Huber, R.; Burggraf, S.; Mayer, T.; Barns, S. M.; Rossnagel, P.; Stetter, K. O., Isolation of a Hyperthermophilic Archaeum Predicted by in-Situ Rna Analysis. *Nature* **1995**, 376, (6535), 57–58.

304. Bronkhorst, P. J. H.; Streekstra, G. J.; Grimbergen, J.; Nijhof, E. J.; Sixma, J. J.; Brakenhoff, G. J., A new method to study shape recovery of red blood cells using multiple optical trapping. *Biophysical Journal* **1995**, 69, (5), 1666–1673.

305. Bayoudh, S.; Mehta, M.; Rubinsztein-Dunlop, H.; Heckenberg, N. R.; Critchley, C., Micromanipulation of chloroplasts using optical tweezers. *Journal of Microscopy-Oxford* **2001**, 203, 214–222.

306. Brandao, M. M.; Fontes, A.; Barjas-Castro, M. L.; Barbosa, L. C.; Costa, F. F.; Cesar, C. L.; Saad, S. T. O., Optical tweezers for measuring red blood cell elasticity: application to the study of drug response in sickle cell disease. *European Journal of Haematology* **2003**, 70, (4), 207–211.

307. Barjas-Castro, M. L.; Brandao, M. M.; Fontes, A.; Costa, F. F.; Cesar, C. L.; Saad, S. T. O., Elastic properties of irradiated RBCs measured by optical tweezers. *Transfusion* **2002**, 42, (9), 1196–1199.

308. Yin, S. H.; Zhang, X. Q.; Zhan, C.; Wu, J. T.; Xu, J. C.; Cheung, J., Measuring single cardiac myocyte contractile force via moving a magnetic bead. *Biophysical Journal* **2005**, 88, (2), 1489–1495.

309. Curtis, J. E.; Grier, D. G., Structure of optical vortices. *Physical Review Letters* 90, (13): Art. No. 133901 APR 4 **2003**.

310. Edelstein, R. L.; Tamanaha, C. R.; Sheehan, P. E.; Miller, M. M.; Baselt, D. R.; Whitman, L. J.; Colton, R. J., The BARC biosensor applied to the detection of biological warfare agents. *Biosensors & Bioelectronics* **2000**, 14, (10–11), 805–813.

3

Chemical Force Microscopy Nanoscale Probing of Fundamental Chemical Interactions

Aleksandr Noy, Dmitry V. Vezenov, and Charles M. Lieber

1   Basic Principles of Chemical Force Microscopy

1.1   Chemical Sensitivity in Scanning Probe Microscopy Measurements

Intermolecular forces impact a wide spectrum of problems in condensed phases: from molecular recognition, self-assembly, and protein folding at the molecular and nanometer scale, to interfacial fracture, friction, and lubrication at a macroscopic length scale. Understanding these phenomena, regardless of the length scale, requires fundamental knowledge of the magnitude and range of underlying weak interactions between basic chemical functionalities in these systems (Figure 1). While the theoretical description has long recognized that intermolecular forces are necessarily microscopic in origin, experimental efforts in direct force measurements at the microscopic level have been lagging behind and have only intensified in the course of the last decade. Atomic force microscopy (AFM)[1,2] is an ideal tool for probing interactions between various chemical groups, since it has pico-Newton force sensitivity (i.e., several orders of magnitude better than the weakest chemical bond[3]) and sub-nanometer spatial resolution (i.e., approaching the length of a chemical bond). These features enable AFM to produce nanometer to micron scale images of surface topography, adhesion, friction, and compliance, and make it an essential characterization technique for fields ranging from materials science to biology.

As the name implies, intermolecular forces are at the center of the AFM operation. However, during the routine use of this technique the specific chemical groups on an AFM probe tip are typically ill-defined. To overcome this inherent limitation of the AFM, Lieber and co-workers introduced the concept of chemical modification of force probes to make them sensitive to specific molecular interactions[4]. By using chemically-functionalized tips, a force microscope can be transformed into a tool that can (i) quantify forces between different molecular groups, (ii) probe surface free energies on a nanometer scale, (iii) determine pK_a values of the surface acid/base groups locally, and (iv) map the spatial distribution of specific functional groups and their ionization state. This ability to discriminate between chemically distinct functional groups has led the Lieber group to name the variation of force microscopy carried out with specifically functionalized tips "chemical force microscopy" (CFM)[4].

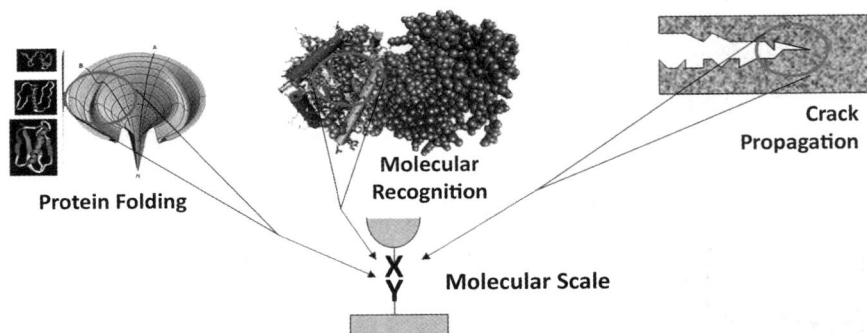

Figure 1. Molecular-scale view of the different examples of condensed phase phenomena illustrating the importance of intermolecular interactions.

1.2   Measuring Interaction Forces with an Atomic Force Microscope

A typical force microscope consists of an integrated cantilever-tip assembly interacting with the sample surface, a detector that measures the displacement of the cantilever and feedback electronics to maintain a constant imaging parameter such as tip-sample separation or force (Figure 2). The integrated cantilever-tip assemblies can have single or V-shaped beams[5] and normal spring constants, k_z, in the range of 0.01–100 N/m. By far, the most popular and versatile detection scheme in AFM is optical lever deflection.[6] In this scheme, the vertical displacement due to normal forces and lateral twist due to friction of the cantilever are measured using a quadrant photodiode, as shown in Figure 2. Force values are determined from the normal displacement, Δz, of the cantilever from its rest position. With an instrumental sensitivity on the order of 0.1 Å, minimal forces in the range of 10^{-13}–10^{-8} N (depending on the cantilever stiffness) can be measured. Hence, AFM can in principle measure molecular interactions ranging from weak van der Waals ($<10^{-12}$ N) to strong covalent (10^{-7} N) bonds.[7–9] In practice, the displacement (and corresponding force) sensitivity is limited by thermally excited cantilever vibrations, optical and electronic noise.[10] If the measurements are conducted in ambient air or liquids, the thermal noise is especially important. For example, cantilever quality factors drop from 10^3-10^5 in vacuum to 10^0-10^2 in fluids due to hydrodynamic damping. The thermal noise limited minimal force is then on the order of 1–20 pN at room temperature. Use of specially designed small cantilevers for AFM can push the force detection threshold to even lower values.[11,12]

AFM measures the magnitude of intermolecular interactions by performing a force-distance measurement, commonly referred to as an F-D curve or simply as a "force curve" (Figure 3). In these measurements, the deflection of the cantilever is recorded during the sample approach-withdrawal cycle.[10] The magnitude of the jump in the retraction trace corresponds to the adhesion between functional groups on the tip and sample surfaces. The observed deflection of the cantilever is converted into a force of adhesion using the cantilever spring constant.

The sphere-on-flat tip-sample geometry of the typical AFM force measurement does not correspond to the interaction between two molecules, as shown in Figure 1. However, the general features of the interaction potential are the same; that is, the potential has a minimum and increases nonlinearly from this minimum (Figure 3). If the cantilever were infinitely stiff, the probe deflection would have simply traced the gradient of the interaction potential (of course, an infinitely stiff cantilever would not have generated any measurable deflection, making such an experiment fairly useless). In practice, the molecular force gradients are higher than the stiffness of the typical cantilevers over a substantial part of the intermolecular force profile; therefore, most AFM cantilevers experience mechanical

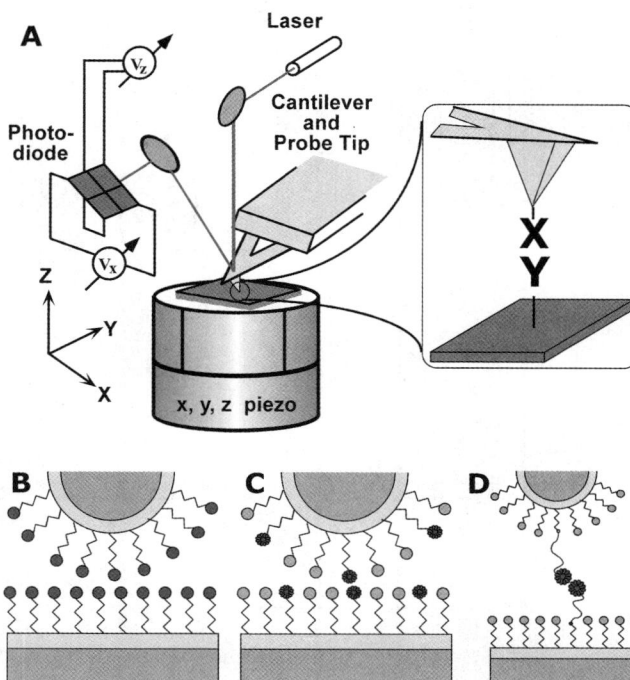

Figure 2. (A) Schematics of the atomic force microscope. The inset shows a close-up of the interactions of the probe and sample surfaces modified with specific functional groups. (B-D) Possible configurations of probe tip and sample functionalization. (B) All terminal functionalities bear the same chemical groups. (C) Active functional groups (red) on the tip are "diluted" with inactive spacer molecules (grey). (D) Interacting groups are attached to the surfaces of tip and sample through long flexible polymer tethers.

instabilities during the force curve cycle. Whenever tip-surface force gradient exceeds the cantilever spring constant, the probe will jump in and out of contact with the sample surface (Figure 3). The magnitude of the adhesion force, which corresponds to the jump out from the potential minimum, is measured precisely in these experiments. Several recent developments have also attempted to use the AFM to map the entire interaction potential (see the chapter by P. Ashby for a detailed discussion of these efforts).

1.2.1 CFM and Other Force Spectroscopy Techniques

CFM is not unique in its ability to measure interactions between molecules and molecular assemblies; other techniques such as the surface forces apparatus (SFA)[13], an elastomer lens-on-plate (or "JKR") apparatus[14,15], colloidal probe microscope[16,17], interfacial force microscope[18], and optical tweezers[19] have also been successful in probing molecular-scale forces. Yet AFM-based techniques hold the distinction of being able to probe specific interactions on surfaces *locally*, thus having the ability to combine quantitative intermolecular force measurements with high-resolution imaging.

1.3 AFM Tip Modification with Chemical Functional Groups

General strategies for modification of the AFM probe tips with different chemical functionalities are covered extensively in the chapter 7; therefore, we will only briefly touch upon the basic strategies of probe functionalization for CFM measurements. To probe

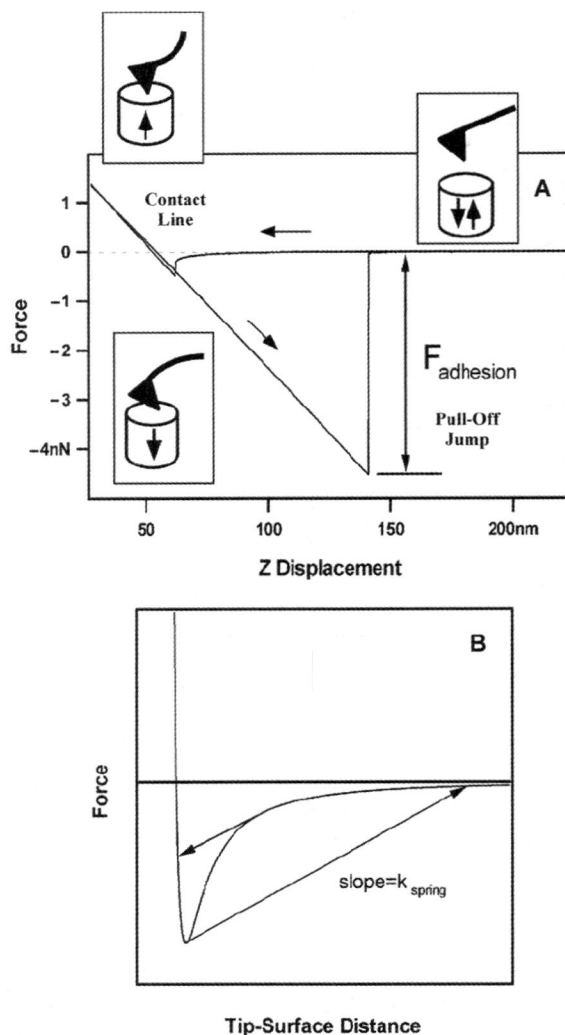

Figure 3. (A) Typical force-vs-Z-piezo displacement curve. At large separations, no force is observed between tip and sample. At short distances, the van der Waals attraction will pull the tip abruptly into contact with the sample (jump into contact point). After that, the deflection of the soft cantilever tracks the movement of the sample (linear compliance regime). Hysteresis in the force between tip and sample is observed when the tip is withdrawn from the sample. The finite force necessary to pull the tip off the sample surface corresponds to the adhesive force between functional groups on the surfaces of the tip and sample. (B) The cycle in (A) is shown as a schematic intermolecular potential between the tip-sample functional groups. Whenever the force gradient exceeds the cantilever spring constant k_{spring} the system becomes mechanically unstable and cantilever jumps occur.

interactions between functional groups in a rationally controlled manner, we need to modify the probes with well defined molecular layers. The simplest way that this modification can be accomplished is by using monolayers of amphiphilic molecules chemisorbed on the surface of the tip. Different types of interactions can then in principle be studied by varying the head group of the amphiphile.

By far the most well developed approach involves self-assembly of monolayers (SAMs) of functionalized organic thiols onto the surfaces of Au-coated Si_3N_4 or Si tips.

Stable, robust, and crystalline monolayers of alkyl thiols or disulfides containing a variety of terminal groups are a staple of the surface science literature and are readily prepared[20–25]. Using systems of thiol SAMs on Au, systematic studies have been carried out on the interactions between basic chemical groups on the probe tip and similarly modified Au substrates.[4, 26–37] Covalent modification of AFM probes with reactive silanes or through direct Si-C bond formation has also been reported.[102–104]

1.4 Role of Environment in CFM Experiments: Experiment Design and Basic Types of Detectable Forces

The environment, in which surfaces interact, plays a crucial role in determining the nature and the magnitude of measured forces (Figure 4). To probe interactions determined solely by solid surface free energies (i.e., bare functional groups), we need to measure adhesion forces in vacuum. These measurements (Figure 4A) would admittedly be the easiest to compare with the available body of the computational literature; but unfortunately more often than not these measurements have very little to do with the real systems in which the functional groups are exposed to the ambient environment or immersed in water or other liquid medium. Force measurements carried out in ambient air (Figure 4B) are much more relevant to the practical applications, yet they are also much more difficult to interpret. The major contribution to the interactions in ambient air (and the major obstacle to consistent data interpretation) comes from capillary forces generated by the meniscus forming between the AFM probe and sample surfaces (Figure 4C). These forces[8, 38, 39] are usually 1–2 orders of magnitude higher than specific chemical interactions, and thus they almost always obscure the relatively small differences in molecular forces between tip and sample functional groups. We also note that capillary condensation will emphasize the relative degree of wetability (hydrophilicity) and can serve as a basis for discriminating between hydrophobic and hydrophilic groups when

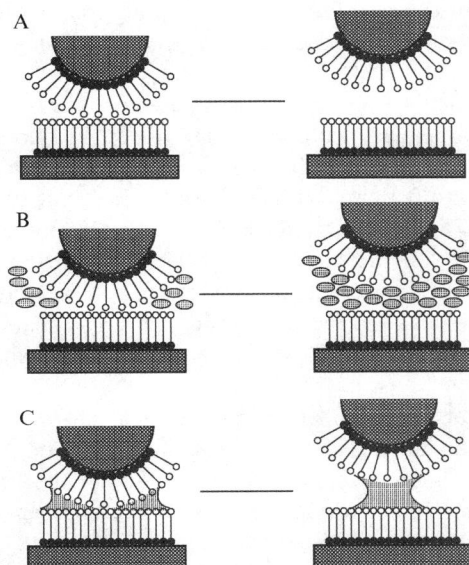

Figure 4. Comparison between force measurements in ultrahigh vacuum, liquids, and ambient air. (A) Interactions under high vacuum conditions are determined by solid surface free energies. (B) Pulling apart the surfaces under liquid will result in their solvation upon separation. The magnitude of the adhesion force is defined by solid-liquid surface free energies. (C) Experiments conducted under ambient conditions reflect the wettability of the surfaces, since the predominant interaction results from capillary forces.

imaging under ambient conditions.[40] Measurements performed in a dry inert gas atmosphere may seem an attractive approach to circumventing the limitations imposed by capillary forces;[41] however, researchers should use extreme caution in interpreting the results of these experiments, since it is difficult to exclude or account for the presence of adsorbed vapor on high energy surfaces.

The only general approach for eliminating the capillary effect is to conduct the experiments in liquid rather than in air.[42] In this case adhesion force measurements with both surfaces immersed in liquid will reflect the interplay between surface free energies of *solvated* functional groups. In addition to introducing the solvation energy components into the overall energy balance of the system, the presence of tightly-bound solvation layers on the interacting surfaces can introduce effects reminiscent of the collective ("hydrophobic") interactions, and further complicate interpretation of the results (the end of this chapter presents a more detailed discussion of these effects). Understanding fundamental interactions in aqueous solutions is especially important, since AFM has become a routine technique for probing and imaging biological systems.[43–54]

1.5   Making Sure that CFM Experiments Work: Characterization and Calibration of Force Probes and the Force Microscope

To obtain absolute force values and make direct comparisons between studies requires knowledge of the cantilever spring constants and the tip radii. Researchers have introduced several approaches for experimental calibration of normal spring constants of the AFM cantilevers, k_z,[55–58] yet the thermal resonance method[56] has become the dominant technique. Current generation commercial AFMs make calibration of the individual tips easier than ever by incorporating automated calibration routines directly into the microscope control software. As the process variations during mass-fabrication of the AFM cantilevers could introduce up to 50% deviations in the cantilever spring constants for nominally identical cantilevers, with additional changes in effective stiffness due to gold coating,[26] *individual calibration of every cantilever used for force spectroscopy measurements is imperative*. The lateral spring constants k_L of cantilevers can also be determined experimentally[59] or derived using the corresponding normal spring constants through the cantilever geometry.[26, 60–62]

In addition, the sensitivity of the optical detector, which measures the cantilever displacement, must be calibrated to obtain absolute forces. The normal (z) sensitivity of the detector can be readily determined from the linear compliance region of the force curve when the tip is pushed against a stiff sample (Figure 3A); the literature also contains examples of noncontact methods.[63, 64] The lateral (xy) sensitivity of the detector in the past was frequently calibrated from atomic stick-slip friction loops,[65] although implicit neglect of the lateral stiffness of the contact could introduce a significant error. A more robust, but more involved, procedure than the use of stick-slip friction loops requires a standard with two different well defined surface slopes. It is also possible to relate the normal sensitivity of a detector to its lateral sensitivity.[63, 66, 67]

Finally, the tip radius is an important parameter to characterize, since it affects the contact area between the tip and the sample; that is, the number of molecular contacts. Three methods for tip characterization have been applied in CFM: 105(i) tip imaging using scanning electron microscopy (SEM), which provides a direct measure of the radius and is the most common approach; (ii) the "adhesion standard" system (e.g., CH_3-H_2O-CH_3), which provides the benchmarked work of adhesion to define effective tip radius using measured adhesion force, known work of adhesion, W_{132}, and a selected contact mechanics model, e.g., $R = F_d/(1.5\pi W)$; (iii) tip shape reconstruction algorithms, which can generate the probe tip shape from the AFM images of a tip characterizer, such as sharp Si spikes on a flat surface, atomically sharp features on the (305) face of $SrTiO_3$, colloidal gold clusters, or even random sharp features on the sample. In at least one test, blind tip shape restoration (method iii) produced tip radius values that were in excellent agreements with the estimates from the direct SEM imaging of the probes (method i).[106]

2 Probing Specific Nanoscale Interactions with CFM

A large number of CFM studies of the interactions between different chemical functional groups covalently linked to tips and samples have appeared since the initial report of CFM.[4, 26] Tables 1 and 2 present some typical results. Adhesion forces between tips and substrates modified with SAMs terminating in CF_3, CH_3, OCH_3, CH_2Br, OH, COOH, $COCH_3$, $CONH_2$, and NH_2 groups have been measured in organic[4, 26, 33] and aqueous solvents,[27] and inert dry atmosphere.[31,68] Even a quick glance through these tables shows that the notion of determining a defined bond strength for a particular functional group pair interaction is too simplistic. Interaction forces measured between the same functionalities in different solvents can differ by almost an order of magnitude; and, even more troubling, measurements performed by different research groups using similar probe functionalization in the same solvent sometimes produce different results. It is clear that the measured interaction force can be influenced by many different parameters. In the following sections we quickly review the observed trends, outline the experimental factors that influence forces in CFM, and present a detailed discussion of the basic concepts that are useful in designing and interpreting CFM experiments.

2.1 CFM in Air and in Gas Environments

Several studies of functional group interactions have been carried out in dry gases[31, 68] and are summarized in Table 1. Several qualitative trends are noticeable. The adhesive forces between tips and samples modified with SAMs that both terminate in nonpolar groups are small[68] or undetectable within the resolution of reported experiment.[31] Observed adhesive forces are also small when one of the SAM surfaces terminates with nonpolar groups and the other with polar groups. In contrast, when both tip and sample SAM surfaces terminate with hydrogen bonding groups (Table 1, entries 1, 4–7) significant adhesion is observed. The relative magnitudes of the adhesive forces are also in accord with the with expected bond strengths; that is, $COOH/NH_2 > COOH/COOH > NH_2/NH_2$. Note that the use of dried gases is not sufficient to completely preclude adsorption of water and other species on the surfaces of SAMs. Hence, to determine unambiguously the magnitudes of the interactions between bare molecular assemblies, it is necessary to carry out CFM studies under ultrahigh vacuum conditions, which can ensure clean surfaces.

Table 1. Adhesion Forces Between Functional Groups in Gaseous Environment.

Functional Group (Tip-Surface)	Monolayer, Chain Length	Medium	Adhesion (nN)	Ref.
COOH-COOH	thiol SAM, C_{16}	dry Ar	62	31
COOH-CH_3	thiol SAM, C_{16}	dry Ar	≈0	31
CH_3-CH_3	thiol SAM, C_{18}	dry Ar	≈0	31
NH_2-COOH	thiol SAM, C_{11}	dry N_2	4.3±0.4	68
COOH-COOH	thiol SAM, C_{11}	dry N_2	1.4±0.3	68
NH_2-NH_2	thiol SAM, C_{11}	dry N_2	0.7±0.2	68
CH_3-CH_3	thiol SAM, C_{11}	dry N_2	0.4±0.2	68

2.2   CFM in Organic Solvents

Initial CFM experiments carried out in organic solvents focused on probing van der Waals and hydrogen bonding interactions, while CFM performed in electrolyte solution assessed hydrophobic and electrostatic forces (Table 2). Representative F-D curves obtained in ethanol using Au-coated tips and samples that were functionalized with SAMs terminating in either CH_3 or COOH groups readily reveal the difference between the individual interactions (Figure 5A). To quantify the differences and uncertainties in the adhesive interactions between different functional groups, however, it is always necessary to record multiple force curves for each type of intermolecular interaction and each tip-sample pair. Histograms of the observed adhesion forces typically exhibit Gaussian distributions (Figure 5B) and yield mean adhesion forces (± its experimental uncertainty) of 2.3 ± 0.8, 1.0 ± 0.4, and 0.3 ± 0.2 nN for interactions between COOH/COOH, CH_3/CH_3, and CH_3/COOH

Table 2. Interaction strength between AFM tips and samples functionalized with specific functional groups measured in different solvents.

Functionality	Probe functionalization monolayer	Solvent	Adhesion (nN)	Reference
CH_3-CH_3	Silane, C_2	EtOH	0.4 ± 0.3	29
CH_3-CH_3	Silane, C_9	EtOH	0.7 ± 0.6	29
CH_3-CH_3	Silane, C_{14}	EtOH	2.4 ± 1.2	29
CH_3-CH_3	Silane, C_{18}	EtOH	3.5 ± 2.3	29
CH_3-CH_3	Thiol, C_{18}	EtOH	1.0 ± 0.4	26
CH_3-CH_3	Thiol, C_{12}	EtOH	2.3 ± 1.1	33
CF_3-CF_3	Silane, C_{10}	EtOH	15.4	30
CH_3-CF_3	Silane, C_{18}, C_{10}	EtOH	*repulsive*	30
CH_3-CH_3	Silane, C_{18}	$CF_3(CF_2)_6CF_3$	52	30
CF_3-CF_3	Silane, C_{10}	$CH_3(CH_2)_6CH_3$	21	30
CH_3-CH_3	Thiol, C_{12}	$CH_3(CH_2)_{14}CH_3$	0.07 ± 0.05	33
COOH-COOH	Thiol, C_{11}	$CH_3(CH_2)_{14}CH_3$	0.11 ± 0.02	33
COOH-COOH	Thiol, C_{11}	Hexane	0.95 ± 0.26	101
COOH-CH3	Thiol, C_{11}, C_{18}	EtOH	0.3 ± 0.2	26
COOH-COOH	Thiol, C_{11}	EtOH	2.3 ± 0.8	26
COOH-COOH	Thiol, C_{11}	EtOH	0.27 ± 0.04	33
COOH-COOH	Thiol, C_{11}	PrOH	1.37 ± 0.26	101
CH_2OH-CH_2OH	Thiol, C_{11}	EtOH	0.18 ± 0.18	33
COOH-COOH	Thiol, C_{11}	Water	2.8 ± 0.2	101
COOH-COOH	Thiol, C_{11}	Water,	7.0 ± 0.2	27
COOH-COOH	Thiol, C_{11}	pH < 5	2.3 ± 1.1	33
COOH-CH_2OH	Thiol, C_{11}	Water, DI	1.1 ± 0.5	27
CH_2OH-CH_2OH	Thiol, C_{11}	Water, pH < 5	1.0 ± 0.2	27
CH_2OH-CH_2OH	Thiol, C_{11}	Water,pH < 5	0.3 ± 0.05	33
CH_3-CH_3	Thiol, C_{18}	Water, DI	60 ± 5	27
CH_3-CH_3	Thiol, C_{12}	Water	12.5 ± 4.4	33

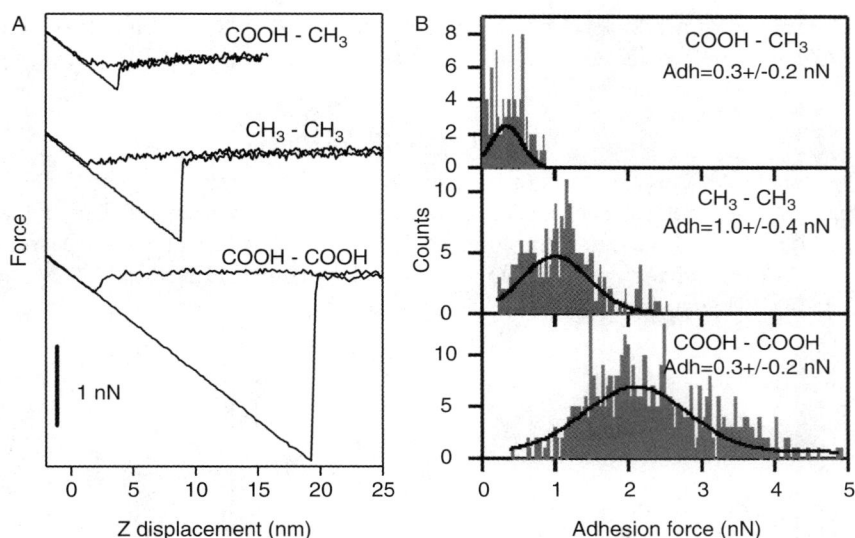

Figure 5. Specific interactions between basic chemical functionalities in ethanol. (A) Representative force vs. distance traces recorded between samples and tips functionalized with simple organic functional groups.[4] (B) Histograms of adhesion forces recorded between similar tip/sample pairs in repeated measurements. Solid lines indicate Gaussian fits to the data.[26]

groups, respectively. Since the mean values do not overlap, it is possible to differentiate between these chemically distinct functional groups by measuring the adhesion forces with a tip that terminates in a defined functionality. The observed trend in the magnitudes of the adhesive interactions between tip/sample functional groups, i.e., COOH/COOH > CH_3/CH_3 > CH_3/COOH, agrees with the qualitative expectation that interactions between hydrogen bonding groups (i.e., COOH) will be greater than between non-hydrogen bonding groups (i.e., CH_3).

When the solvent is chemically similar to the tip and sample terminal functional groups (for example, CH_3 groups in hexadecane or CH_2OH groups in ethanol), the forces required to separate the tip and surface are small (Table 2). Cross-interactions between immiscible components were found to be either very small (e.g., CH_3/COOH in ethanol) or even repulsive (e.g., CH_3/CF_3 in ethanol). In contrast, when the solvent is immiscible with the functional groups that terminate the SAMs on the tip and sample, the adhesive forces are extremely large. Both van der Waals (Table 2 CH_3 groups in perfluorohexane or CF_3 groups in hexane) and hydrophobic (Table 2 CH_3 groups in water) interactions can be responsible for this latter behavior.

3 Interpretation of CFM Experiments

3.1 Thermodynamic Model: Surface Free Energies and Surface Tension Components Models

3.1.1 Contact Mechanics Approach to Tip-Surface Contact in CFM

Although force microscopy with sharp probes approaches the limit of point contact, in practice the number of interacting molecular species in CFM experiments remains on the order of tens to hundreds. Energies of intermolecular interactions and the number of functional groups contributing to experimentally observed forces can be derived from adhesion data by considering the contact deformation between the probe and a surface.

The problem of adhesive contact between the tip of radius R_1 and sample of local curvatures R_2 can be treated in terms of elastic contact between a flat surface and a sphere of effective radius R:

$$R = (1/R_1 + 1/R_2) \tag{1}$$

The sphere deforms due to both repulsive forces (Born repulsion) within the area of contact and attractive forces near the edge of the contact zone and outside it. The interatomic distances within the contact zone differ little, and this area can be considered to be a flat circle of radius a. For repulsive-only interactions (hard-wall potential, Hertz model), the dependence of the contact area size on external load P is well known:

$$a^3 = \frac{RP}{K}, \tag{2}$$

$$\text{where } K = \frac{4}{3}\left(\frac{1-v_1^2}{E_1} + \frac{1-v_2^2}{E_2}\right)^{-1} \tag{3}$$

is the effective elastic constant of the system (v is the Poisson ratio and E is the Young's modulus). Realistic potentials are not straightforward to implement in the models of contact mechanics, because interaction force depends on the intermolecular separation, in other words, it is defined by the surface profile of the deformed sphere, which in turn depends on the interaction force. One can decouple the force-surface profile dependency by assuming that i) the profile is not changed because of the presence of attractive forces outside the contact zone, or ii) attractive forces act only within the contact area (zero range forces).

The first option means that the forces are based on the Hertz result and leads to the DMT (Derjaguin, Muler, and Toporov) model of adhesion:[69] the radius a, the surface profile, and the stress distribution are given by Hertz equations with the external load substituted by the total force $P + F_a$, which includes adhesion F_a. The force of adhesion can then be related to the thermodynamic work of adhesion by summing up interactions in the gap between the surfaces outside the contact zone,

$$a^3 = \frac{RP_1}{K} \qquad P_1 = P + 2\pi W_{132} \qquad F_a = 2\pi R W_{132} \tag{4}$$

The work of adhesion for separating the sample and tip gives the balance of interfacial free energies $W_{132} = \gamma_{13} + \gamma_{23} - \gamma_{12}$. The second choice, implemented in the JKR (Johnson, Kendall, and Roberts) model[70], results in a different stress distribution: compressive in the center of the contact zone, changing to tensile when approaching the boundary (and zero outside the contact circle). The total energy of the system is minimized when the external load P is substituted by an apparent load P_1:

$$a^3 = \frac{RP_1}{K} \qquad P_1 = P + 3\pi W_{132}R + \sqrt{6\pi W_{132}RP + \left(3\pi W_{132}R\right)^2} \qquad F_a = \frac{3}{2}\pi R W_{132} \tag{5}$$

The two models differ substantially in predicted contact area, force of adhesion, and surface profile. JKR theory predicts a finite radius of contact under zero external load and when surfaces separate: $a_{0(JKR)} = \left(\frac{6\pi W_{132}R^2}{K}\right)^{1/3}$ and $a_{s(JKR)} = \frac{a_{0(JKR)}}{4^{1/3}} \approx 0.63 a_{0(JKR)}$, respectively. One can estimate the number of molecular contacts in adhesive interactions by dividing the contact area at pull-off, a_s, by the area occupied by a single functional group. Corresponding quantities for DMT theory are $a_{0(DMT)} = \left(\dfrac{2\pi W_{132} R^2}{K}\right)^{1/3}$ and $a_{s(DMT)} = 0$. The estimate of the number of molecular contacts in the DMT model must consider the range of intermolecular forces z_0.

A self-consistent approach to the contact problem typically requires numerical solutions. Such calculations based on the Lennard-Jones potential showed[71] that the DMT and JKR results correspond to the opposite ends of a spectrum of a non-dimensional parameter (so-called Tabor elasticity parameter):

$$\mu = \left(\frac{16}{9}\frac{RW^2}{K^2 z_0^3}\right)^{1/3} \tag{6}$$

This parameter asserts the relative importance of the deformation under surface forces: for $\mu < 0.1$ the DMT model is appropriate, for $\mu > 5$ the JKR model applies. Although the JKR model predicts infinite stresses at the perimeter of the contact zone, whereas DMT model predicts discontinuous stress, numerical results did not display any abnormal stress distributions.

To avoid self-consistent calculations based on a specific potential, Maugis derived an analytical solution to the adhesive contact problem[72] by using the Dugdale approximation that the adhesive stress has a constant value σ_0 (theoretical stress) until a separation $h_0 = W/\sigma_0$ is reached at radius c, whereupon it falls to zero (Figure 6). The net normalized force is given by (m = c/a):

$$\bar{P} = \bar{a}^3 - \lambda \bar{a}^2 \left(\sqrt{m^2 - 1} + m^2 \arccos(1/m)\right), \tag{7}$$

where parameter λ is a measure of the ratio of the elastic deformation to the range of surface forces and $\bar{a}$ and $\bar{P}$ are the scaled radius of the contact zone and external load:

$$\lambda = 2\sigma_0 \left(\frac{RK^2}{\pi W}\right)^{1/3} = 1.16\mu \quad \bar{a} = \frac{a}{\left(\pi WR^2/K\right)^{1/3}} \quad \bar{P} = \frac{P}{\pi WR} \tag{8}$$

The elasticity parameter λ is related to m through:

$$\frac{1}{2}\lambda\,\bar{a}^2\left[(m^2 - 2)\arccos(1/m) + \sqrt{m^2 - 1}\right] + \frac{4}{3}\lambda^2 \bar{a}$$
$$\left[\sqrt{m^2 - 1}\arccos(1/m) - m + 1\right] = 1. \tag{9}$$

When λ is increased, m→1 (i.e., contact forces only, c = a) and the JKR limit is recovered:

$$\bar{P} = \bar{a}^3 - \sqrt{6\bar{a}^3} = \bar{P}_1 - \sqrt{6\bar{P}_1}. \tag{10}$$

When λ is decreased, m→∞ (i.e., long-ranged forces, c >> a) and the DMT limit is achieved:

$$\bar{P} = \bar{a}^3 - 2. \tag{11}$$

The Maugis-Dugdale (MD) model is an accurate representation of the adhesion in the presence of a liquid meniscus (constant pressure inside the meniscus).

The choice of a functional form of the contact area dependence on the external load is crucial to interpretation of the friction forces, which will depend on both the interfacial shear strength of the contact and its size. The difficulty with MD theory is that it does not easily lend itself to fitting the experimental data. This issue was addressed by Carpick et al.,[73] who demonstrated that a simple general equation:

$$\frac{a}{a_{0(\alpha)}} = \left(\frac{\alpha + \sqrt{1 - P/F_{a(\alpha)}}}{1 + \alpha} \right)^{2/3} \tag{12}$$

closely approximates Maugis' solution and can be used to fit experimental data onto a contact area (friction). Numerical results are then used to obtain the Tabor parameter from the fitting parameter α ($\alpha = 0$ corresponds to DMT case, $\alpha = 1$ corresponds to JKR case).

The exact value of the numerical coefficient in Equation 4 and 5 for F_a (3/2 vs. 2) is beyond the experimental uncertainty of CFM experiments (widths of adhesion force distributions plus errors in determination of k and R, see Tables 1 and 2); and both approaches can and have been used to derive surface free energies from adhesion data. On can construct an "adhesion map" of the applicability of particular models depending on the parameter μ and applied force.[74] In the CFM, high modulus materials ($\approx 100\,GPa$) and tip radii $\approx 100\,nm$ result in $\mu < 1$; thus CFM experiments fall into a "transition" zone. Numerical calculations for $\mu < 0.3$ indicate, however, that the compliance of adhesive contact and computed contact radii are still well represented by the JKR equations. Therefore, it appears that JKR equations give good predictions even under conditions well outside the expected JKR zone. We note that it is impossible to place a given system in the respective contact mechanics regime based on the adhesion force measurement alone, since contact area is not measured independently in CFM. An estimate of the Tabor elasticity parameter is preferred in this case, although this approach still needs to make assumptions about effective elastic constant of the monolayer/substrate system.

3.1.3 Intermolecular Force Components Theory

The sphere-on-flat tip-sample geometry of the AFM does not correspond to the interaction between two atoms or molecules. The Lennard-Jones potential typically used to represent interaction between molecular species (with minimum energy ε_0):

$$\frac{F_{molec}(z)}{\varepsilon_0/z_0} = 12 \left[-\left(\frac{z}{z_0} \right)^{-7} + \left(\frac{z}{z_0} \right)^{-13} \right] \tag{13}$$

has to be modified for the CFM geometry to account for multiple intermolecular pairs. The fundamental $1/z^7$ law of attraction for dispersion forces between molecules was first derived by London based on second order perturbation theory. An alternative approach by Dzyaloshinskii, Lifshitz, and Pitaevskii[75] uses quantum field theory to relate van der Waals (vdW) forces to spectroscopic properties of materials, but produces an analytically useful result only in the case of vdW pressure between two semi-infinite dielectric slabs separated by a third medium:

$$f(z) = -\frac{A}{6\pi} \frac{1}{z^3} \tag{14}$$

where A is the (non-retarded) Hamaker coefficient, which relates to a detailed dielectric behavior of the materials through a complete electromagnetic spectrum.

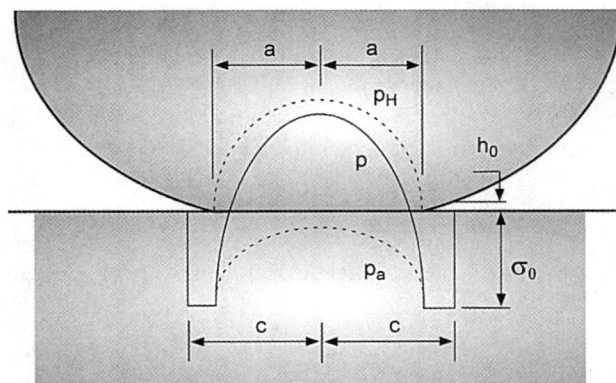

Figure 6. The Maugis-Dugdale stress distribution is a sum of two terms: the Hertz pressure p_H acting on the area of radius a, and adhesive tension p_a acting on the area confined by radius c.

The two scales (macro and molecular) converge in Hamaker's approach: from the power law for intermolecular interactions, $F_{molec}(z) = -C/z^7$, the macroscopic van der Waals force between bodies 1 and 2 with molecular densities ρ_1 and ρ_2 can be obtained by volume integration of the pair-wise interactions:

$$f(z) = -\frac{\pi \rho_1 \rho_2 C}{36 z^3}, \text{ where } \rho_1 \rho_2 C = 6A/\pi^2 \qquad (15)$$

Thus, London's result for interactions across vacuum (α is atomic polarizability)[76]:

$$U_{molec} = -\frac{3}{4\pi} h \frac{\nu_{e1}\nu_{e2}}{\nu_{e1} + \nu_{e2}} \frac{\alpha_1 \alpha_2}{z^6} \qquad (16)$$

leads to a geometrical mean combining rule for the corresponding Hamaker coefficients ($\nu_{e1} \approx \nu_{e2}$, $\rho_1 \approx \rho_2$) and for the surface free energies in the case of Lifshitz–van der Waals (LW) forces:

$$A_{12} = \pm\sqrt{A_{11}A_{22}} \text{ and } \gamma_{12}^{LW} = \sqrt{\gamma_1^{LW}\gamma_2^{LW}} \qquad (17)$$

The pull-off forces, as described in the contact mechanics models, depend on the surface free energies through the thermodynamic work of adhesion. In associating solvents and for polar surfaces, it is important to take a proper account of both additive (symmetric) van der Waals interactions and complementary (asymmetric) electron donor-acceptor interactions (hydrogen bonding). Donor-acceptor or Lewis acid-base interactions are short-ranged (contact) forces and are not accounted for in London's or Lifshitz's treatments of weak intermolecular forces. These interactions, however, influence thermodynamic data, and thus must be included in computational models for analysis and prediction of physicochemical properties—for example, in solvation models and the linear free energy relationship approaches.

Hydrogen bonding interactions operate over the background of the omnipresent dispersion forces. While the dispersion forces for dielectrics are often very similar in magnitude, the hydrogen bonding interaction differentiates various classes of organic functional groups quite dramatically. One can compare, for example, boiling points of homologous hydrocarbons and alcohols or adhesion forces between CH_3/CH_3 and $COOH/COOH$ pairs.

A successful scale of the strength of acid-base interactions was proposed by Fowkes, van Oss, Chaudhury, and Good (FOCG),[77] who recognized the need for having two values to describe the polar surface tension component: one representing electron accepting and the other representing electron donating abilities in combining relations. In this model, the total surface tension of a polar system is separated into van der Waals, γ^{vW}, and Lewis acid, γ^+; and Lewis base, γ^-, components:

$$\gamma_{total} = \gamma^{LW} + \gamma^{AB} \tag{18}$$

where $\gamma^{AB} = 2\sqrt{\gamma^+ \gamma^-}$. For cross-interactions, the following combing rule applies:

$$\gamma_{12} = \sqrt{\gamma_1^{LW}\gamma_2^{LW}} + \sqrt{\gamma_1^+\gamma_2^-} + \sqrt{\gamma_2^+\gamma_1^-} \tag{19}$$

and the solid-liquid interfacial tension is then given by:

$$
\begin{aligned}
\gamma_{SL} &= \gamma_{SV} + \gamma_{LV} - 2\left(\sqrt{\gamma_{SV}^{LW}\gamma_{LV}^{LW}} + \sqrt{\gamma_{LV}^+\gamma_{SV}^-} + \sqrt{\gamma_{SV}^+\gamma_{LV}^-}\right) \\
&= \left(\sqrt{\gamma_{SV}^{LW}} - \sqrt{\gamma_{LV}^{LW}}\right)^2 + 2\left(\sqrt{\gamma_{LV}^+\gamma_{LV}^-} + \sqrt{\gamma_{SV}^+\gamma_{SV}^-}\right. \\
&\quad \left. - \sqrt{\gamma_{SV}^+\gamma_{LV}^-} - \sqrt{\gamma_{SV}^-\gamma_{LV}^+}\right)
\end{aligned}
\tag{20}
$$

The interpretation of adhesion measurements in liquids can be complicated if all components are involved in acid-base interactions. Although there is no solid fundamental theoretical basis for applying the geometric mean combining rule for interactions of acid-base type (unlike for van der Waals interactions), the FOCG model is conceptually simple and can be applied successfully to rationalize the trends in adhesion forces observed between organic functional groups of model SAMs or polymer systems in organic liquids.

3.2 Scaling Relationships in CFM Experiments

3.2.1 Experimental Geometry: Surface Roughness and Probe Size Effects

One of the advantages of using nanoscale point probes in adhesion studies is that they drastically lower the requirements for the surface quality necessary to achieve molecularly smooth contact, e.g., the sample does not have to display macroscopic atomically smooth areas. Quasi-equilibrium pull-off forces predicted by the contact mechanics equations discussed in the previous sections should be directly proportional to the effective radius defined by Equation 1. We can immediately recognize, however, that substrate roughness, e.g., local variations in the substrate radius of curvature will affect the magnitude of adhesion. For example, analysis of force and topography maps from the AFM measurements on a chemically homogeneous, hydrophobic sample-silanized etched silicon[78] showed an unambiguous direct correlation between the substrate's local curvature and the force of adhesion.

The width of the adhesion force distribution (σ_F) for a $HS(CH_2)_{15}COOH/HS(CH_2)_{15}COOH$ pair in ethanol was halved when the polycrystalline Au substrates were replaced by the single crystalline, annealed Au on mica.[79] Similar observations[80] were reported for the adhesion forces in water (Figure 7) using the same $HS(CH_2)_{19}CH_3$ terminated tip on three $HS(CH_2)_{15}CH_3$ modified substrates: (i) 11-nm sputtered Au film, (ii) 110-nm sputtered Au film, and (iii) annealed Au(111) on mica. Remarkably, Au substrates presenting large areas of the (111) face produced the same mean value of adhesion, while the distribution width was a factor of five smaller. These results suggest that the spread of the local curvature distribution of the substrate could be a primary factor responsible for the width of adhesion force distributions for chemically identical tip-sample combinations.

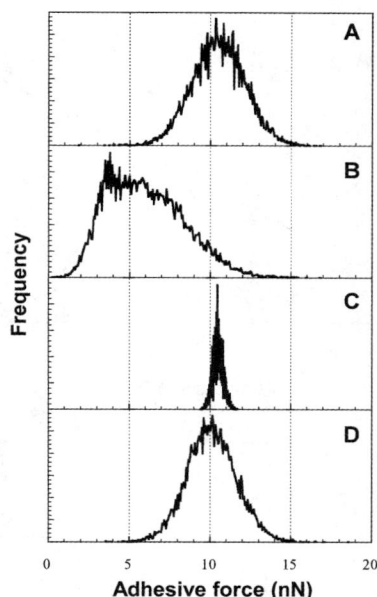

Figure 7. Histograms of adhesive force mappings using the same CH_3-terminated tip on $HS(CH_2)_{19}CH_3$ SAMs formed on: (A) 11 nm thick sputtered gold, (B) 110 nm thick sputtered gold, (C) 100 nm thick thermally evaporated Au(111) on mica, and (D) same as (A). The sequence of CFM measurements was (A)-(B)-(C)-(D).[80]

The most serious practical consequence of varying sample flatness for the interpretation of the chemical force microscopy results is that the measured distributions almost never reflect the inherent statistical distribution width for a given chemical functionality. Thus, statistical treatments of force fluctuations in CFM should be used with caution—a significant contribution to σ_F almost always arises from the fluctuations in the tip-sample contact area due to the differences in local surface curvature.

Scaling of the measured adhesion force values with the size of the contact area and, correspondingly, with the number of bonds comprising the interactions is generally a complicated topic, especially at high loading rates when the kinetic effects are strong[81]. In the quasi-equilibrium limit, some generalizations are possible on the scaling of the adhesion forces with the probe size. The relationship between the tip radius, R, and force of adhesion, F_a, was probed systematically[82] for hydrophobic contacts (CH_3/CH_3) in water. For tip radii ranging from R = 15 nm to R = 125 nm, a remarkably good linear correlation (Figure 8A) existed between the measured adhesion forces and the probe radii.

Moreover, when researchers accounted for substrate roughness (radius of curvature of Au grains) by defining and effective probe radius, the linear fit showed a zero intercept (Figure 8B), as predicted by the contact mechanics models.[82] If we use the value of the thermodynamic work of adhesion in the CH_3-H_2O-CH_3 system of W = 103 mJ/m^2, data in Figure 8 produce a proportionality coefficient of $(1.59 \pm 0.15)\pi W$ between the pull-off force and the effective radius. This value is very close to the $1.5\pi W$ value predicted by the JKR model.

One of the possible solutions to mitigating the effects of the surface roughness on the observed interactions is to use CFM probes of extremely small size. Lieber and co-workers demonstrated an important development in this direction by using chemically-modified carbon nanotube AFM probes for force measurements.[83–85] These probes (especially single-wall carbon nanotube probes) offer a versatile setup for measuring basic chemical

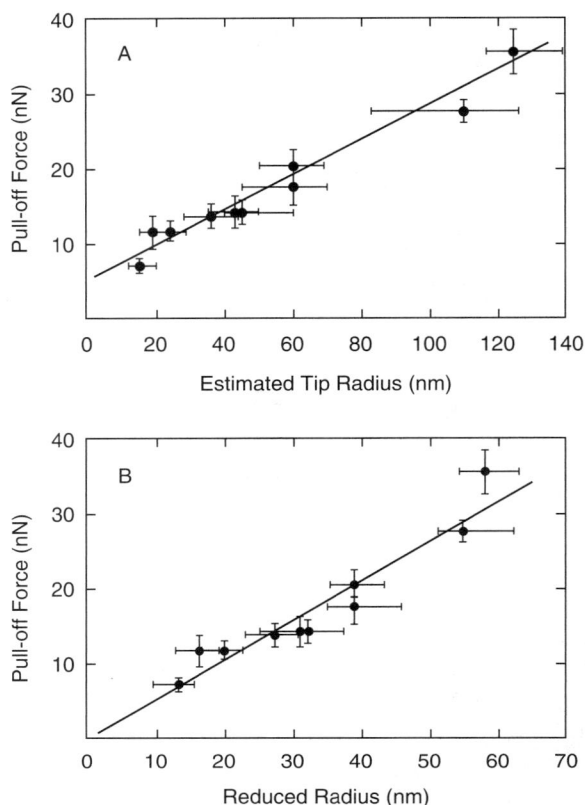

Figure 8. (A) Plot of mean pull-off force vs estimated tip radius (circles) for 10 tips. The solid line shows a linear least squares fit. The vertical error bars represent the standard deviation of the average pull-off force while the horizontal error bars are estimated maximum and minimum radius. (B) Plot of mean pull-off force vs reduced tip radius, taking into account roughness of the substrate. Reproduced from reference.[82]

interactions on a nearly single molecule basis. Unfortunately, the full potential of these probes has yet to be realized, mostly due to the experimental challenges in the mass production of such probes.

3.2.2 Applied Load and Measured Adhesion Values

The main weakness of CFM, as well as all other AFM-based adhesion measurement techniques, is its inability to obtain an independent measure of the contact area. Frequently, a tip-sample friction force is assumed to provide a simultaneous measurement of the contact area on the assumption that friction is proportional to the actual contact area and interfacial shear stress for corresponding functional group pairs.[86]

$$F = \tau_0 \pi a^2 \qquad (21)$$

Note that the general relationship between the radius of the area of contact and the external load (Equation 12) predicts a nonlinear relationship between friction force and applied load. Surprisingly, most of the reported experimental friction-load curves in chemical force microscopy showed approximately linear behavior.[26, 87–90] Vezenov et al.[35] attributed the apparent linear shape of these curves to the averaging of multiple single-asperity contacts that occurs when a friction force is averaged over a path of several micrometers in a typical

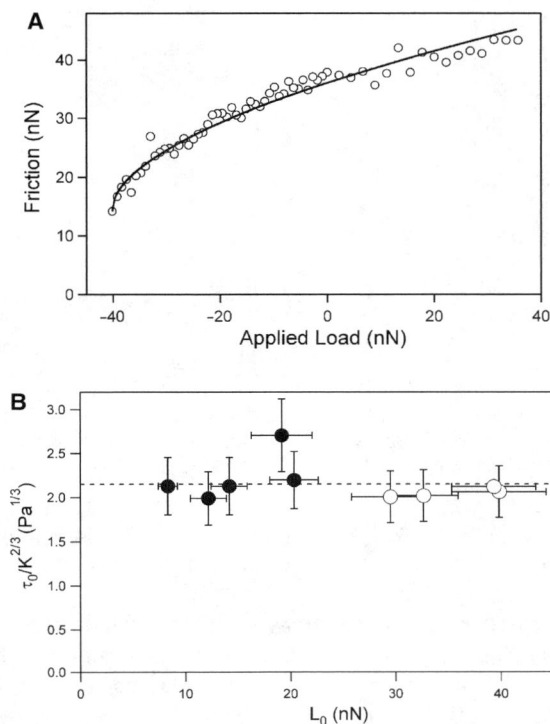

Figure 9. (A) Friction force versus applied load curve (o) for a CH_3 terminated tip and sample in water. The shape of the curve is consistent with the nonlinear dependence of the contact area on external load predicted by the JKR model (fit). (B) Interfacial shear strength, τ_0, of methyl-methyl contacts in methanol-water mixtures normalized by the elastic constant, $K^{2/3}$, plotted versus the force of adhesion measured simultaneously. Closed and open symbols represent different experiments.[91]

friction measurement. The nonlinear JKR-like behavior was detectable only for relatively blunt tips when large forces for hydrophobic SAMs in water turned substrate imperfections into secondary effects (Figure 9A). Nonlinear relationships were also observed between friction and applied load in methanol-water mixtures using methyl terminated siloxane SAMs on smooth Si substrates and silicon nitride tips (Figure 9B).[35] For the same tip-sample pair, the adhesion increased with higher water content; however, the interfacial shear stress, determined by fitting to a contact mechanics model, remained constant.

3.2.3 Adhesion and the Surface Free Energy: Systematic Variations of Adhesion in Mixed Solvent and Mixed SAM Systems

A fruitful strategy to test the scaling relationship between the adhesion forces and the interfacial free energies predicted by the contact mechanics theories for quasi-equilibrium unbinding is to use the same tip and sample pair and vary either the solvent composition or the SAM composition on the probe. This arrangement preserves the geometrical parameters of the system while varying the interfacial free energies of the interacting system in a smooth and predictable manner. For example, Vezenov et al. used the same tip-sample pair terminating in CH_3 groups to determine adhesion forces in a series of methanol-water mixtures.[91] These measurements produced an unambiguous linear correlation between adhesion forces and corresponding surface free energy values determined from independent contact angle

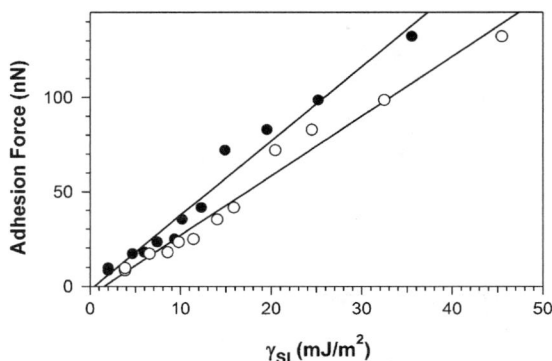

Figure 10. Adhesion between CH_3-terminated tips and samples versus solid-liquid surface free energy determined from advancing (open symbols) and receding (closed symbols) contact angles for CH_3-terminated SAM (γ_{SV} = 19.3 mJ/m^2).[91]

measurements, thus providing an additional corroboration of the scaling relationships predicted in such systems by the contact mechanics models (Figure 10).

3.2.4 Mechanical Properties of Sams and Effect of Chain Packing

The degree of packing in the SAM is affected by the anchoring mechanism and defect density and is generally different for thiolates on Au, trichrolosilanes on Si, and trimethoxysilanes on Si. The tilt angle of alkyl chains with respect to substrate is influenced by the anchoring group density in these SAMs. Thicknesses of these monolayers measured by ellipsometry (2.2 nm, 2.4 nm, and 1.9 nm, respectively) is consistent with an all-trans conformation and an expected trend in tilt angles of 30°, 10°, and 40°, respectively. Most important for the CFM measurements, the orientation of the terminal group is therefore different for these three commonly encountered types of the SAMs. Klein and co-workers performed elegant experiments exploring this degree of freedom in the CFM experiments by measuring adhesion forces between $HS(CH_2)_{17}CH_3$/Au tips and $HS(CH_2)_{17}CH_3$, $Cl_3Si(CH_2)_{17}CH_3$, and $(CH_3O)_3Si(CH_2)_{17}CH_3$ monolayers in water. Adhesion forces were sensitive to SAM internal organization.[92] The adhesion results obtained with three different probes, while nominally reflecting methyl-methyl interactions in water, displayed a trend that paralleled the quality of organization of the SAMs: adhesion forces ratios were 1/(0.797±0.005)/(0.687±0.006) for HS-, Cl_3Si-, and $(CH_3O)_3Si$-anchor groups, correspondingly (the error is the standard deviation of the mean for different tips). Interestingly, the contact angles, measured on the same series of samples, were much less sensitive to the type of SAMs (111°, 110°, and 108°). Thus, different chain packing and orientation of the terminal group of the self-assembled monolayer can lead to measurable differences in the pull-off forces and thus can serve as a sensitive probe of the conformation of the surface chemical groups. The sensitivity to conformation was also demonstrated by the discrimination of chiral isomers on the basis of adhesion forces detected in CFM experiments.[93] On the other hand, the use of the work of adhesion of a "reference" system, such as methyl-methyl contacts in water (W = 103 mJ/m^2), to determine effective tip radius[33,94] should be used with caution, because nominally the same interface can produce a different apparent work of adhesion depending on the packing of the underlying alkane chain.

3.2.5 Role of the Solvent and Hydrogen Bonding Interactions

Surface free energy arguments explain the magnitudes of adhesion forces measured between organic functional groups in ethanol. Lieber and co-workers observed[91] that adhesion measured by CFM did not correlate with solvent polarity (dipole moment or dielectric constant) or cohesion energy. On the other hand, they found[91] that the surface tension component (STC) theory (Section 3.1.3) is generally useful in explaining CFM data in various solvents.

For CH_3-terminated SAMs, the last three terms in Equation 20 are all zero (i.e., $\gamma^{AB} = \gamma^+ = \gamma^- = 0$), while the van der Waals component γ^{LW} determined from contact angle measurements with liquid hydrocarbons (19.3 mJ/m^2) is essentially the same as the corresponding γ^{LW} values for alcohols and water (21.8, 18.5, and 20.1 mJ/m^2 for water, MeOH, and EtOH, respectively). Thus, the first term in Equation 20 is also negligible and the value of the adhesion force between two methyl surfaces in alcohols (or water) is essentially a measure of the strength of the hydrogen bonding interaction (in a free energy sense) in liquid alcohol (water):

$$W_{CH_3/ROH/CH_3} = 2\,\gamma^{AB}_{ROH}\ \ (R = CH_3CH_2, CH_3 \text{ or } H) \tag{22}$$

Ethanol and methanol have indistinguishable values of surface tension (22.8 and 22.6 mJ/m^2, respectively) but different hydrogen bonding components ($\gamma^{AB} = 2.7$ and 4.1 mJ/m^2, respectively), both of which are much smaller than the corresponding value for water ($\gamma^{AB} = 51$ mJ/m^2). When the same tip and sample are used in the adhesion experiment, forces between CH_3 groups in MeOH were consistently 1.5–2 times greater than those in EtOH, in agreement with the ratio of 1.6 between corresponding γ^{AB} values. Clear and Nealey measured adhesion forces between $HS(CH_2)_{17}CH_3$ modified Au-coated probes and CH_3- and COOH-terminated siloxane monolayers on Si substrates.[94] For a number of solvents (hexadecane, ethanol, propanediols, and water), they also found a good agreement between the work of adhesion for CH_3/CH_3 interface (i) obtained experimentally with CFM, (ii) predicted form contact angle measurements, and (iii) the values calculated from the STC model.

Ethylene glycol (EG) and dimethylsulfoxide (DMSO) present another example of two solvents with a similar surface tension (48 and 44 mJ/m^2) but quite different adhesion forces between apolar surfaces. These solvents have significant acid-base components (19 and 8 mJ/m^2, respectively) and dispersion contributions that differ from that of the CH_3-terminated SAM (29 and 36 mJ/m^2, respectively). Adhesive forces between methyl surfaces in EG are predicted (Equation 20) to be greater than those in DMSO by a factor of 1.8, while experiments yielded a factor of 1.8–2.0 difference.[91] These data independently confirm the validity of the intermolecular force components approach for treating adhesive interactions between organic groups in liquid medium.

The methanol-water mixture provided a simple way to generate similar solvents that spanned a large range of hydrogen bonding ability ($\gamma^{AB} = 4$ to 51 mJ/m^2). With a nonpolar SAM, the force of adhesion increased monotonically in mixed solvents of higher water content; whereas, with a SAM having a hydrogen bonding component (COOH groups), higher water content led to decreased adhesion compared to the nonpolar counterpart. Overall, surface tension component interpretation of the CFM data in solvents showed that competition between hydrogen bonding *within the solvent* and hydrogen bonding *between surface groups and the solvent* provided the main contribution to adhesion forces between organic functional groups in liquids. Water, with its exceptional hydrogen bonding properties (both as acceptor and donor), used as a solvent in CFM measurements will therefore discriminate best between various organic groups.

Liquids that form strong interfacial bonds can display two adhesion minima: (i) one corresponding to contact between the groups of the tip and the first solvation shell of the surface when the maximum load on the tip in contact is small, and (ii) the second, deeper, minimum corresponding to contact between the tip and the surface groups when the maximum load is high and the solvation shell is fully penetrated. For OH-OH pair in octanol, the CFM gives $\gamma_{SL} = 0.16$ mJ/m^2 and 0.60 mJ/m^2, respectively, for the two minima.[95]

A set of interesting experiments involves COOH and OH groups in both nonpolar and hydrogen bonding solvents. The acid-base components of surface free energy of these SAMs are not readily available from contact angle measurements, because most test liquids will completely wet such surfaces. The CFM in these systems can potentially provide, along with a dispersion γ^{LW} component, the values of γ^+ and γ^- which together will completely characterize adhesion between these SAMs and other organic surfaces. For example, STC treatment[95] of $W_{OH/HD/OH}$ and $W_{OH/HD/CH3}$ values (HD is hexadecane) using Equation 20 yields $\gamma^{AB}_{OH} \approx 1.1\text{-}1.5\,mJ/m^2$. On the other hand, one would obtain $\gamma^{AB}_{OH} \approx 24.6\,mJ/m^2$ for the same monolayers when using values of the work of adhesion $W_{OH/HOH/OH}$ and $W_{OH/HOH/CH3}$ found with CFM in water. Warszynski et al. argued[95] that the discrepancy can be resolved if one assumes that in a combining relationship (Equation 19) one needs to take the values for surface free energy of solids saturated with the respective liquids. Rearrangement of surface groups in response to different environments is another possible factor responsible for changing the γ^{AB} component of the SAM. In addition, the STC model has been shown to have internal inconsistencies, and CFM could be a powerful tool to provide estimates of surface tension components directly for each individual solvent/surface pair. Clearly, the γ^{AB} values for high surface energy groups are not available by other means; in these situations, CFM can be used as an independent method to construct the respective acid-base scales.

4 Kinetic Effects in CFM Measurements

4.1 Chemical Force Microscopy in the Kinetic Regime

So far we have concentrated on the scaling behaviors described by the thermodynamic models of chemical force microscopy. Yet, since CFM typically probes the interactions of a very small number of individual molecular contacts, the thermal fluctuations are still significant, and thus they should exhibit some elements of the bond rupture kinetics typical for non-equilibrium forced unbinding of single bonds. These effects are discussed in detail in the other chapters, so here we provide only a very cursory sketch, mostly for the sake of consistency. If the interacting system is loaded at a very high rate, such that the bond rupture occurs far from equilibrium, the applied force would exponentially amplify the kinetic rate of dissociation. As solved by Evans and Ritchie,[96] the kinetic equations for the case of linear loading (as encountered in the AFM experiments) predict that the pull-off force at constant loading rate r_f is:[97]

$$f_{\text{pull-off}} = \frac{k_B T}{x_\beta} \ln\left(\frac{r_f}{r_0}\right) \tag{23}$$

Characteristic loading rate r_0 is defined as:

$$r_0 = \frac{k_B T}{x_\beta} \cdot \frac{1}{\tau_D \exp\left(\dfrac{E_0}{k_B T}\right)}$$

where τ_D represents the inverse of the diffusion-limited attempt frequency, E_0 is the depth of the energy well, and x_β is the distance to the transition state.

Direct application of this formalism to the CFM experiment presents several significant challenges. CFM almost always involves the rupture of multiple individual bonds connected in parallel. Williams and Evans considered the kinetics of the bond rupture in such systems[98, 99] and showed that the kinetics of bond rupture is nontrivial in the general case of uncorrelated bonds. However, the analysis could be simplified by assuming that all the bonds are correlated, i.e., they share a single reaction coordinate. Then the system could represent a single "macro-bond" with the total potential equal to the sum of the potentials of individual components.[99] For the serial loading of N identical bonds Williams and Evans obtained the following expression for the unbinding force:

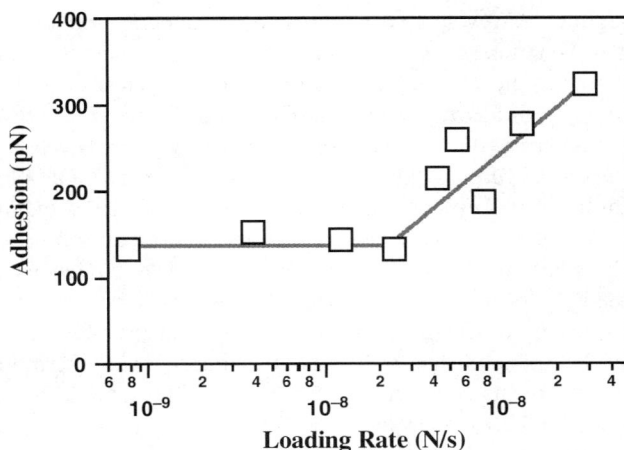

Figure 11. Binding forces between COOH-modified probe and sample in ethanol plotted as a function of loading rate.[34] Lines indicate fits according to Equation 25 in the equilibrium and nonequilibrium unbinding regimes.

$$f_{pull-off} = \frac{k_B T}{x_\beta} \ln\left(\frac{r_f}{r_0} N \exp\left(\frac{(N-1)E}{k_B T} \right) \right)$$ (24)

Qualitatively, in the case of such parallel loading of N bonds, the binding force is only slightly smaller than N times single bond strength (yet the scaling is not exact!). Another important feature of this case is that the distance scale of the interactions is unchanged, i.e., the width of the potential for the "macro-bond" is still equal to the width of the potential for a single bond. Moreover, the scaling of the bond strength with the loading rate predicted by the kinetic model for a single bond case is still valid.

Configuration constraints imposed by the rigid self-assembled monolayers provide an almost ideal parallel loading case; therefore, we can use the CFM measurements in the kinetic regime to determine the width of the interaction potential. Noy and coworkers[34] tested this regime for the interactions of COOH-terminated surfaces (Figure 11) and observed the behavior predicted by the kinetic analysis. When the tip-sample junction was loaded slowly, the unbinding force was virtually independent of the loading rate, indicating unbinding in the equilibrium regime. As the loading rate increased further, the system transitioned to the non-equilibrium unbinding and started to exhibit a characteristic exponential increase in the binding force with the loading rate. The measured slope of this increase estimates the distance to the transition state as 0.6 Å for the interactions between COOH functionalities. The distance to the transition state is significantly shorter than the values typically observed for interactions between biological macromolecules (see the Chapters 5 and 8), which is reasonable to expect considering the size difference. Interestingly, this value is comparable to the barrier width *per base pair* (0.7 Å) obtained by Strunz et al. in the DNA unbinding experiments.[100] It is tempting to assign the value obtained in these measurements to the hydrogen bond potential, but, as we will show in the next section, the real picture is more complicated.

4.2 Probing Entropic Barriers in Nanoscale Adhesion with CFM

As we have discussed in Section 2, solvation plays a very important role in shaping the interactions between chemically-modified probes and surfaces. The surface tension component model provides the quantitative framework for predicting binding forces in the near-equilibrium regime. However, we can gain a deeper understanding of the origin of such forces if we consider the temperature dependence of the tip-sample interaction strength. Noy and co-workers[34, 36] studied the strength of the interactions between several different surfaces as a function of temperature. Intuitively, we expect the binding force to decrease as the temperature increases and the thermal fluctuations gain more energy to break the bond. Surprisingly, the researchers observed (Figure 12) that for interactions between COOH-terminated surfaces in a polar, hydrogen-bonding solvent the interaction strength *increased* with the temperature.[36] This behavior was also present for interactions of other hydrophilic functionalities in polar solvents. Conversely, when the liquid medium was switched to a nonpolar solvent (hexane) the temperature trend reversed. The researchers attributed this behavior to the large negative entropy accompanying ordering of solvent molecules at the interfaces.[36] This negative entropy destabilizes the unbound state and leads to the observed counterintuitive temperature dependence. Nonpolar solvents do not tend to form ordered layers and thus do not contribute to these *entropic* solvation barriers.

The kinetic model outlined in the previous section can provide a quantitative interpretation of this phenomenological picture. If we separate the energy barrier into enthalpic and entropic components, $E_0 = \Delta H - T\Delta S$, and substitute Equation into Equation we can represent the temperature dependence of pull-off forces in the following form:

$$f_{pull-off} = \frac{\Delta H}{x_\beta} - \frac{\Delta S}{x_\beta} T - \frac{k_B T}{x_\beta} \ln \left[\frac{k_B T}{r_f \tau_D x_\beta} \right]. \tag{25}$$

The first two terms in Equation 25 describe the enthalpic and the entropic contribution to the bond strength, and the third term describes the contribution of thermal motion to the bond strength. In other words, the first two components describe the true energy-barrier

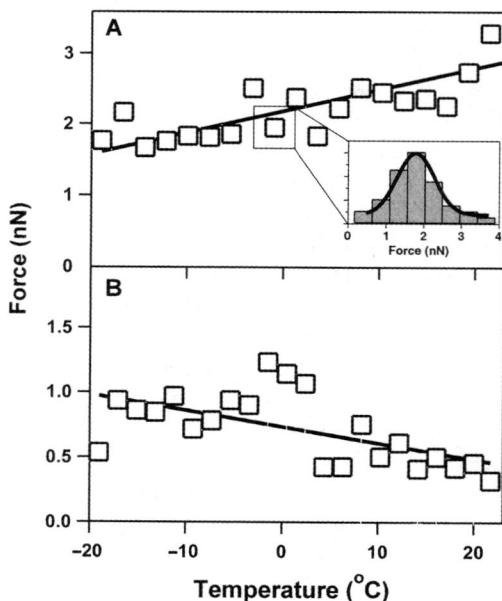

Figure 12. (A) Binding force as a function of temperature for interactions of COOH-modified probe and sample in ethanol. Inset shows a binding force histogram at one temperature point. (B) Binding force as a function of temperature for the interactions of COOH-modified probe and sample in hexane.

contribution and the (always negative) third component describes the "thermal weakening" of a bond caused by the thermal fluctuations helping the system to get over the activation barrier.

Equation 25 highlights another bit of nontrivial physics of the behavior of intermolecular bonds under external load. The third term in Equation 25 ("thermal weakening") always increases in magnitude as the temperature increases, leading to the overall decrease in the observed force, in full agreement with the intuitive picture. Yet the entropic term can lead to either increase or decrease in the overall interaction force depending on the sign on the entropy change for the unbinding process. Therefore, for the cases when the energy barrier has a large entropy component (i.e., in cases of associating solvents and high surface energy groups), we expect the bond strength to increase with the temperature. The relative magnitude of the entropic and the kinetic terms in Equation 25 defines two regimes of bond rupture: (i) thermally-dominated kinetics, where the kinetic weakening leads to a decrease in the observed bond strength with the increase in temperature; and (ii) barrier-dominated kinetics, where the entropic term overwhelms the kinetic term and leads to an increase in interaction strength with the increase in temperature. Furthermore, Equation 25 also indicates that the entropic regime of unbinding must exist only over a limited range of temperatures. As the temperature increases, the kinetic term, which increases as $T \cdot lnT$, will overwhelm the entropic term, which increases only linearly. For the entropic forces caused by the ordering of the solvent molecules at the surface, this crossover point simply corresponds to the situation when the thermal motion overwhelms molecular ordering in the solvent layers in the vicinity of the surface.

5 Conclusions

The chemical force microscopy approach provides several remarkable opportunities for force spectroscopy of specific intermolecular interactions. First, this technique creates a nanoscale tool that can address directly the roles that the different chemical functionalities, the nature of the solvent, and environmental variables play in shaping the strength of intermolecular interactions. Second, CFM studies go beyond the naïve notion that intermolecular interaction strength is determined solely by the nature of the interacting groups, and thus can be used as a universal tool for chemical identification. Indeed, CFM studies show that the interaction strength between two chemical species always reflects not only the nature of the species, but also the context of the environment surrounding these species, and in particular the solvent medium, which always plays a critical role in shaping intermolecular interactions in condensed phases.

Despite these complications, practical realizations of chemical force microscopy have shown tremendous progress towards understanding intermolecular interactions on the molecular level. The studies described in this chapter reveal the central role interfacial free energies play in determining the magnitudes of the observed adhesion (and friction) forces. Continuum contact mechanics models also provide a robust framework for quantitative understanding of the CFM results. We want to emphasize that a quantitative approach to the CFM experiments is almost always necessary to realize the full potential of the technique. Thus it is imperative to use the versatile arsenal of calibration and characterization tools that has been developed for the CFM experiments since the time of the inception of this technique.

Recently, an emerging kinetic view of intermolecular interactions has led to another paradigm shift for understanding these interactions. The kinetic model shows that the measured interaction strength depends not only on the energy landscape of the system, but also on the loading history prior to the bond breakup. This new paradigm refocuses our attention on the energy landscape as a fundamental characteristic of the interaction. Moreover, force spectroscopy approaches derived from the kinetic model allow direct characterization of the potential energy barrier geometry. These developments are only an indication of the rich opportunities that lie ahead, and chemical force microscopy studies will continue to reveal a true picture of the energy landscapes in complex chemical and biological systems.

References

1. Binnig, G.; Quate, C. F.; Gerber, C. *Phys. Rev. Lett.* **1986**, *56*, 930–3.
2. Quate, C. *Surf. Sci.* **1994**, *299/300*, 980.
3. Smith, D. P. E. *Rev. Sci. Instr.* **1995**, *66*, 3191–5.
4. Frisbie, C. D.; Rozsnyai, L. F.; Noy, A.; Wrighton, M. S.; Lieber, C. M. *Science* **1994**, *265*, 2071.
5. Albrecht, T. R.; Akamine, S.; Carver, T. E.; Quate, C. F. *J. Vac. Sci. Technol. A, Vac. Surf. Films* **1990**, *8*, 3386–96.
6. Meyer, G.; Amer, N. M. *Appl. Phys. Lett.* **1988**, *53*, 1045–7.
7. Cleveland, J. P.; Schaffer, T. E.; Hansma, P. K. *Phys. Rev. B* **1995**, *52*, R8692–R8695.
8. Israelachvili, J. *Intermolecular and Surface Forces*. Academic Press: New York, 1992.
9. Grandbois, M.; Beyer, M.; Rief, M.; Clausen-Schaumann, H.; Gaub, H. E. *Science* **1999**, *283*, 1727–1730.
10. Burnham, N. A.; Colton, R. J.; Pollock, H. M. *J. Vac. Sci. Technol. A, Vac. Surf. Films* **1991**, *9*, 2548–56.
11. Schaffer, T.; Viani, M.; Walters, D.; Drake, B.; Runge, E.; Cleveland, J.; Wendman, M.; Hansma, P. In *Micromachining and Imaging*; SPIE: San Jose, CA, 1997; Vol. 3009, p 48–52.
12. Viani, M. B.; Schaffer, T. E.; Chand, A.; Rief, M.; Gaub, H. E.; Hansma, P. K. *J. Appl. Phys.* **1999**, *86*, 2258–62.
13. Israelachvili, J. *Acc. Chem. Res.* **1987**, *20*, 415–21.
14. Chaudhury, M. K.; Whitesides, G. M. *Langmuir* **1991**, *7*, 1013–25.
15. Chaudhury, M. K.; Whitesides, G. M. *Science* **1992**, *255(5049)*, 1230–2.
16. Kappl, M.; Butt, H. J. *Particle & Particle Systems Characterization* **2002**, *19*, 129.
17. Ducker, W. A.; Senden, T. J.; Pashley, R. M. *Nature* **1991**, *353*, 239–241.
18. Houston, J. E.; Michalske, T. A. *Nature* **1992**, *356*, 266–267.
19. Bustamante, C.; Macosko, J. C.; Wuite, G. J. L. *Nature Reviews Molec. Cell Biol.* **2000**, *1*, 130–136.
20. Nuzzo, R. G.; Allara, D. L. *J. Am. Chem. Soc.* **1983**, *105*, 4481.
21. Porter, M. D.; Bright, T. B.; Allara, D. L.; Chidsey, C. E. D. *J. Am. Chem. Soc.* **1987**, *109*, 3559–3568.
22. Bain, C. D.; Troughton, E. B.; Tao, Y.-T.; Evall, J.; Whitesides, G. M.; Nuzzo, R. G. *J. Am. Chem. Soc.* **1989**, *111*.
23. Bain, C. D.; Evall, J.; Whitesides, G. M. *J. Am. Chem. Soc* **1989**, *111*, 7155.
24. Bain, C.; Whitesides, G. *Angew. Chem. Int. Ed. Engl.* **1989**, *28*, 506–512.
25. Whitesides, G. M.; Laibinis, P. E. *Langmuir* **1990**, *6*, 87.
26. Noy, A.; Frisbie, C. D.; Rozsnyai, L. F.; Wrighton, M. S.; Lieber, C. M. *J. Am. Chem. Soc.* **1995**, *117*, 7943–7951.
27. Vezenov, D. V.; Noy, A.; Rosznyai, L. F.; Lieber, C. M. *J. Am.Chem. Soc.* **1997**, *119*, 2006–2015.
28. Noy, A.; Sanders, C. H.; Vezenov, D. V.; Wong, S. S.; Lieber, C. M. *Langmuir* **1998**, *14*, 1508–1511.
29. Nakagawa, T.; Ogawa, K.; Kurumizawa, T.; Ozaki, S. *Japanese J. Appl. Phys., Part 2 (Letters)* **1993**, *32*, L294–6.
30. Nakagawa, T.; Ogawa, K.; Kurumizawa, T. *J. Vac. Sci. Technol. B, Microelectron. Nanometer Struct.* **1994**, *12*, 2215–18.
31. Green, J.-B. D.; McDermott, M. T.; Porter, M. D.; Siperko, L. M. *J. Phys.Chem.* **1995**, *99*, 10960.
32. Thomas, R. C.; Tangyunyong, P.; Houston, J. E.; Michalske, T. A.; Crooks, R. M. *J. Phys. Chem.* **1994**, *98*, 4493.
33. Sinniah, S. K.; Steel, A. B.; Miller, C. J.; Reutt-Robey, J. E. *J. Am. Chem. Soc.* **1996**, *118*, 8925–8931.
34. Zepeda, S.; Yeh, Y.; Noy, A. *Langmuir* **2003**, *19*, 1457–1461.
35. Vezenov, D. V.; Noy, A.; Lieber, C. M. *J. Adhes. Sci.Tech.* **2003**, *17*, 1385–1401.
36. Noy, A.; Zepeda, S.; Orme, C. A.; Yeh, Y.; De Yoreo, J. J. *J. Am. Chem. Soc.* **2003**, *125*, 1356 – 1362.
37. Noy, A.; Huser, T. R. *Rev. Sci. Instr.***2003**, *74*, 1217–1221.
38. Xiao, X.; Qian, L. *Langmuir* **2000**, *16*, 8153–8158.
39. Jang, J.; Schatz, G. C.; Ratner, M. A. *J. Chem. Phys.* **2004**, *120*, 1157–1160.
40. Wilbur, J. L.; Biebuyck, H. A.; MacDonald, J. C.; Whitesides, G. M. *Langmuir* **1995**, *11*, 825–831.
41. Poggi, M. A.; Bottomley, L. A.; Lillehei, P. T. *Nano Lett.* **2004**, *4*, 61–64.
42. Drake, B.; Prater, C. B.; Weisenhorn, A. L.; Gould, S. A. C.; Albrecht, T. R.; Quate, C. F.; Cannell, D. S.; Hansma, H. G.; Hansma, P. K. *Science* **1989**, *243*, 1586–9.
43. Fotiadis, D.; Scheuring, S.; Muller, S. A.; Engel, A.; Muller, D. J. *Micron* **2002**, *33*, 385–397.
44. Malkin, A. J.; Land, T. A.; Kuznetsov, Y. G.; McPherson, A.; Deyoreo, J. J. *Phys.l Rev. Lett.* **1995**, *75*, 2778–2781.

45. Land, T. A.; Malkin, A. J.; Kuznetsov, Y. G.; McPherson, A.; Deyoreo, J. J. *Phys.l Rev. Lett.* **1995**, *75*, 2774–2777.
46. Kuznetsov, Y. G.; Malkin, A. J.; McPherson, A. *J. Cryst. Growth* **1999**, *196*, 489–502.
47. Kuznetsov, Y. G.; Malkin, A. J.; Land, T. A.; DeYoreo, J. J.; Barba, A. P.; Konnert, J.; McPherson, A. *Biophys. J.* **1997**, *72*, 2357–2364.
48. Malkin, A. J.; Kuznetsov, Y. G.; Lucas, R. W.; McPherson, A. *J. Struct. Biol.* **1999**, *127*, 35–43.
49. Li, M. Q. *Appl. Phys. A:* **1999**, *68*, 255–258.
50. Woolley, A. T.; Cheung, C. L.; Hafner, J. H.; Lieber, C. M. *Chemistry & Biology* **2000**, *7*, 193–204.
51. Noy, A.; Vezenov, D.; Kayyem, J.; Meade, T.; Lieber, C. *Chemistry & Biology* **1997**, *4*, 519–527.
52. Friddle, R.; Klare, J. E.; Martin, S.; Corzett, M.; Balhorn, R.; Baldwin, E. P.; Baskin, R.; Noy, A. *Biophys. J.* **2004**, *86*, 1632–1639.
53. Sulchek, T. A.; Friddle, R. W.; Langry, K.; Lau, E. Y.; Albrecht, H.; Ratto, T. V.; DeNardo, S. J.; Colvin, M. E.; Noy, A. *Proc. Natl. Acad. Sci. USA* **2005**, *102*, 16638–16643.
54. Kienberger, F.; Ebner, A.; Gruber, H. J.; Hinterdorfer, P. *Acc. Chem. Res* **2006**, *39*, 29–36.
55. Cleveland, J. P.; Manne, S.; Bocek, D.; Hansma, P. K. *Rev. Sci. Instr.* **1993**, *64*, 403–5.
56. Hutter, J. L.; Bechhoefer, J. *Rev. Sci. Instr.* **1993**, *64*, 1868–73.
57. Sader, J. E.; Larson, I.; Mulvaney, P.; White, L. R. *Rev. Sci. Instr.* **1995**, *66*, 3789–98.
58. Sader, J. E.; Chon, J. W. M.; Mulvaney, P. *Rev. Sci. Instr.* **1999**, *70*, 3967–9.
59. Ogletree, D. F.; Carprick, R. W.; Salmeron, M. *Rev. Sci. Instr.* **1996**, *67*, 3298.
60. Warmack, R. J.; Zheng, X. Y.; Thundat, T.; Allison, D. P. *Rev. Sci. Instr.* **1994**, *65*, 394–9.
61. Neumeister, J. M.; Ducker, W. A. *Rev. Sci. Instr.* **1994**, *65*, 2527–31.
62. Hazel, J. L.; Tsukruk, V. V. *Thin Solid Films* **1999**, *339*, 249–57.
63. Marti, O. *Phys. Scr. Vol. T (Sweden), Physica Scripta Volume T* **1993**, *T49B*, 599–604.
64. D'Costa, N. P.; Hoh, J. H. *Rev.f Sci. Instr.* **1995**, *66*, 5096–7.
65. Liu, Y.; Wu, T.; Fennel-Evans, D. *Langmuir* **1994**, *10*, 2241.
66. Liu, Y. H.; Evans, D. F.; Song, Q.; Grainger, D. W. *Langmuir* **1996**, *12*, 1235–1244.
67. Schwarz, U. D.; Koster, P.; Wiesendanger, R. *Rev. Sci. Instr.* **1996**, *67*, 2560–7.
68. Thomas, R. C.; Houston, J. E.; Crooks, R. M.; Kim, T.; Michalske, T. A. *J. Am. Chem. Soc.* **1995**, *117*, 117.
69. Derjaguin, B. V.; Muller, V. M.; Toporov, Y. P. *J. Coll. Interf. Sci.* **1975**, *53*, 314–26.
70. Carpick, R. W.; Agrait, N.; Ogletree, D. F.; Salmeron, M. *J. Vac. Sci. Technol. B, Microelectron. Nanometer Struct.* **1996**, *14*, 1289–95.
71. Muller, V. M.; Yushchenko, V. S.; Derjaguin, B. V. *J. Coll. Interf. Sci.* **1980**, *77*, 91–101.
72. Maugis, D. *Langmuir* **1995**, *11*, 679–682.
73. Carpick, R. W.; Sasaki, D. Y.; Burns, A. R. *Tribology Letters* **1999**, *7*, 79–85.
74. Johnson, K. L. *Proc. Roy.l Soc., London, A* **1997**, *453*, 163–179.
75. Dzyaloshinskii, I. E.; Lifshitz, E. M.; Pitaevskii, L. P. *Physics-Uspekhi* **1961**, *4*, 153–176.
76. London, F. *Trans. Faraday Soc* **1937**, *33*, 10.
77. van Oss, C. J.; Chaudhury, M. K.; Good, R. J. *Chem. Rev.* **1988**, *88*, 927–41.
78. Segeren, L.; Siebum, B.; Karssenberg, F. G.; Van Den Berg, J. W. A.; Vancso, G. J. *J. Adh. Sci. Technol.* **2002**, *16*, 793–828.
79. McKendry, R.; Theoclitou, M.; Abell, C.; Rayment, T. *Langmuir* **1998**, *14*, 2846–2849.
80. Sato, F.; Okui, H.; Akiba, U.; Suga, K.; Fujihira, M. *Ultramicroscopy* **2003**, *97*, 303–14.
81. Williams, P. M. *Analytica Chim. Acta* **2003**, *479*, 107–115.
82. Skulason, H.; Frisbie, C. D. *Langmuir* **2000**, *16*, 6294–6297.
83. Wong, S. S.; Joselevich, E.; Woolley, A. T.; Cheung, C. L.; Lieber, C. M. *Nature* **1998**, *394*, 52–55.
84. Hafner, J. H.; Cheung, C. L.; Lieber, C. M. *J. Am. Chem. Soc.* **1999**, *121*, 9750–9751.
85. Hafner, J. H.; Cheung, C. L.; Oosterkamp, T. H.; Lieber, C. M. *J. Phys. Chem. B* **2001**, *105*, 743–746.
86. Caprick, R.; Salmeron, M. *Chem. Rev.* **1997**, *97*, 1163–1194.
87. Beake, B. D.; Leggett, G. J.; Shipway, P. H. *Surf. Interf. Analysis* **1999**, *27*, 1084–1091.
88. Brewer, N. J.; Leggett, G. J. *Langmuir* **2004**, *20*, 4109–4115.
89. Leggett, G. J.; Brewer, N. J.; Chong, K. S. L.; Matter, S. *Phys. Chem. Chem. Phys* **2005**, *7*, 1107–1120.
90. Brewer, N. J.; Foster, T. T.; Leggett, G. J.; Alexander, M. R.; McAlpine, E. *J Phys. Chem. B,* **2004**, *108*, 4723–4728.
91. Vezenov, D. V.; Zhuk, A. V.; Whitesides, G. M.; Lieber, C. M. *J. Am. Chem. Soc.* **2002**, *124*, 10578–10588.
92. Duwez, A. S.; Jonas, U.; Klein, H. *Chem. Phys. Chem.* **2003**, *4*, 1107–1111.
93. Mckendry, R.; Theoclitou, M.; Rayment, T.; Abell, C. *Nature* **1998**, *391*, 566–568.
94. Clear, S. C.; Nealey, P. F. *J. Coll. Interf. Sci.* **1999**, *213*, 238–250.
95. Warszynski, P.; Papastavrou, G.; Wantke, K. D.; Mohwald, H. *Colloids and Surfaces A: Physicochemical and Engineering Aspects* **2003**, *214*, 61–75.
96. Evans, E.; Ritchie, K. *Biophys. J.* **1997**, *72*, 1541–1555.

97. Evans, E. *Faraday Discussions* **1999**, 1–16.

98. Evans, E.; Williams, P. In *Physics of Bio-Molecules and Cells*; Flyvbjerg, H., Jülicher, F., Ormos, P., David, F., Eds.; Springer and EDP Sciences: Heidelberg, 2002; Vol. 75, p 147–185.

99. Evans, E.; Williams, P. In *Physics of Bio-Molecules and Cells*; Flyvbjerg, H., Jülicher, F., Ormos, P., David, F., Eds.; Springer and EDP Sciences: Heidelberg, 2002; Vol. 75, p 187–203.

100. Strunz, T.; Oroszlan, K.; Schafer, R.; Guntherodt, H. J. *Proc. Natl. Acad. Sci. U.S.A.* **1999**, *96*, 11277–11282.

101. Han, T.; Williams, J. M.; Beebe, T. P., Jr. *Anal. Chim. Acta* **1995**, *307*, 365–376.

102. Ito, T.; Namba, M.; Buehlmann, P. and Umezawa, Y. Langmuir, **1997**, *13*, 4323–4332.

103. Headrick, J. E. and Berrie, C. L. Langmuir, **2004**, *20*, 4124–4131.

104. Ara, M. and Tada, H. *Appl. Phys. Lett.* **2003**, *83*, 578–580.

105. Vezenov, D. V.; Noy, A.; Ashby, P. J. *Adh. Sci. Tech.* **2005**, *19*, 313–364.

106. Tormoen, G. W.; Drelich, J.; Beach, E. R., J. *Adh. Sci. Tech.* **2004**, *18*, 1–17.

Chemical Force Microscopy: Force Spectroscopy and Imaging of Complex Interactions in Molecular Assemblies

Dmitry V. Vezenov, Aleksandr Noy, and Charles M. Lieber

1 Introduction

Many macroscopic processes and events in condensed phases are driven by intermolecular interactions that extend beyond the basic pair-wise adhesive interactions of functional groups that were the focus of the previous chapter. The spectrum of these interactions ranges from the long-range forces that by necessity involve large ensembles of functional groups, to friction and wear processes that involve constant breaking and reforming of noncovalent bonds at different length and time scales. Often these interactions are accompanied by substantial energy dissipation, and thus rigorous quantitative treatment of these events becomes a considerable challenge.

Despite such difficulties, Chemical Force Microscopy (CFM) provides an excellent opportunity to probe these events with an unprecedented degree of chemical specificity. We first describe the use of CFM for probing ionization processes at surfaces. The long-range nature of Coulombic interactions and double-layer formation in aqueous solutions of electrolytes leads to the involvement of a large number of species both in solution and on the surfaces, and the interplay between these species gives rise to complex processes such as charge regulation. We show how rational control over the chemical functionality of the atomic force microscope (AFM) probe allows local probing of these processes and provides information about the ionization state of the surface functional groups.

Monitoring friction at the local scale provides an excellent opportunity to study the role of specific interactions in this complex dissipative process. Extreme curvature of the AFM tip provides a convenient system for studying single asperity constants, and CFM measurements provide an ideal vehicle for probing the fundamentals of the friction forces. On a molecular level, friction process involves repeated breaking of noncovalent bonds; therefore, we need to control the chemistry at the sheared interface to probe the link between adhesion and friction forces in these idealized model systems.

An important extension of the friction force studies is the use of chemically-functionalized AFM tips for rational imaging of surface functional groups. We describe how the observations of chemical sensitivity in friction forces translate into the chemically-sensitive imaging of surface patterns of functional groups. Besides representing an immensely important subset of the AFM-based imaging techniques, tapping (intermittent-contact) mode imaging provides an opportunity for direct probing of the energy dissipation in the tip-sample junction; and we show that chemical modification of the AFM probes provides us with an opportunity to tune these interactions. Overall, chemical modification of the AFM probes gives a versatile and powerful method for probing the dynamic of complex intermolecular interactions in condensed phases.

2   Probing Adhesion Forces in Molecular Assemblies Bearing Charges

2.1   Chemical Force Titrations: Operating Principle

The acid-base chemistry of ionizable sites immobilized at interfaces is central to understanding a number of complex dispersed systems, such as micelles, nanoparticles, polymers, polyelectrolytes, and biological macromolecules, thus making water an important medium for CFM experiments. The charged state of the surfaces is of fundamental interest in colloidal stability, particle adhesion to surfaces, and adsorption of macromolecules.[1, 2] An important consideration for these systems is that the dissociation constants of the surface groups can differ from those of their monomer analogs in solution. Several factors can cause these differences: (i) a low dielectric permittivity of an adjacent hydrocarbon region; (ii) fewer degrees of freedom for the immobilized species; (iii) an excess electrostatic free energy of the supporting surface; and (iv) changes in the solution dielectric constant in the vicinity of charged surfaces.[3] While a theoretical description must make estimates of these considerations, CFM includes these factors by design.

The surface free energy depends on the state and the degree of ionization of the functional groups, which can be monitored by measuring a contact angle of a buffered droplet versus pH of the drop solution. A large change in the contact angle occurs at a pH equal to the pKa of the surface groups, when half of the maximum number of ionizable groups have dissociated.[4] Chemically-modified AFM tips provide an alternative way to probe such changes in solid-liquid surface free energies with pH. The surface charge induced by the dissociation of acidic/basic groups can be detected by monitoring the adhesive force with an AFM probe sensitive to electrochemical interactions, and variations in the sign and magnitude of the interactions will indicate changes in the surface charge: a transition from adhesion to repulsion will occur at a pH$\approx$pK of the functional group on the surface. This AFM-based approach to determining local pKs has been termed "chemical force titration" (CFT) by Lieber and co-workers.[5]

CFT probes pH dependence of adhesion forces between basic or acidic functional groups, or more generally, dependence of adhesion forces between ionizable groups as a function of the concentration of counterions. Several aspects of ionization of surface groups have been investigated with CFM: (i) large pK_a shifts of amine-terminated SAMs; (ii) observation of inverted CFT curves for aromatic amines; (iii) effect of ionic strength (IS) on the shape of the CFT curves and peak positions; and (iv) ion adsorption by neutral groups.

Adhesion force values measured at different solution pH for tips and samples functionalized with 3-aminopropyltriethoxysilane (APTES) SAMs terminating with amine groups show a sharp drop to zero (indicating a repulsive interaction) below a pH of 4 (Figure 1A). This decrease and elimination of an attractive force between the tip and sample at low pH is consistent with the protonation of the amine groups on these two surfaces. Contact angle values measured using buffered solution droplets on this surface (Figure 1B) also show a sharp transition (an increase in wetability) as the droplet pH is reduced below 4.5. Hence local force microscopy measurements using a modified probe tip and macroscopic wetting studies provide very similar values for the pK of the surface amine group in the APTES-derived SAMs.

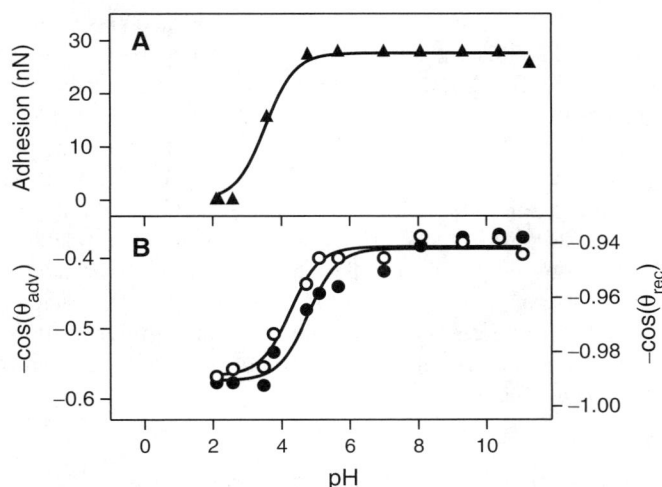

Figure 1. (A) Adhesion force between sample and tip functionalized with amino groups versus pH (B) Negative cosines of the advancing (open circles) and receding (filled circles) contact angles of phosphate buffer drops on a Si sample modified with APTES as a function of pH.

The apparent pKa obtained from force microscopy (pK = 3.9) and contact angle wetting (pK = 4.3) for the surface amine group is 6–7 pK units lower than bulk solution values.[6] Similarly, large pK_a shifts and strong adhesion at high pH were found for thiolate long-chain $(-(CH_2)_{11}-)$ SAM pairs bearing amine functionality.[7] Large shifts in dissociation constants observed in CA titrations for mixed acid-methyl monolayers were attributed to unfavorable solvation of the carboxylate anion at the monolayer interface.[8] Large pK shifts relative to solution were also observed in studies of amino groups grafted onto the surface of a hydrophobic polymer.[9] In addition, simulations of the titration of surface amine groups showed large negative shifts in pK when the amine was poorly solvated.[10] Since the interfacial tension in alkylamine-water system is negligibly small (< 0.1 mJ/m^2),[11] the high contact angles and large adhesion forces at pH>4 indicate that the APTES-derived monolayers are relatively hydrophobic. The hydrophobic nature of the SAM likely arises from a partially disordered structure that exposes methylene groups at the surface. Hence one can attribute the large observed pK shift to a hydrophobic environment surrounding the amine groups. A vapor-phase deposition procedure[12] improves the quality of the APTES modification of SiO_2/Si substrates, and this silanization method produces reproducible monolayers with a higher $pK_{1/2}$ (7.4) and a factor of three lower adhesion forces. On the other hand, increasing the chain length from 2 to 11 methylene units of analogous amine-terminated thiol SAM did not result in significant $pK_{1/2}$ changes. Notably, the measured adhesion force, F_a, was larger for a short-chain pair than for a long-chain pair as expected from the degree of ordering in these SAMs.[13]

The ability to detect such pK changes locally by AFM should be of significant utility to studies of biological and polymer systems. Several studies have already demonstrated the potential of the technique: He et al. demonstrated the local character of pK_a measurements with CFT by observing $pK_{1/2}$ of 5.6 for COOH SAM on micropatterned substrates (1–2 μm regions terminating in COOH or CH$_3$), for which contact angle titration showed a transition at $pK_{1/2} = 11.0$,[14] and the Vancso group reported a remarkable spatial resolution of ca. 50 nm for CFT of plasma treated polymer substrates.[15, 16]

CFT of basic functional groups, in which a nitrogen atom is involved with aromatic functionality (either being a part of the aromatic ring, as in 4-mercaptopyridine HS-Py, or through conjugation, as in 4-aminothiophenol HS-Ph-NH$_2$), produced high adhesion force at low pH and low/repulsive forces at high pH—contrary to expectations based on their charged state.[17, 18] To explain this phenomenon, *ab initio* calculations of surface charges were carried out for APTES, HS-Ph-NH$_2$, and HS-Py molecules.[19] Calculated surface charges (in units of electron charge) for these groups were, respectively, 0.606, 0.607, and 0.035 in the protonated state, and −0.09, −0.298, and −0.909 in the neutral state. A highly polarized state of Py groups together with efficient charge dispersion in Py-H$^+$ is consistent with repulsion between neutral Py groups and attraction between charged Py-H$^+$ groups observed in CFTs. A significant surface charge of Ph-NH$_2$ groups is responsible for low adhesion at high pH. Protonated Ph-NH$_3^+$ groups, however, carry the same charge as primary amine APTES. Since some fraction of neutral groups carrying comparable and opposite surface charge (−0.3 vs 0.6) is available at low pH, the observed adhesion reflects the interaction between surfaces with reduced overall charge and having a distribution of positive (due to Ph-NH$_3^+$) and negative (due to Ph-NH$_2$) local charge.

2.2   Chemical Force Titrations on Hydrophilic Surfaces

Contact angle goniometry is limited to sufficiently hydrophobic surfaces that do not show complete wetting in both non-ionized and ionized states. In the case of high free energy surfaces, such as COOH terminated SAMs, it is necessary to dilute the hydrophilic groups with a hydrophobic surface component and either pretreat the surface[8] or use another liquid (vs. vapor phase)[20] to perform a contact angle experiment. As CFT results on the APTES system demonstrate, the incorporation of a hydrophobic component must be used with considerable caution, since it can produce very large pK shifts. Such shifts in the pK$_a$ of surface COOH relative to bulk solution have been observed in cases of mixed COOH and CH$_3$3 SAMs.[8, 20] Hence it has not been possible to determine the pK of a homogeneous COOH-terminated surface by the contact angle approach.

In contrast, force titrations provide a direct measure of the solid-liquid interfacial free energy and completely bypass the limitations discussed in the previous paragraph. A prominent feature in the force titration curve obtained for COOH-terminated sample and tip (Figure 2A) is the sharp transition from positive adhesion forces at low pH to zero (indicating repulsion) at high pH. The observed repulsion at pH>6 can be attributed to interaction between electrical double layers formed in the presence of charged carboxylate groups, while the adhesive interaction at low pH values originates from hydrogen bonding between uncharged carboxyl groups. The force versus separation curves become fully reversible and practically identical at all pH values higher than 7. This indicates that the surface charge density is saturated under these conditions. Based on these CFT data, the pK$_a$ of the surface-confined carboxylic acid is 5.5 ± 0.5. This value lies within 0.75 pK units of the pK$_a$ for COOH functionality in aqueous solution.[6] The similarity of surface-confined and solution pK$_a$s strongly indicates that solvation effects do not play a significant role in determining the ionization behavior of pure COOH-terminated SAMs.

Control experiments with both hydrophilic and hydrophobic groups that do not dissociate in aqueous solutions do not display pH-dependent transitions. CFM titration curves for tip/sample SAMs terminating in OH/OH and CH$_3$/CH$_3$ functionalities show an approximately constant, finite adhesive interaction throughout the whole pH range studied (Figure 2B). To probe pKs on unknown surfaces, we need to use tips functionalized with SAMs that (i) do not exhibit a pH-dependent change in ionization and (ii) are hydrophilic. The hydroxyl-terminated SAM meets these requirements, and has been used to determine the pK$_a$ of carboxyl-terminated SAM as shown in Figure 2C. Remarkably, the dissociation constant obtained from that titration curve is the same as that determined from the data in Figure 2A using COOH-terminated SAMs on both sample and tip.

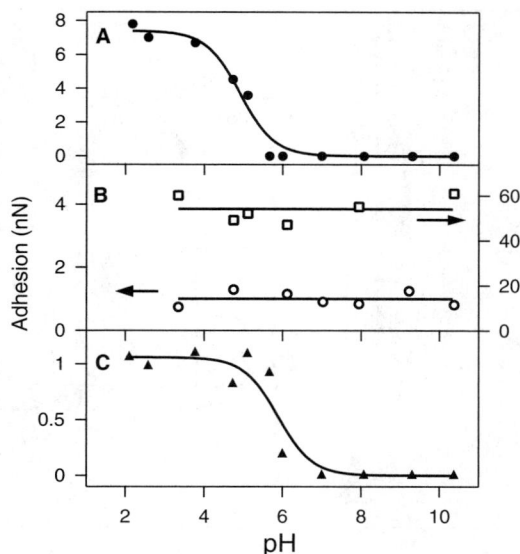

Figure 2. CFT curves recorded in buffered solutions (ionic strength of 0.01 M) for (A) COOH-COOH; (B) CH$_3$-CH$_3$ (squares) and OH-OH (circles); and (C) COOH-OH contacts 5.

Wong et al. achieved the ultimate resolution of CFT by applying it to identify the functional groups at the terminus of chemically modified carbon nanotubes (CNTs).[21, 22] Shortening the CNT by an electrical arc discharged in the presence of oxygen resulted in termination of the CNT with carboxylic acid groups. CFT of such CNT showed adhesion curves (Figure 3) similar to those of analogous SAMs systems (Figure 2A). Subsequent chemical modification of the nanotube ends to amine functionality resulted in titration curves characteristic of basic functional groups. Alternatively, a COOH functional group at the CNT end could be converted chemically into neutral benzyl moiety, for which CFTs no longer displayed the transitions as expected when dissociation of surface groups cannot occur.

2.3 Peaks in CFT Curves Due to Hydrogen Bonding and Ion Bridges

CFT at low ionic strength ($<10^{-4}$ M) resulted in another type of unusual behavior: instead of a sigmoidal transition from the high adhesion between neutral groups to low adhesion (or repulsion) between charged groups, Smith et al. observed a peak in adhesion force at intermediate pH for COOH,[23–25] PO(OH)$_2$,[24–26] and NH$_2$[13] groups. Their interpretation of 10- to 20-fold increase of adhesion between hydrophilic groups centers on a hypothesis of formation of "strong hydrogen bonds" between neutral and charged groups representing a conjugated acid/base pair, e.g., carboxyl and carboxylate. For a given fraction β of dissociated groups, the total adhesion force is comprised of two contributions from a total of N groups: (i) "weak" hydrogen bonds between neutral groups (e.g., COOH/COOH) with a single force value of f_{hb} and (ii) "strong" ionic hydrogen bonds (e.g., COOH/COO$^-$) that are a factor of m stronger:

$$F_a = N f_{hb} \left[2 \left(1-\beta\right) \beta \, m + \left(1-\beta\right)\left(1-\beta\right) m \right] \tag{1}$$

Smith and co-workers were able to reproduce the shape of the CFT peak with the equation above for HS(CH$_2$)$_{15}$COOH SAMs resulting in a fitted value m$\approx$16 (Figure 4), although for shorter HS(CH$_2$)$_{10}$COOH SAM, the observed peak was wider than predicted.[23]

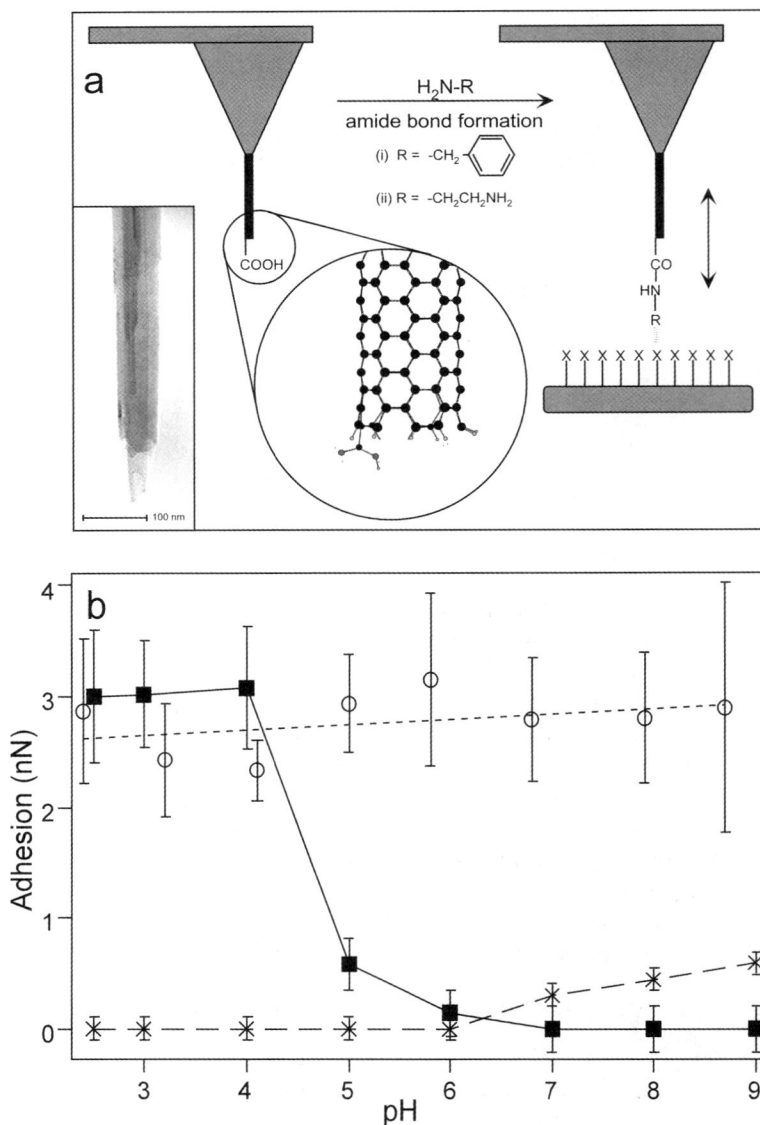

Figure 3. (A) Diagram illustrating the modification of a nanotube tip by coupling an amine (RNH_2) to a pendant carboxyl group, and the application of this probe to sense specific interactions with functional groups (X) of a substrate. The circular inset is a molecular model of a single nanotube wall with one carboxyl group at the tip end. Inset, TEM image showing the open end of a shortened nanotube tip. (B) Adhesion force as a function of pH between the nanotube tips and a hydroxy-terminated SAM (11-thioundecanol on gold-coated mica): filled squares, carboxyl (unmodified); open circles, phenyl (modified with benzylamine); and crosses, amine (modified with ethylenediamine). Each data point corresponds to the mean of 50–100 adhesion measurements, and the error bars represent one standard deviation. From Ref. 21.

Since the relevant forces reflect the energy balance for interactions between groups on the tip and sample as well as these same surface groups and the solvent, one needs to take these arguments as reflecting the difference in the strength of the respective hydrogen bonds between carboxylate anion and carboxylic acid group on the one hand, and carboxylate anion and water on the other hand. Differences in hydrogen bond strength could partly reflect differences in the dielectric constant of the environment surrounding charged groups

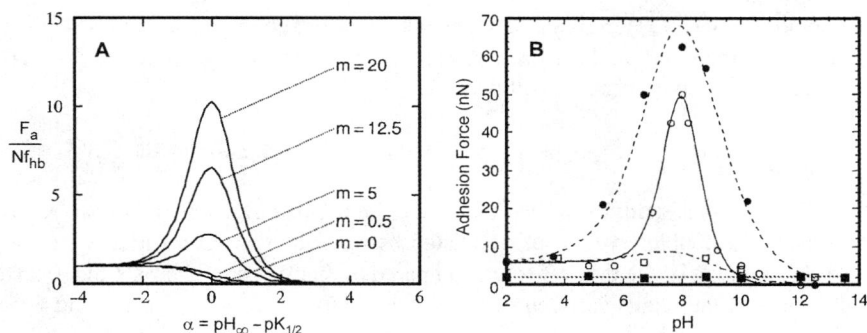

Figure 4. (A) The CFT curves are modeled as a linear combination of two types of hydrogen bonds—strong ionic and weak neutral bonds—and plotted for different values of m (the ratio of the strength of these two bonds). pH∞ is the pH of the bulk solution, and $pK_{1/2}$ is defined as the pH at which half of the surface groups are ionized. The peak in CFT observed for the low electrolyte concentration is reproduced with values of m = 15–20, whereas the sigmoidal step at high ionic strength is reproduced by m~0, i.e., no strong hydrogen bonds formed (e.g., because of the interaction of ions in the buffer with the ionized acid groups in the SAMs). (B) Effect of SAM packing on CFT curves in a buffer of very low electrolyte concentration (10^{-7} M) for tips and substrates modified with $HS(CH_2)_{15}COOH$ (16:16, open circles), $HS(CH_2)_{10}COOH$ (11:11, solid circles), $HS(CH_2)_{10}COOH$ and $HS(CH_2)_3COOH$ (11:3, open squares), and $HS(CH_2)_3COOH$ (3:3, solid squares). The 16:16 peak is accurately fitted by Equation 1, yielding a value of m = 16), while the rest of the curves (11:11, 11:3, and 3:3 data) are only guides to the eye.

when exposed to water, or when sandwiched between organic SAMs. Similarly shaped CFT curves were observed for $(HS(CH_2)_{11}O)_2PO(OH)$ modified tips and surfaces.[27] One caveat of this model is that while the peak in adhesion force is predicted for $\beta = \frac{1}{2}$, the $pK_{1/2}$ values should be more appropriately treated as pH values where adhesion force takes $\frac{1}{2}$ its maximum value, rather than pH for the degree of dissociation of 0.5 for surface groups. An analysis of the noncontact regime of CFT with force feedback showed that the surface potential of COOH groups at pH = 7 reflects the maximum β value of about 15%,[28] while surface charge calculation gives the pK_a of 7.7. Simultaneous analysis of contact (adhesion) and noncontact (double layer) forces appears to be the most consistent way for determining the pK_a of surface functional groups.

CFT of neutral surfaces that can form complexes with ions from solution can result in a sandwich-type interfacial bond: adhesion between methylsulfanyl groups on the tip and sample in a solution of $AgNO_3$ (IS = 0.1 M, KNO_3) showed a peak at $p[Ag^+] = 2$, while adhesion between similarly prepared hexyl monolayers was insensitive to variations in Ag^+ concentration.[29] A model of binding was proposed that involved formation of 1:2 interfacial complexes, $R(CH_3)S-Ag^+-S(CH_3)R$, that promoted strong interfacial bonds at a low concentration of silver ions. At a high concentration ($>10^{-2}$ M) of Ag^+, the surface coverage increased (approached saturation): competitive repulsion from $R(CH_3)S-Ag^+/Ag^+-S(CH_3)R$ interactions resulted in a sharp drop in adhesion. Interestingly, detailed analysis of a noncontact regime in CFTs on nominally neutral surfaces—hydrophobic, methyl-terminated, and hydrophilic, ethylene oxide-terminated—revealed that both interfaces can undergo complex charging as a function of solution pH via adsorption of hydronium and hydroxyl ions.[30]

2.4 Modeling of Adhesion Forces in the Presence Charge

The JKR theory of contact mechanics (see chapter 3) can serve as a reasonable basis for understanding adhesion data in an aqueous medium. To interpret pH dependent adhesion data in electrolyte solutions it is also important to consider long-range electrostatic forces between the tip and sample surface. Since the JKR theory is based on energy balance, it is reasonable to expect no adhesion (i.e., Hertzian behavior) when the free energy of a double layer per unit area w_{DL} balances the interfacial surface tension γ_{SL}. Quantitatively, the pull-off force, $P_{pull-off}$, can be related to these two terms as follows:

$$P_{pull-off} - P_{DL} = -\frac{3}{2}\pi R(W_{SLS} - 2w_{DL}), \; or \; P_{pull-off} = -\frac{3}{2}\pi RW_{SLS} + \frac{5}{2}P_{DL} \quad (2)$$

where $P_{DL} \approx 2\pi Rw_{DL}$ is an additional load that has to be applied to a spherically shaped tip due to the presence of a double layer. Thus, repulsion between like-charged surfaces ($P_{DL} > 0$) will decrease the magnitude of the pull-off force compared to that given by the JKR theory. There is a threshold value of the repulsive electrical double layer force $P_{DL} = P_{pull-off}$, beyond which the deformation of a spherical tip should be fully reversible with a contact radius going monotonically to zero (no pull-off force) as the load is reduced. This condition is equivalent to $\gamma_{SL} \approx w_{DL}$; that is, the attractive surface free energy component is canceled by the repulsive double layer term. The corresponding surface potential

$$\Psi = \sqrt{\frac{\lambda}{\varepsilon\varepsilon_0}} \gamma_{SL} \quad (3)$$

is independent of the tip radius for $\lambda \ll R$ (λ is the Debye length, ε is the dielectric constant, and ε_0 is the permittivity of vacuum). Therefore, the change from adhesive to repulsive behavior is characteristic of the ionization state of the interacting surfaces and can be used to estimate the surface potential.

This model allows us to calculate the surface free energies and surface potentials of hydrophilic SAMs. The values of γ_{SL} determined from adhesion data for OH and COOH (fully protonated) terminated surfaces were $8 \, mJ/m^2$ and $16 \, mJ/m^2$, respectively, and are in good agreement with the measurements of interfacial tension for the two-phase systems consisting of water and melts of either long-chained alcohols (7-8 mJ/m^2)[31] or carboxylic acids (10-11 mJ/m^2).[32] The surface potential of the carboxylate SAM at pH>6 and ionic strength of 0.01 M, 65 mV, calculated from Equation 3, is substantiated by analysis of F-D curves discussed in the next section.

2.5   Probing Double Layer Forces

The electrochemical origin of the pH-dependent repulsive forces can be verified by changing the Debye screening length λ; (i.e., by changing the solution ionic strength). The repulsive interaction becomes progressively longer-ranged (Figure 5, left panel) as the solution ionic strength decreases. A detailed analysis of the electrical double layer forces must take into account the surface charge-potential regulation imposed by the potential dependent binding of counterions (H^+ and Na^+ in the case of COOH SAMs) at the interface; for example, by using a model with a linearized condition for charge-potential regulation.[33] For carboxylate surfaces, this model fitted the experimental data very well down to separations of 1 nm. The condition of constant charge (vs. constant potential) was approached in these experiments independently of the ionic strength. The values of the Debye screening length extracted from experimental F-D curves agree well with the values calculated from the solution ionic strengths. Significantly, the surface potential calculated using this analysis of the carboxyl terminated surface at pH = 7.2 and 0.01 M ionic strength, 60 mV, is close to the value estimated from adhesion measurements.

Thus, it is possible to determine double layer parameters of systems being imaged by simultaneously recording F-D curves, using a tip bearing a functional group with a predetermined ionization behavior. Force-distance curves obtained with magnetic force feedback removed the jump-in instability and allowed recording of the full force profile for both neutral OH and charged COOH groups.[28] The F-D curves in the case of OH SAMs showed only

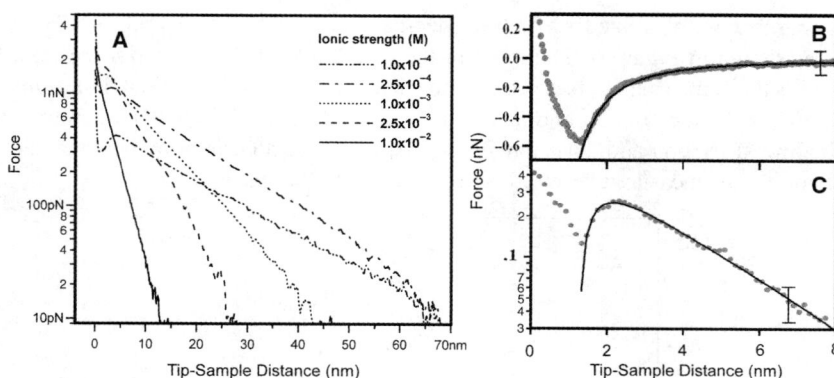

Figure 5. (A) Repulsive double layer forces versus tip-sample separation recorded on approach between COOH modified tips and samples at different ionic strengths of the pH 7.2 buffer. (B, C) Force profiles obtained using magnetic force feedback for (B) OH-terminated SAMs in deionized water and (C) COOH-terminated SAMs in 0.01 M, pH 7 phosphate buffer. Circles: Experimental data; Lines: model fits. The error bars are representative of the respective data sets 5.

attraction, whereas in the case of COOH SAMs double layer repulsion dominated van der Waals attraction at long (> 2 nm) distances.

3 Chemical Effects in Friction Experiments

CFM experiments on multiple organic monolayer systems demonstrated that friction and adhesion forces correlate at the nanometer scale. The friction force between functional groups on a sample surface and the tip is usually determined by recording the lateral deflection of the cantilever as the sample is scanned in a forwards/backwards cycle along the direction perpendicular to the cantilever axis to produce a friction loop.[34] The externally applied load is controlled independently through the cantilever normal deflection. The numerical values of the normal and lateral forces are determined using the cantilever spring constants, and friction coefficients are obtained from the slopes of the corresponding friction versus load (F-L) curves.

CFM uses chemically modified tips bearing well-defined functional groups to assess unambiguously chemical contributions to friction. As mentioned in the section on the mechanics of tip-substrate contact in the preceding chapter, one has to be careful in selecting conditions for quantitative analysis of chemical effects on frictions force. Ideally, such characterization comes from comparison of values of the shear stress for each functional group pair. These values are rarely available, because they require accurate knowledge of mechanical and geometrical parameters of the system (cantilever constants, tip radii, effective elastic constants of monolayer-substrate contact). Quantitative *comparison* of different functional groups in CFM is frequently achieved by ensuring that the systems are structurally similar by using (i) the same monolayer system (i.e., thiol SAMs on polycrystalline gold) and (ii) CFM tips of similar characteristics (or the same tip with different surfaces/media).

In ambient conditions, it is difficult to distinguish clearly true chemical contributions to friction from other factors primarily due to the large magnitude of capillary adhesion. This effect can be eliminated by performing measurements either in liquids or ultrahigh vacuum. Friction forces between tips and samples modified with different functional groups have been measured in ethanol,[35] water,[5] and dry argon[36] as a function of an applied load. In all of these measurements, the friction forces were found to increase linearly with the applied load. For a fixed external load the absolute friction force decreased as COOH/COOH > CH$_3$/CH$_3$ > COOH/CH$_3$[35, 36], as shown in Figure 6. The trend in the magnitude of the friction forces and friction coefficients is the same as that observed for the adhesion forces: COOH/COOH-terminated tips and samples yield large friction and adhesion forces, while the COOH/CH$_3$ combination displayed the lowest friction and adhesion. Thus, there is a direct correlation between the friction and adhesion forces measured between well-defined SAM surfaces.

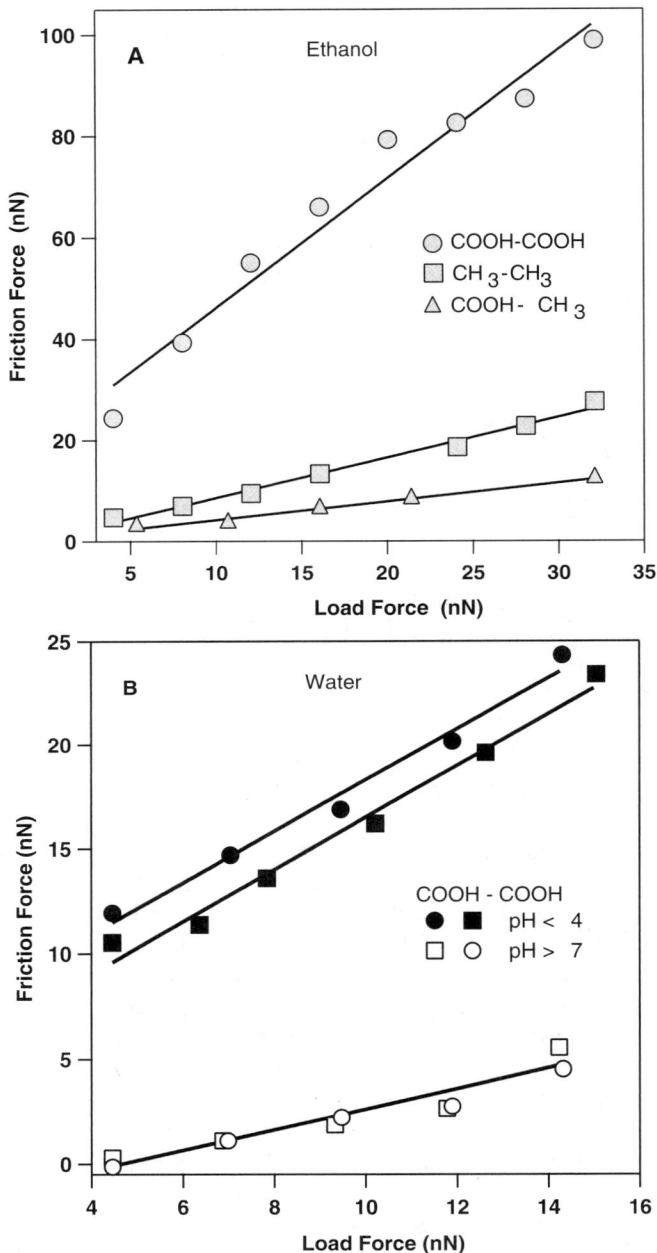

Figure 6. (A) Summary of the friction force versus applied load data recorded for functionalized samples and tips terminating in COOH/COOH (○), CH$_3$/CH$_3$ (□) and COOH/CH$_3$ (△) in EtOH (B) Friction force versus applied load curves for COOH-COOH functionalized samples and tips recorded at pH below and above pK$_a$ of surface carboxyl group.

The change in the ionization state of functional groups can also be detected by recording frictional forces at different pH. F-L curves for COOH-terminated tips and samples are linear, but fall into two distinct categories: larger friction forces and friction coefficients are found in solutions at pH < 6 compared to pH > 6 (Figure 6). This crossover in behavior occurs at the same region of pH where the normal forces exhibit a transition from attraction to repulsion. There is also a finite load (~4 nN, for the data in Figure 6) necessary to achieve nonzero friction at high pH. This additional load is required to overcome the double layer repulsion (between charged surfaces) and bring the tip into physical contact with the sample surface.

The frictional behavior of tips and samples functionalized with ionizable and non-ionizable SAMs can be summarized in plots of the friction coefficient versus pH. The friction coefficients determined for OH- and CH_3-terminated SAMs are independent of pH, as expected for neutral, non-ionizable functional groups. In contrast, the friction coefficients determined for cases in which one or both SAM surfaces terminate in carboxyl groups show significant decreases at pH above the pK_a of the surface COOH group. The friction coefficients therefore exhibit similar pH dependencies to those observed in adhesion measurements shown in Figure 2. These results suggest that the drop in friction coefficient with changes in pH is also associated with the ionization of surface groups.

The magnitudes of the friction force and friction coefficient for hydroxyl-terminated surfaces were the lowest among hydrophilic group pairs (COOH-COOH, COOH-OH, OH-OH) in aqueous solution. Analysis of F-L curves for methyl terminated tip/sample combinations also yields low friction coefficients; however, for comparable tip radii the magnitude of the friction force at low external loads is almost an order of magnitude greater than for either carboxyl or hydroxyl functionalized surfaces. The large magnitude of the friction force between methyl surfaces in aqueous media originates from the large contact area between hydrophobic surfaces.

Other monolayer properties can affect adhesive and frictional behavior of organic layers. For example, AFM studies of phase-segregated Langmuir-Blodgett films have indicated that elasticity and crystallinity correlate with observed friction[37] and adhesion.[38] Chain-length dependence was observed for friction on silane SAMs on mica[39] and LB films of saturated carboxylic acids.[40] This chain-length dependence may in part reflect structural differences in the different monolayers versus intrinsic viscoelastic effects.[39] Therefore, some additional information on mechanical properties of these SAMs may be required to ascribe observed differences in adhesion and friction to chemical effects in structurally well-defined systems.

4 Chemical Imaging Contrast in CFM Experiments

Scanning force microscopy (SFM) has transformed the fields of nanoscience and surface science by providing visualization in real space of structures and features of molecular and nanometer sizes. Initially, the images available with SFM could be interpreted as ultra-high resolution profilometry, which provides very limited information about sample properties and composition. Thus one clear goal for the development of the AFM imaging capabilities has always been achieving some degree of chemical sensitivity. Such sensitivity to the surface composition combined with unparalleled resolution can make force microscopy a versatile surface analysis tool. The contrast in AFM images originates from the interactions between the probe tip and the surface. In general, these interactions depend on the surface chemistry, morphology, mechanical properties, and on the nature of the surrounding medium. Thus, if we want chemical contrast to be the dominant contribution to the AFM image, it is necessary to identify and enhance forces that are chemically specific and eliminate interactions that are not correlated with the chemical identity of the functional groups at the interface. If the origin of the chemical contrast is understood, one can envision fine-tuning the probe-surface interactions, or even the operating conditions of the instrument, as a way of enhancing the imaging sensitivity and specificity.

The CFM approach, which focuses on the chemical modification of the AFM tips as a way of probing and enhancing specific interactions, represents a logical first step for such an approach. We note that CFM, like any force spectroscopy approach, fundamentally relies on intermolecular, typically non-covalent, interactions to discriminate between different functionalities. Therefore, it may not achieve the identification power of other surface analysis techniques, such as secondary ion mass spectroscopy, or infrared spectroscopy. The spatial resolution of CFM, on the other hand, is orders of magnitude better than the lateral resolution of these other techniques. This high resolution, taken together with the applicability of AFM to a wide range of samples, offsets these inherent drawbacks. We now describe several CFM approaches that researchers have used to obtain maps of the surface that could be rationally interpreted on the basis of the tip-sample interactions.

4.1 Lateral Force Imaging

The results presented in the previous sections show that chemical modification of probe tips is often sufficient to influence normal and lateral forces in a rational way. Specifically, the observed differences in friction forces could readily produce lateral force images of heterogeneous surfaces with predictable contrast.[5, 35, 36, 41] Patterned SAMs of alkane thiols on Au surfaces present a convenient model system for such studies, because they represent an ideal model surface that exhibits chemical contrast. Patterned SAMs have homogeneous mechanical properties throughout the sample, are readily prepared, and can incorporate a variety of different terminal functionalities. In addition, patterned surfaces can be made relatively flat, thus eliminating unwanted topographic contributions to lateral force images. Indeed, a topographical image (Figure 7A) of a photochemically patterned SAM[42] that has $10 \mu m$- $\times 10 \mu m$-square regions terminated with -COOH groups and surrounded with CH_3-terminated groups is almost completely flat (certainly, a topographical image fails to distinguish the pattern!). Yet the friction maps of these samples taken with different tip functionality readily show contrast that reveals chemical information about the surfaces (Figures 7B, 7C). Friction maps recorded with COOH tips display high friction on the area of the sample that terminates in COOH groups, and low friction on the CH_3 terminated regions. Images recorded with CH_3 tips exhibit a reversal in the friction contrast: low friction is found in the area of the sample that contains the COOH-terminated SAM, and higher friction is observed in the surrounding CH_3 regions.

As expected, this reversal in friction contrast occurs only with changes in the probe tip functionality and is consistent with the friction forces obtained on the homogenous SAM surfaces. As typical for the friction-based measurements, the image resolution is not that of a single functional group but rather an ensemble of groups defined by the tip contact area. The apparent resolution in the images shown in Figure 7 is limited by the resolution of the photopatterning method used in preparing the sample (~200 nm). Employing a microcontact printing technique for pattern generation does improve the resolution by a factor of 4.[43] Researchers have achieved much higher resolution (on the order of 10–20 nm) using a monolayer-bilayer COOH/CH_3 system with functionalized tips in dry Ar atmosphere.[36]

The imaging contrast could be enhanced further by changing the solvent characteristics, such as composition[44] or pH[5]. We have demonstrated the validity of this approach by mapping changes in functional group ionization states with varying solution pH values. Images of COOH/OH patterned surfaces obtained with a COOH-terminated tip in different pH solutions show that the friction contrast between COOH/OH regions can be reversibly inverted, with the change in contrast always occurring near the pK_a of the surface carboxyl (Figures 7E, 7F). The reversals in friction contrast presented in these examples occur only with changes in the probe tip functionality in organic media or a change in the probe tip ionization state in aqueous

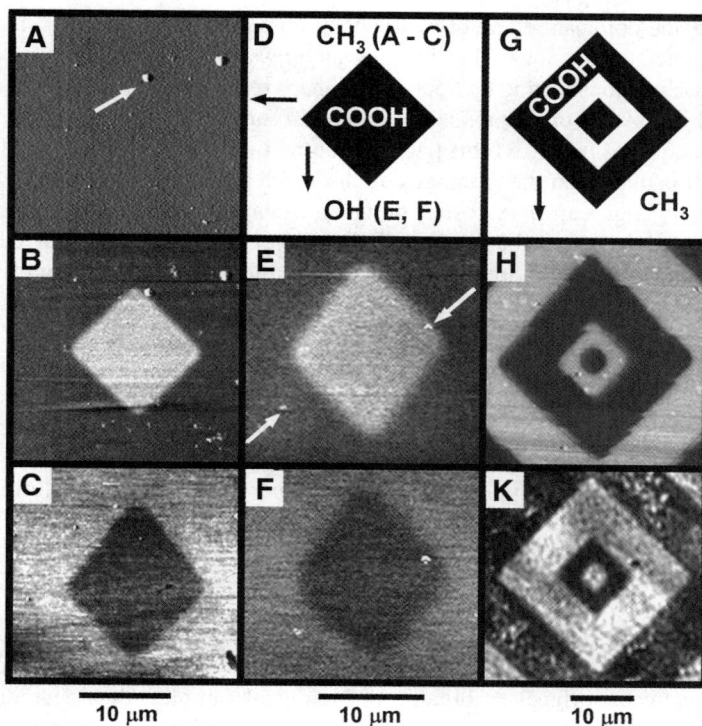

Figure 7. CFM images (dark gray represents areas of low friction, light gray represents areas of high friction) of photochemically generated patterns: COOH/CH$_3$ groups, image taken with COOH (A – topography, B – friction) or CH$_3$ (C) tips in ethanol; COOH/OH groups, image taken with COOH tips in 0.01 M phosphate buffer at low (E) and high (F) pH; COOH/CH$_3$ groups, image taken with CH$_3$ tips in water (H) or ambient air (K). Arrangement of the chemical functionality on the surfaces is outlined in (D) and (G). The same area of the sample is readily identified by the particles contaminating the surface (arrows) in images shown in (A) and (B), (E) and (F).

solvents. These results show that AFM imaging with chemically-modified probes in a controlled environment can produce rationally-interpretable image contrast.

Guntherodt and co-workers pioneered use of friction forces to map different domains in phase-segregated LB films.[37] The image contrast in these investigations, however, arose due to the differences in the elastic properties of the LB film domains and not the chemical functionality at the film surface. It is important to point out that researchers need to use caution when interpreting AFM images as pure maps of surface composition, even when the image is taken with a chemically-modified tip; many systems exhibit complex interactions between the tip and the surface. Only when viscoelastic effects (or site-to-site differences due to them) are eliminated, can the chemical components of the tip-surface interaction dominate and consequently determine the image contrast. Conversely, if the interfacial chemistry is the same, or the probe functionality is chosen to minimize the differences in chemical interactions, then the second order effects (e.g., viscoelastic properties, contact stiffness, etc.) could become dominant.[45]

The conditions under which the imaging is performed are important, since dominant interactions depend on the media even for the same tip-surface system as illustrated by lateral force images of a COOH/CH$_3$ pattern acquired in water and air using a CH$_3$-functionalized tip in Figures 7H, 7K. The friction contrast between COOH and CH$_3$ functional groups is relatively large in both images. In water, this pattern shows high friction on the methyl-terminated regions and low friction over the carboxyl-terminated regions of the sample. This result is readily understood on the basis of the friction force measurement that we have discussed in the previous sections. The contrast reflects the dominant effect of hydrophobic forces (leading to large contact areas) that mask other chemical interactions. Thus imaging under water with hydrophobic CH_3-terminated tips is an approach to constructing hydrophobicity maps of sample surfaces and could be used to map hydrophobic domains on membranes of biological surfaces.

Images acquired in air exhibit friction contrast opposite to those obtained in aqueous solution. It is clear that these images do not reflect chemical interactions directly, but rather are due to large capillary forces between the sample and tip over the hydrophilic COOH-terminated areas of the surface that are wet more readily than CH_3 regions.[35] Although chemical modification of the tip does not influence the contrast controlled by capillary forces, a treatment of AFM probes that produces hydrophobic surfaces can reduce capillary forces and enhance the image resolution.[44, 46, 47] Perhaps the most elegant variation of this approach has been demonstrated by Lieber and co-workers in a series of publications using carbon nanotube AFM probes for high-resolution imaging of colloids and biomolecules. Small size, high aspect ratio, and hydrophobic carbon nanotube sidewalls all contributed to reducing the capillary forces, and, as a result, the CNT probes exhibited minimal adhesion in air, providing extremely stable high-resolution imaging of sample surfaces in ambient conditions.[22, 48–50]

4.2 Tapping Mode Imaging and Energy Dissipation Imaging

Despite conceptual simplicity of the chemical contrast in friction force imaging, practical drawbacks of contact mode imaging are well-known—it exerts unnecessarily high forces on the delicate surfaces of organic or biological samples, thus degrading the imaging resolution significantly, or worse, leading to the tip-induced sample damage. It is not surprising then that AFM has been widely adopted for imaging of such "soft" samples in the intermittent-contact, or "tapping", mode of imaging. In this mode, the AFM probe vibrates near its resonance frequency with substantial amplitude and strikes or "taps" the surface for a brief portion of each oscillation cycle. The resulting images typically exhibit much crisper contrast and the imaging process causes much less surface damage than contact mode imaging. Tapping mode imaging had been broadly adopted for imaging of organic matter, and the origins of the impressive image contrast of the phase component of the tapping mode images were soon established by J. Cleveland and co-workers.[51, 52] They reported the key observation that the phase image contrast reflected the energy dissipation in the tip-sample junction. Specifically, the power dissipated in the tip-sample junction, P_t, is related to the observed phase lag ϕ and cantilever parameters by a simple expression:

$$P_t = \frac{1}{2} \frac{k\omega_0}{Q_{cant}} \left(A_0 A \sin\varphi - A^2 \right) \tag{4}$$

or for constant amplitude imaging:

$$\Delta P_t \propto \frac{k\omega_0}{Q_{cant}} \Delta(\sin\varphi) \tag{5}$$

where k is the cantilever spring constant, Q_{cant} is the quality factor, ω_0 is the resonant frequency, and A_0 and A are the cantilever oscillation amplitudes far away from the surface (free vibration) and during imaging (intermittent contact), respectively.

AFM experiments typically feature quite fast loading rates; therefore the tip pull-off from the surface is almost never an equilibrium process. From the point of view of the cantilever dynamics, adhesion interactions measured with soft cantilevers are dissipative, because the elastic energy stored in the cantilever is dissipated by the viscous damping in the interaction medium after adhesion contact is broken. Therefore, samples presenting areas of large differences in adhesion forces ($\Delta F \propto \Delta P_t$) should produce stronger phase contrast ($\Delta \sin \varphi \approx \Delta \varphi$).

The ability to probe the energy dissipated due to the tip-sample interactions in a direct measurement opened up an opportunity for chemical force microscopy to use tapping mode for probing specific interactions of the functional groups on the AFM probe and the surface of a sample. With the experimental conditions adjusted to minimize the contributions of other interactions (e.g., viscoelastic energy dissipation), the energy dissipated at the probe-surface interface should directly reflect interactions of the chemical functionalities. The first demonstration by Noy et al.[53] of the coupling between adhesion and phase lag in tapping mode images used a familiar model system of interacting self-assembled monolayers on gold surfaces. To keep the variations in the experimental conditions to the minimum, they adjusted the strength of the interactions by varying the composition of the water/methanol medium filling the AFM fluid cell.

Image contrast between sample regions of different composition observed in these measurements was clearly increasing with the increase in the water content (Figure 8). This trend can easily be rationalized by noting that increase in the water content accentuates the differences in the interaction strength between the CH_3-functionalized probe and hydrophobic regions versus hydrophilic regions of the sample.[54] Indeed, a detailed comparison of the observed phase lag with the measured tip-surface interfacial free energies in different solvent mixtures showed a clear correlation between the two (Figure 9). Similarly, Wong et al. observed chemical sensitivity of phase images of $COOH/CH_3$ SAM patterns obtained using nanotube tips—image contrast inverted when COOH groups at the open ends of CNT were converted to benzyl functionality.[22]

Ashby and Lieber have later refined this approach and developed the Energy Dissipation Chemical Force Microscopy (ED-CFM) approach, where they used an equation similar to Equation 4 construct direct maps of the tip-sample energy dissipation and correlate them to the chemical composition of the tip and surface functional groups.[55] Remarkably, they observed that this approach not only distinguished between the different functionalities on the sample surface, but also aided in removing topography-generated artifacts. A comparison of the amplitude, phase lag, and energy dissipation images of the same region of the sample surface that contains variation in the chemical group functionality and topographic roughness highlights the fact that ED-CFM imaging is considerably more successful in rejecting topographic artifacts than the conventional phase imaging (Figure 10).

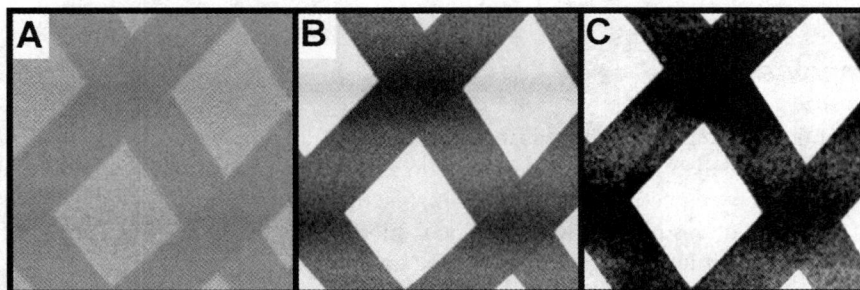

Figure 8. Phase lag maps of a SAM sample patterned with COOH-terminated square regions surrounded by a CH_3-terminated background recorded with the same CH_3-terminated tip in a series of methanol-water solvents. The water content, phase contrast and differences in tip-sample work of adhesion are (A) 20%, (B) 60%, and (C) 80%. The gray scale in each of these 25 μm × 25 μm images represents phase variation of 50°. The imaging set point (A/A_0 ratio) was maintained at 0.6 in all three images.

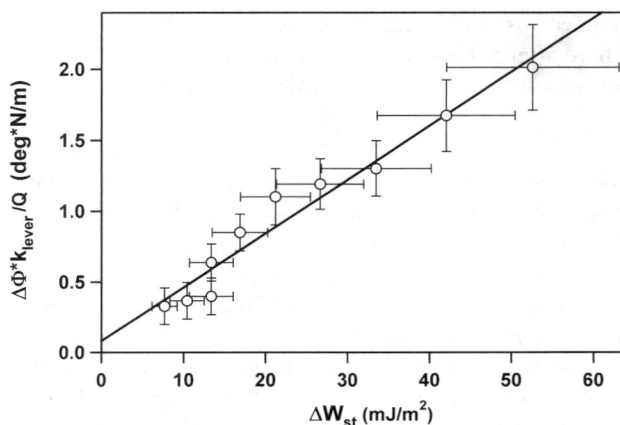

Figure 9. Plot of the product of the cantilever spring constant and phase contrast divided by the cantilever quality factor versus the difference in tip-sample interaction free energies for the two distinct sample regions shown in Figure 8. The data were obtained with an increasing percentage of water from 0 to 80%. The straight line is a linear fit to the data. The plot combines data obtained with two cantilevers having different spring constants and quality factors. Both cantilevers had tips terminating in CH_3 groups. Error bars represent standard deviations determined by bearing analysis of phase lag images and errors in the work of adhesion values originating from uncertainties in the tip radii.

L. Chen et al. have performed further experimental and computational investigations of the AFM probe interactions with the surface in tapping mode imaging and highlighted the importance of controlling the cantilever damping in achieving the highest image resolution.[56] Incorporation of the active methods of controlling such damping (i.e., "Q-control" functionality) directly into the commercial force microscopy instrumentation have already resulted in a further refinement of imaging techniques. It will be interesting to see if it can produce an enhanced ability to separate and isolate the contributions of chemical interactions in chemical force microscopy imaging. Another intriguing possibility arising from the efforts to improve imaging resolution is the use of higher harmonic imaging (i.e., Dual-AC™ mode imaging) that could potentially help to obtain well-defined *quantitative* images of the strength of the tip-sample interactions.

5 Conclusions

Chemically-modified tips produced by covalently linking molecules to force microscope probes proved to be an excellent tool to explore, quantify, and map adhesion and friction forces between complex assemblies of functional groups on a tip and sample. This methodology could also be used to determine the local pK's of surface ionizable functional groups using force titrations with probes bearing groups that undergo dissociation. The interactions observed between modified tip and sample surfaces in aqueous solutions agree well with the predictions of double layer and contact mechanics models. These models can be used to extract surface free energies and double layer parameters that are essential to understanding interactions in aqueous media.

The friction forces between modified tips and samples are also chemically specific; moreover, the magnitudes of the friction forces parallel the trends found for adhesion forces.

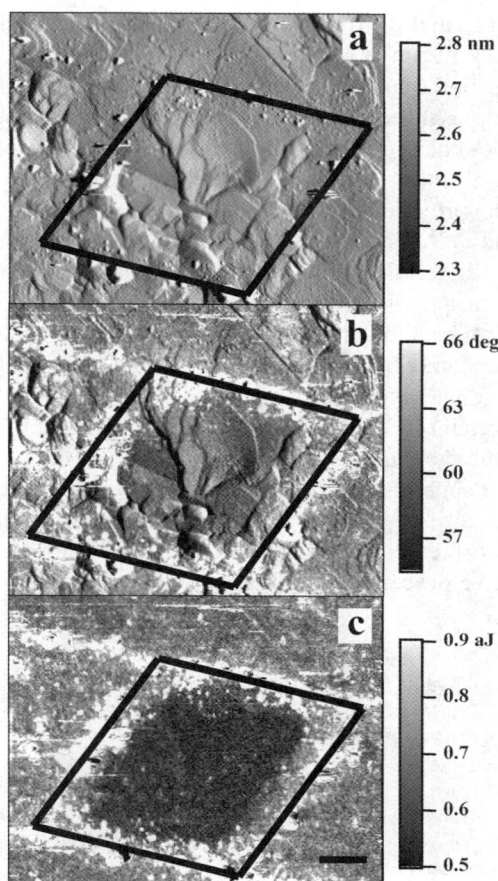

Figure 10. (A) Amplitude, (B) phase lag, and (C) energy dissipation images of a patterned SAM surface of hydroxyl surrounding a carboxyl square. The black square highlights the edges of the pattern. The tip is functionalized with hydroxyl terminated SAM. The images are obtained in 0.01MpH2 phosphate buffer. The topography is coupled into the amplitude and phase images, but compensated in the energy dissipation image. The scale bar at the lower right is 200 nm 55.

This observation underscores the fundamental role of the basic non-covalent interactions in shaping microscopic mechanisms of surface phenomena such as adhesion and friction. The parallels between the two phenomena extend also to force titrations: friction between ionizable groups changes upon dissociation of interacting functionalities; for example, the friction forces for COOH-terminated surfaces decreased significantly at a pH corresponding to the pK_a determined from the adhesion measurements.

Significantly, this predictable dependence of friction forces on the tip and sample functionality could form the basis for chemical force microscopy where lateral force images are interpreted in terms of the strength of adhesive and shear interactions between functional groups. Therefore, in conjunction with adhesion data, CFM can distinguish different functional group domains in organic and aqueous solvents. When present, the hydrophobic effect dominates both adhesion and friction forces, and hence lateral force images taken with methyl-terminated tips in aqueous solutions can map hydrophobic regions on a sample. On hydrophilic surfaces, observed pH-dependent changes in friction forces of ionizable groups can be exploited to map spatial distribution of hydrophilic functional groups and to define their ionization state as a function of pH.

CFM imaging can find many uses in basic and applied research: from quantifying the strength of interactions in host-guest systems to identifying surface distribution of chemical modifications or chemically distinct domains in soft materials such co-polymers. A wide range of intermolecular interactions can be studied by the CFM technique, and analysis of these data can provide basic thermodynamic information relevant to chemical, biological, polymer, and colloidal systems. CFM imaging of systems such as polymers, biomolecules, and other materials could produce direct images of the spatial distribution of functional groups, hydrophobic versus hydrophilic domains, and the time course of the surface reorganization.[57] The approach of using force titrations to determine the local pK of acidic and basic groups may be applicable to probing the local electrostatic properties of protein surfaces in their native environments and the ionization of colloidal particles at the nanoscale. Assignment by CFT of the chemical functionality at the terminus of the shortened carbon nanotube to carboxylic acid groups proves the potential for unparalleled spatial resolution for such identification. Studies using chemically-modified tips also provide new approaches to imaging energy dissipation processes during dynamic (intermittent or sliding) tip-surface contact and will lead to new insights into the molecular mechanisms of dissipative processes relevant to tribology.

References

1. *Surface and Colloid Chemistry*; Birdi, K. S., Ed.; CRC Press: Boca Raton, 1997.
2. Myers, D. *Surfaces, Interfaces, and Colloids*; John Wiley & Sons: New York, 1999.
3. Zhmud, B. V.; Golub, A. A. *J. Colloid Interface Sci.* **1994**, *167*, 186.
4. Holmes-Farley, S. R.; Reamey, R. H.; McCarthy, T. J.; Deutch, J.; Whitesides, G. M. *Langmuir* **1988**, *4*, 921.
5. Vezenov, D. V.; Noy, A.; Rosznyai, L. F.; Lieber, C. M. *J. Am.Chem. Soc.* **1997**, *119*, 2006–2015.
6. *CRC Handbook of Chemistry and Physics*; 72 ed.; Lide, D. R., Ed.; CRC Press: Boca Raton, FL, 1991.
7. van der Vegte, E.; Hadziioannou, G. *J. Phys. Chem. B* **1997**, *101*, 9563–9569.
8. Creager, S. E.; Clark, J. *Langmuir* **1994**, *10*, 3675.
9. Chatelier, R.; Drummond, C.; Chan, D.; Vasic, Z.; Gengenbach, T.; Griesser, H. *Langmuir* **1995**, *11*, 4122.
10. Smart, J. L.; McCammon, J. A. *J. Am Chem. Soc.* **1996**, *118*, 2283–2284.
11. Glinski, G. C.; Platten, J. K.; De Saedeleer, C. *J. Colloid Interface Sci.* **1993**, *162*, 129.
12. Zhang, H.; He, H. X.; Wang, J.; Mu, T.; Liu, Z. F. *Applied Physics A:* **1998**, *A66*, S269–S271.
13. Wallwork, M. L.; Smith, D. A.; Zhang, J.; Kirkham, J.; Robinson, C. Langmuir 2001, 17, 1126–1131.
14. He, H.-X.; Huang, W.; Zhang, H.; Li, Q. G.; Li, S. F. Y.; Liu, Z. F. *Langmuir* **2000**, *16*, 517–521.
15. Schoenherr, H.; van Os, M. T.; Foerch, R.; Timmons, R. B.; Knoll, W.; Vancso, G. J. *Chemistr. Mater.* **2000**, *12*, 3689–3694.
16. Schonherr, H.; van Os, M. T.; Vancso, G. J.; Forch, R.; Knoll, W.; Hruska, Z.; Kurdi, J.; Arefi-Khonsari, F. *Chem. Comm.* **2000**, 1303–1304.
17. Zhang, H.; He, H.-X.; Mu, T.; Liu, Z.-F. *Thin Solid Films* **1998**, *327–329*, 778–780.
18. Zhang, H.; Zhang, H.-L.; He, H.-X.; Zhu, T.; Liu, Z.-F. *Mater. Sci. & Engin., C.* **1999**, *C8–C9*, 191–194.
19. Wang, J.; Zhang, H.; He, H.; Hou, T.; Liu, Z.; Xu, X. *Theochem* **1998**, *451*, 295–303.
20. Lee, T. R.; Carey, R. I.; Biebuyck, H. A.; Whitesides, G. M. *Langmuir* **1994**, *10*, 741.
21. Wong, S. S.; Joselevich, E.; Woolley, A. T.; Cheung, C. L.; Lieber, C. M. *Nature* **1998**, *394*, 52–55.
22. Wong, S. S.; Woolley, A. T.; Joselevich, E.; Cheung, C. L.; Lieber, C. M. *J. Am. Chem. Soc.* **1998**, *120*, 8557–8558.
23. Smith, D. A.; Wallwork, M. L.; Zhang, J.; Kirkham, J.; Robinson, C.; Marsh, A.; Wong, M. *J. Phys. Chem. B* **2000**, *104*, 8862–8870.
24. Smith, D. A.; Robinson, C.; Kirkham, J.; Zhang, J.; Wallwork, M. L. *Rev. Analyt. Chem.* **2001**, *20*, 1–27.
25. Smith, D. A.; Connell, S. D.; Robinson, C.; Kirkham, J. *Anal. Chim. Acta* **2003**, *479*, 39–57.
26. Zhang, J.; Kirkham, J.; Robinson, C.; Wallwork, M. L.; Smith, D. A.; Marsh, A.; Wong, M. *Analyt. Chemi.* **2000**, *72*, 1973–1978.
27. Kreller, D. I.; Gibson, G.; vanLoon, G. W.; Horton, J. H. *J. Colloid Interf. Sci.* **2002**, *254*, 205–213.
28. Ashby, P. D.; Chen, L. W.; Lieber, C. M. *J. Am. Chem. Soc.* **2000**, *122*, 9467–9472.
29. Ito, T.; Citterio, D.; Buehlmann, P.; Umezawa, Y. *Langmuir* **1999**, *15*, 2788–2793.
30. Dicke, C.; Hahner, G. *J. Am. Chem. Soc.* **2002**, *124*, 12619–25.

31. Glinski, J.; Chavepeyer, G.; Platten, J. K.; De Saedeleer, C. *J. Colloid Interface Sci.* **1993**, *158*, 382.
32. Chavepeyer, G.; De Saedeleer, C.; Platten, J. *J. Colloid Interface Sci.* **1994**, *167*, 464.
33. Reiner, E. S.; Radke, C. J. *Adv. Colloid Interface Sci.* **1993**, *58*, 87.
34. Overney, R. M.; Takano, H.; Fujihira, M.; Paulus, W.; Ringsdorf, H. *Phys. Rev. Lett.* **1994**, *72*, 3546–9.
35. Noy, A.; Frisbie, C. D.; Rozsnyai, L. F.; Wrighton, M. S.; Lieber, C. M. *J. Am. Chem. Soc.* **1995**, *117*, 7943–7951.
36. Green, J.-B. D.; McDermott, M. T.; Porter, M. D.; Siperko, L. M. *J. Phys. Chem.* **1995**, *99*, 10960.
37. Overney, R. M.; Meyer, E.; Frommer, J.; Brodbeck, D.; Luethi, R.; Howald, L.; Giintherodt, H. J.; Fujihira, M.; Takano, H.; Gotoh, Y. *Nature* **1992**, *359*, 133–135.
38. Berger, C. E. H.; van der Werf, K. O.; Kooyman, R. P. H.; de Grooth, B. G.; Greve, J. *Langmuir* **1995**, *11*, 4188.
39. Xiao, X.; Hu, J.; Charych, D. H.; Salmeron, M. *Langmuir* **1996**, *12*, 235.
40. Lee, G. U.; Chrisey, L. A.; Ferrall, C. E.; Pilloff, D. E.; Turner, N. H.; Colton, R. J. *Israel J. Chem.* **1996**, *36*, 81.
41. Frisbie, C. D.; Rozsnyai, L. F.; Noy, A.; Wrighton, M. S.; Lieber, C. M. *Science* **1994**, *265*, 2071.
42. Wollman, E. W.; Kang, D.; Frisbie, C. D.; Lorkovic, I. M.; Wrighton, M. S. *J. Am. Chem. Soc.* **1994**, *116*, 4395–4404.
43. Wilbur, J. L.; Biebuyck, H. A.; MacDonald, J. C.; Whitesides, G. M. *Langmuir* **1995**, *11*, 825–831.
44. Sinniah, S. K.; Steel, A. B.; Miller, C. J.; Reutt-Robey, J. E. *J. Am. Chem. Soc.* **1996**, *118*, 8925–8931.
45. Dufrene, Y. F.; Barger, W. R.; Green, J. B. D.; Lee, G. U. *Langmuir* **1997**, *13*, 4779–4784.
46. Alley, R. L.; Komvopoulos, K.; Howe, R. T. *J. Appl. Phys.* **1994**, *76*, 5731–7.
47. Knapp, H. F.; Wiegrabe, W.; Heim, M.; Eschrich, R.; Guckenberger, R. *Biophys. J.* **1995**, *69*, 708–15.
48. Wong, S. S.; Woolley, A. T.; Odom, T. W.; Huang, J. L.; Kim, P.; Vezenov, D. V.; Lieber, C. M. *Appl. Phys. Lett.* **1998**, *73*, 3465–3467.
49. Wong, S. S.; Joselevich, E.; Woolley, A. T.; Cheung, C. L.; Lieber, C. M. *Nature* **1998**, *394*, 52–55.
50. Wong, S. S.; Harper, J. D.; Lansbury, P. T.; Lieber, C. M. *J. Am. Chem. Soc.* **1998**, *120*, 603–604.
51. Cleveland, J.; Anczykowski, B.; Schmidt, A.; Elings, V. *Appl. Phys. Lett.* **1997**, *72*, 2613–2615.
52. Anczykowski, B.; Gotsmann, B.; Fuchs, H.; Cleveland, J. P.; Elings, V. B. *Appl. Surf. Sci* **1999**, *140*, 376.
53. Noy, A.; Sanders, C. H.; Vezenov, D. V.; Wong, S. S.; Lieber, C. M. *Langmuir* **1998**, *14*, 1508–1511.
54. Vezenov, D. V.; Zhuk, A. V.; Whitesides, G. M.; Lieber, C. M. *J. Am. Chem. Soc.* **2002**, *124*, 10578–10588.
55. Ashby, P. D.; Lieber, C. M. *J. Am. Chem. Soc.* **2005**, *127*, 6814–6818.
56. Chen, L.; Cheung, C. L.; Ashby, P. D.; Lieber, C. M. *Nano Lett* **2004**, *4*, 1725–1731.
57. Hillborg, H.; Tomczak, N.; Olah, A.; Schonherr, H.; Vancso, G. J. *Langmuir* **2004**, *20*, 785–794.

Dynamic Force Spectroscopy with the Atomic Force Microscope

Phil Williams

The dynamic force spectroscopy (DFS) experiment has been with us for nearly a decade [1]. By studying the effect of force on the dissociation kinetics of molecular interactions, hitherto hidden information about physics, chemistry, and biology is gained. Since the statement of the theory and the first demonstration of the experiment [2], we have seen developments in theory, experimental practice, and data analysis. Advances in theory have suggested the possibility of measuring more than dissociation rates over transition states and their displacements, such as the change in energy of the system at the transition state [3–5], the roughness of the dissociation landscape [6–8], and equilibrium phenomena [9]. Today, the instrumentation used to undertake DFS that is most prevalent in the literature is the atomic force microscope (AFM). The significant advances we have seen in both theory and experiment have sometimes taken place in isolation, and here I believe it is worth considering the application of DFS with current AFM technology. How accurately can we do DFS with an AFM? What exactly can we measure with the AFM, and what advances are needed?

Development of a Theory

The theory behind DFS has been stated many times and will not be repeated in its entirety here [1, 9–11]. It is worth, however, remembering the basis and assumptions on which it has been developed.

A molecular interaction is favourable if the free energy of the complex is less than the combined free energy of the separated components. This difference in free energy, ΔG, is a quantity described by biologists as *affinity*. Two molecules with an affinity for each other will, when separated, associate over time and hence will have moved together. Their movement is through diffusion; they explore the space around them via thermal motion along the energy landscape in which they reside. The shape of this landscape affects the rate of movement. It is possible to calculate the rate at which the molecules will associate (and dissociate) over a landscape, involving the integration of the (Bolzmann-weighted) free energy surface. Considering molecular dissociation, for that is what we wish to measure, one can collapse all the degrees of freedom of the system into a single *reaction coordinate* of molecule separation. This reaction coordinate represents the free energy of the system with one degree of freedom (molecular separation) removed. We arrive at our first approximation by considering molecular interactions with significant affinity (ΔG greater than a few thermal energy units, $k_B T$) and separated by a similarly significant barrier. Here, the integration is dominated by the shape of the landscape at the bound state and at the barrier. The 'tightness' of the landscape at the bound state provides an 'entropic pressure' forcing the molecules to separate, and the magnitude and curvature of the transition state control how quickly this pressure can force the molecules apart. With a harmonic approximation to the shape of the landscape at both the bound and transition states, the rate of escape can be shown to be equal to

$$k_{off} = \left(\frac{\omega_C^{\frac{1}{2}} \omega_{TS}^{\frac{1}{2}}}{2\pi\zeta} \right) \exp\left(-\frac{\Delta G^{\ddagger}}{k_B T} \right) \tag{1}$$

where ω_C is the curvature of the landscape at the bound state, ω_{TS} is the curvature at the transition state, ζ is the frictional coefficient (defines the rate of diffusion) and $\Delta G^{\ddagger}$ is the free energy difference between bound and transition states (the height of the barrier). For molecular interactions on the nanometre length scale the exponential prefactor will be an order of magnitude greater than the field-free diffusion ($k_B T/\zeta$). The prefactor for dissociation of a ligand with a diffusion constant of 10^{-5} cm^2 sec^{-1} bound to protein (biotin has a constant of approximately 6×10^{-6} cm^2 sec^{-1} [12, 13]), for example, will be of the order of 10^{10} sec^{-1}.

Application of a persistent force to the molecular interaction will tilt the energy landscape in the direction of the force, and the component of the force along the dissociation reaction coordinate will lower $\Delta G^{\ddagger}$. Providing that the unbonding transition state is sharp and the binding potential deep, then ω_C and ω_{TS} are unaffected by force. This assumption of sharp barrier and deep potential gives rise to an approximate expression of the dissociation rate under persistent force f, as

$$k_{off}(f) = \left(\frac{\omega_C^{\frac{1}{2}} \omega_{TS}^{\frac{1}{2}}}{2\pi\zeta} \right) \exp\left(-\frac{\left(\Delta G^{\ddagger} - fx^{\ddagger} \right)}{k_B T} \right) = k_{off} \exp\left(\frac{f}{f_{\beta}} \right) \tag{2}$$

where $f_{\beta} = k_B T/x^{\ddagger}$ with $x^{\ddagger}$ equal to the displacement of the unbonding transition state along the reaction coordinate. This dissociation rate defines the probability that the interaction will break at a given force.

In the AFM, force is applied to a bond by attaching one half of the interacting pair to a surface, the other half to the AFM cantilever tip, and separating the two. The speed at which the force can be changed is limited by the mechanical movement of the cantilever and substrate. Force cannot be applied instantaneously in the AFM, and measurement of the dissociation probability as a function of force is better measured indirectly through the ramp-of-force protocol.

For a population of bonds (that do not recombine once broken), the number remaining falls in time as a simple kinetic expression

$$\frac{dS}{dt} = -k_{off} \exp\left(\frac{f}{f_{\beta}} \right) S(t) \tag{3}$$

where force varies linearly in time

$$\frac{df}{dt} = r_f \tag{4}$$

then

$$\frac{dS}{df} = -\frac{1}{r_f} k_{off} \exp\left(\frac{f}{f_{\beta}} \right) S(f) \tag{5}$$

The expression for dissociation in Eqn. 2, however, was for a persistent force, and here it is varying. Providing the force increases slowly so the energy landscape does not change significantly on the timescale of the molecular motion (see above), then this approximation is valid. If, however, force changes rapidly on this timescale, or the stiffness of the pulling potential is comparable to that of the unbonding potential, then we can go no further with this description (discussed later is an alternative analysis proposed by Hummer and Szabo [3]). The number of bonds that remain when force is increased from 0 to f is

$$S(f) = \exp\left\{\frac{f_\beta k_{off}}{r_f}\left[1 - \exp\left(\frac{f}{f_\beta}\right)\right]\right\} \tag{6}$$

Hence the probability of a bond breaking in the ramp of force experiment at a given force is the product

$$P(f) = \left[\frac{1}{r_f}k_{off}\exp\left(\frac{f}{f_\beta}\right)\right]\exp\left\{\frac{f_\beta k_{off}}{r_f}\left[1 - \exp\left(\frac{f}{f_\beta}\right)\right]\right\} \tag{7}$$

The mode of this distribution, f^*, is

$$f^* = f_\beta \ln\left(\frac{r_f}{f_\beta k_{off}}\right) \tag{8}$$

Remember there have been several assumptions and approximations made thus far: The binding potential is deep. The landscape is dominated by sharp barriers. Force inhibits rebinding. The transition state does not move under force. The time required for the force to lower the barrier(s) by a significant amount is very much longer than the timescale of molecular diffusion across the landscape.

A plot of f^* versus the logarithm of the loading rate should be linear of slope f_β and intercept with the loading rate axis at $r_f = f_\beta k_{off}$. This is the tenet of dynamic force spectroscopy (see, for example, Figure 1). If any of our assumptions above were incorrect, then f^* would not increase linearly with $\ln(r_f)$. If the binding potential were shallow, then the exponential prefactor would depend on the level of force applied; thus k_{off} would scale with f, in addition to f_β producing curvature in the DFS spectrum. Similarly, if the barriers were not sharp, then their maximum is likely to move with force and f_β would vary with force, again producing curvature in the spectrum. If the landscape were rough, f_β would vary at each level of force and the f^* values would show scatter. If the loading was perturbing the barrier significantly on the diffusion timescale of the system, then the force will be affected by the viscous drag of the system, which scales linearly with rate and not logarithmically. The spectrum would again be curved. Measuring curvature in a dynamic force spectrum would provide valuable information about the molecular system in addition to f_β and k_{off} values, as one could estimate the curvature of the landscape at the bound and transition states. One could also measure the roughness of the energy landscape by measuring scatter in the data, and determine the molecular diffusion and friction constants by pulling quickly. Hummer and Szabo suggest that even in the current experimental regime molecular friction needs to be considered [3]. But can we do this with the AFM? How well can we measure a force distribution? How well can we locate f^*? Over what range of loading rate can we measure f^*? Below I dissect some of the salient issues around measuring dynamic forces with the AFM.

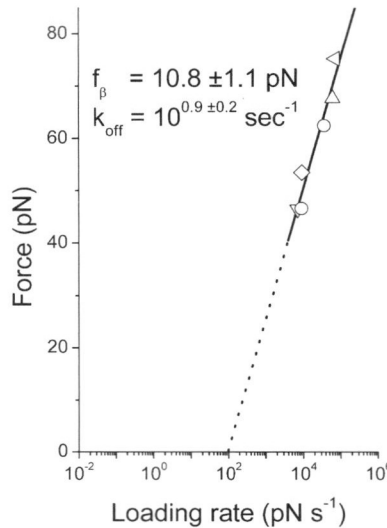

Figure 1. Dynamic force spectrum of the forced dissociation of an RNA duplex of 12 base-pairs in length containing a central 3 bulge helix [14]. Five different cantilevers were used and nearly two orders of magnitude in loading rate were achieved. A single force scale of slope $f_\beta = 10.8\,\mathrm{pN}$ was observed relating to a transition state displaced by 0.4 nm (see [14] for full details).

Why Measure the Mode and Not the Mean?

The simple expression of Eqn. 8 shows how the most probable rupture force varies with loading rate, barrier position, and off-rate derived from a somewhat phenomenological model. Since $p(f)$ isn't symmetric, the mean and mode are different; and as the mode of a distribution is more difficult to estimate than the mean, why isn't the mean force used? There are several reasons for this. The analytical expression for the mean of $p(f)$ is

$$\bar{f} = \frac{\int_0^\infty f p(f)\,df}{\int_0^\infty p(f)\,df} = \int_0^\infty f p(f)\,df \qquad (9)$$

For a single transition state the mean force can be approximated to the series expansion of the exponential integral [15–17]

$$\bar{f} \approx f_\beta \left[\ln\left(\frac{r_f}{f_\beta k_{off}}\right) - \gamma - \sum_{n=1}^\infty \frac{\left(-f_\beta k_{off}/r_f\right)^n}{nn!} \right] \exp\left(\frac{f_\beta k_{off}}{r_f}\right) \qquad (10)$$

where $\gamma = 0.5772\ldots$ is Euler's constant. When $r_f \gg f_\beta k_{off}$ the average force approximates to

$$\bar{f} \approx f_\beta \left[\ln\left(\frac{r_f}{f_\beta k_{off}}\right) - \gamma \right] \approx f^* - 0.6 f_\beta \qquad (11)$$

However, this is only an approximation; so even if we can measure the mean force accurately, we do not have an analytical expression to analyze the data. Secondly, the mean of a distribution is a function of all of the data. In an experiment there may well be instances where multiple connections are made and missed, occasional interactions between the tip and underlying substrate may be misinterpreted and included in the data, and nonlinear loading through polymer linkages all affect the distribution and will change the mean. The mode is less sensitive to such effects. Application of Eqn. 11 can be a useful check, and the quality of the force distributions measured, since

$$f_\beta \approx \frac{f^* - \bar{f}}{\gamma} \tag{12}$$

Testing the Phenomenological Model

Given certain caveats and assumptions, the model expressed by Eqn. 8 shows that the dynamic force spectrum of an interaction will reveal the barrier location and force-free off-rate over the barrier. Dynamic force spectra with multiple slopes indicate the presence of multiple barriers [10, 11, 18]. DFS of the streptavidin/biotin interaction undertaken in a number of laboratories [1, 2, 19, 20] by AFM indicates two barriers; $f_\beta = 7.7$ pN (= 0.5 nm), $k_{off} = 0.006$ sec^{-1} and $f_\beta = 44$ pN (0.1 nm), $k_{off} = 40$ sec^{-1} (Figure 2). Several molecular simulations have been undertaken of this dissociation [21–25]. An adiabatic mapping study revealed potential energy barriers at 0.12 and 0.48 nm; molecular dynamics studies indicate the presence of a barrier too at 0.1 and 0.4 nm [25]. Cross validation of the dissociation rates estimated from DFS with other experiments is more difficult, since there may be barriers not seen in the force experiment that dominate the dissociation kinetics. For the streptavidin/biotin case, for instance, a third barrier is present with force scale $f_\beta \approx 3$ pN unseen by the AFM [25]. Agreement can be found in other systems; recent experiments of an antibody/antigen interaction suggest dissociation rates of similar order to SPR measurements [26, 27].

It is possible to test whether the dissociation rate of the bond does influence the dynamic force spectrum described by Eqn. 8 by varying the architecture of the molecular system. Having two identical molecular interactions in a chain between tip and substrate will double the probability of seeing a rupture event [16]. Our experiments comparing the dynamic force spectrum of the streptavidin/iminobiotin interaction with the spectrum of two of these bonds loaded in series showed the latter spectrum to be displaced by (almost) a factor of two in rate [28]. As expected, by doubling the dissociation rate of the bond, we measured a doubling of the extrapolated off-rates.

A further test of the model was recently undertaken by Noy [29, 30]. It was proposed that multiple bonds loaded simultaneously in a patch, where the force is distributed evenly across the bonds, should produce a dynamic force spectrum with both predictable changes in force scale and extrapolated off-rate, depending on the number of bonds in the patch [16]. Through an ingenious analysis of the AFM force traces, Noy was able to count the number of bonds in the patch. The force spectra for single, two, and three bonds behaved as predicted from use of the simple models above.

The simple model of Eqn. 8 appears to hold for the few systems on which it has been tested. Analysis of the force spectrum using Eqn. 8 provides two pieces of information: the force scale f_β and the extrapolated off-rate k_{off}. These data require at least the measurement of the most probable rupture force at two different loading rates. The accuracy with which we can measure the data depends on the accuracy with which we can measure force, the accuracy of the loading rate, and the difference in loading rate that we can apply. Measuring force with the AFM and

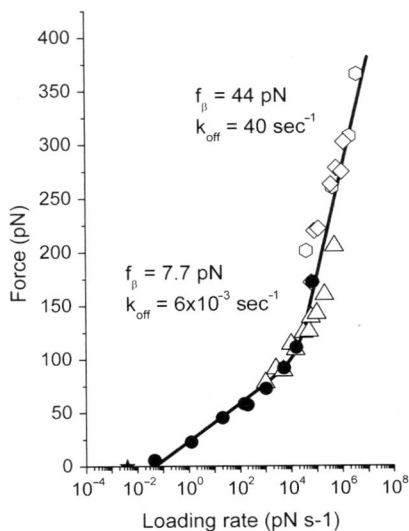

Figure 2. Dynamic force spectrum of the biotin/streptavidin interaction studied by AFM (open symbols) [19, 20] and biomembrane force probe (closed symbols) [2]. Two slopes of force in the semilogarithmic plot are recorded, indicating the presence of two barriers at 0.1 and 0.5 nm. Barriers at these locations are recorded in molecular simulations of this dissociation.

knowing the loading rate require measurement of the stiffness of the cantilever. The cantilever sensitivity also has to be measured, and the photodetector and piezos calibrated. The most probable force also has to be determined from a finite number of measurements.

Spring Constant and Optical Lever Sensitivity Errors

There are a variety of methods used to measure the stiffness of AFM cantilevers, most of which are accurate to within 10% of the true value [31–37]. Errors in the determination of the stiffness K will lead to errors in force spectroscopy measurements taken with them. Similarly, errors in measurement of the optical lever sensitivity (the change in signal from the position-sensitive photodiode for a unit deflection of the lever [typically V nm^{-1}]) S will affect measurements made from the resulting force spectrum [38].

If a single cantilever is used for all of the measurements of a dynamic force study, the forces and loading rates are systematically affected by stiffness error. A 10% overestimation of the stiffness produces a 10% overestimation of the force and a 10% overestimation of the loading rate. The apparent force scale determined from two points (f_1, f_2 at r_{f1}, r_{f2}) on a dynamic force spectrum where both the forces and the loading rates are subject to cantilever stiffness error ε_K and lever deflection error ε_s is

$$f_\beta' = \frac{\Delta \varepsilon_K \varepsilon_s f}{\Delta \ln\left(\varepsilon_K r_f\right)} = \frac{\varepsilon_K \varepsilon_s \left(f_2 - f_1\right)}{\ln\left(r_{f2}/r_{f1}\right)}, \tag{13}$$

and hence the error in the force scale f_β is the same as the error in lever calibration. The 10% overestimation of lever stiffness leads to a 10% overestimation of the force scale. Since the off-rate of the interaction is extrapolated from the force spectrum with

$$k'_{off} = \frac{\varepsilon_K r_f^{f=0}}{\varepsilon_K \varepsilon_S f_\beta} = \frac{r_f^{f=0}}{\varepsilon_S f_\beta} \tag{14}$$

at zero force, and both the loading rate and force scale are equally scaled by the lever stiffness error, the calculated off-rate is unaffected. Unlike error in the lever stiffness, error in the measurement of the cantilever sensitivity affects both the measured force scale and off-rate. As above, a 10% overestimation of the sensitivity produces a 10% overestimation of the force scale, but now also a 10% underestimation of the off-rate (since the apparent loading rate does not involve the lever sensitivity value).

With a dynamic force spectrum requiring many thousands of force measurements, it is certain that more than one cantilever will have to be used. If all the measurements are pooled, the uncertainties in calibration reduce with the root of the number of levers used [39]. Such pooling of data is commonplace, but the analysis above suggests that it may be better to keep data from different levers separate. From a set of force spectra, the variance in force scale will be

$$\langle f_\beta \rangle^2 = \langle K \rangle^2 + \langle S \rangle^2 = \langle \varepsilon_K \rangle^2 + \langle \varepsilon_S \rangle^2, \tag{15}$$

and in off rate

$$\langle k_{off} \rangle^2 = \langle S \rangle^2 = \langle \varepsilon_S \rangle^2, \tag{16}$$

As mentioned, $<\varepsilon_K>^2$ is of order 0.01. As far as I know, studies of $<\varepsilon_S>^2$ have not been published.

Polymer Tethers

The worm-like chain describes the force versus extension behaviour of an inextensible flexible rod and represents reasonably well that measured for unfolded proteins and polymer tethers. The configurational space available to a polymer is reduced when its ends are held, and reduces further as the ends are held at increasing separations. As the ends of a polymer are separated, therefore, the entropy of the polymer is reduced, and this requires work. The work done in separating the ends is equal to the loss of entropy of the system; the polymer behaves as an entropic spring.

The equation

$$f_{WLC}(x) = \frac{k_B T}{b} \left(\frac{1}{4\left(1-\frac{x}{L}\right)^2} - \frac{1}{4} + \frac{x}{l} \right) \tag{17}$$

is used to describe the force required to hold the ends of a polymer of contour length l and persistence length b at separation x.

Nonlinear Loading

The loading rate applied to a bond can be determined from the *fd* trace by the product of the slope (change in force with piezo displacement [pN nm^{-1}]) and the piezo retract velocity [nm sec^{-1}]. Noise in the trace means that either prior filtering (using a median filter) or fitting of an appropriate curve, such as the worm-like chain model, is required. This approach has been used several times, where it is argued that the range in stiffness measured at rupture provides an extended range of loading rates. Wojcikiewicz et al., for example, measured bond strengths between 25 and 325 pN, and by using the difference in system stiffness at rupture were able to plot a dynamic force spectrum covering over three orders of loading rate [40]. Their analysis, however, is problematic, as the loading rates for individual bond rupture events cannot be pooled and used in this fashion.

Consider the worm-like chain model for polymer stretching. The stiffness of the polymer increases with extension, as

$$\frac{dF}{dx} = \frac{k_B T}{b} \left(\frac{1}{2l\left(1 - \frac{x}{l}\right)^3} + \frac{1}{l} \right) \tag{18}$$

and again is nonlinear, showing a transition from Hookian behaviour (scaling with $1/l$) to entropic (scaling with $1/x^3$). The distribution of forces measured at a single retract velocity will exhibit a distribution of loading rates at rupture. A semilog plot of WLC force against chain stiffness appears similar to a dynamic force spectrum of two transition states (two f_β values). As a demonstration, Figure 3a shows the result of a Monte Carlo simulation of bond rupture, where a cantilever (K = 10 pN nm^{-1}) pulls on a single bond (k_{off} = 1 sec^{-1}, f_β = 10 pN) with a polymer tether (l = 32 nm, b = 0.34 nm) 100 times at a single retract velocity of 1000 nm sec^{-1}. Each force is plotted against the logarithm of the slope of the simulated force trace at rupture multiplied by the retract velocity (loading rate at rupture). Figure 3b shows the results of similar calculations of the same bond rupture, this time performed at retract velocities of 100, 1000, and 5000 nm sec^{-1} (50 recordings at each speed), and with 6.4 pN of Gaussian (to mimic thermal) noise added. The transition in stiffness from weak at low forces to strong at high forces produces

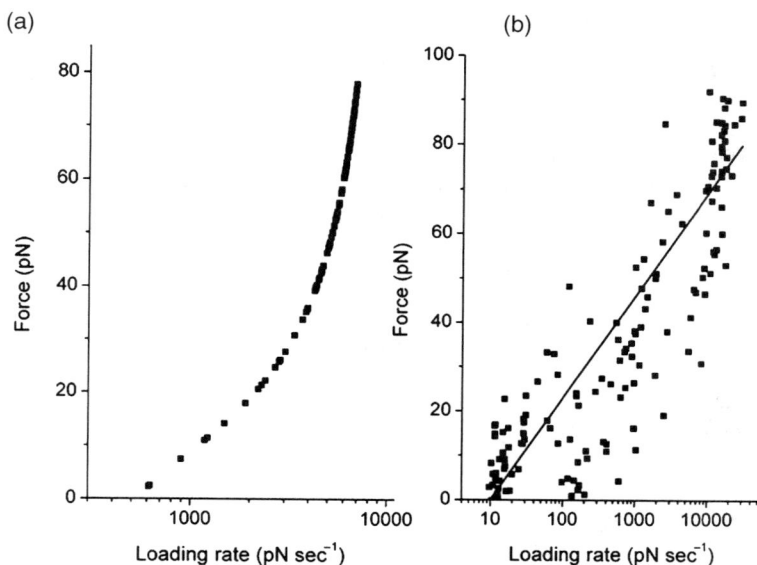

Figure 3. (A) Results of a Monte Carlo simulation of bond rupture using a polymer tether (k_{off} = 1 sec^{-1}, f_β = 10 pN), where the force at rupture is plotted against the instantaneous loading rate at the point of rupture. (B) A force spectrum of Monte Carlo data obtained at three speeds. The force at rupture is plotted against the loading rate at rupture. The true force scale is plotted.

a transition in the 'force-spectrum', which can easily be misinterpreted as the presence of two unbonding transition states.

How can a dynamic force spectrum be produced when flexible polymer linkages are used? A simple method is to calculate or measure the stiffness of the molecular system at each of the modal forces determined from a number of tests at different velocities. For example, the mode rupture force from the simulations above at 100, 1000, and 5000 nm sec^{-1} were 22, 52, and 83 pN, respectively. For the known polymer linker, Eqn. 17 reveals that the polymer is stretched to 18.9, 23.5, and 25.4 nm at these three forces, respectively; and Eqn. 18 gives the stiffness of the polymer at these rupture events of 2.4, 5.1, and 6.9 pN nm^{-1}, respectively. In fact, a rough approximation for the stiffness of a WLC at a high force is

$$K_{WLC}(f) \approx 4 \frac{k_B T}{bl} \left(\frac{bf}{k_B T} \right)^{3/2}. \tag{19}$$

The stiffness of the whole system (polymer and cantilever in series) is calculated, and multiplying by the three velocities gives effective loading rates of 226, 5145, and 34329 pN sec^{-1}, respectively. The resulting dynamic force spectrum (Figure 4a) implies a force scale $f_\beta = 12$ pN and off-rate $k_{off} = 3$ sec^{-1}.

An alternative method was proposed by Evans and Ritchie where the effect of the linker on the loading rate was incorporated into the unbonding probability distribution [41]. For the approximation that unbonding occurs in the entropic regime of the polymer stiffness, the most probable rupture force using a WLC-like polymer linker was shown to be

$$f^* \approx f_\beta \ln\left(\frac{v}{v_\beta}\right) + f_\beta \left[\ln\left(\frac{f^*}{f_\beta} - \frac{3}{2} \right) + \frac{1}{2}\ln\left(\frac{f^*}{f_\beta}\right) \right]$$

$$v_\beta = \frac{lk_{off}}{4}\left(\frac{k_B T}{f_\beta b} \right)^{1/2}, \tag{20}$$

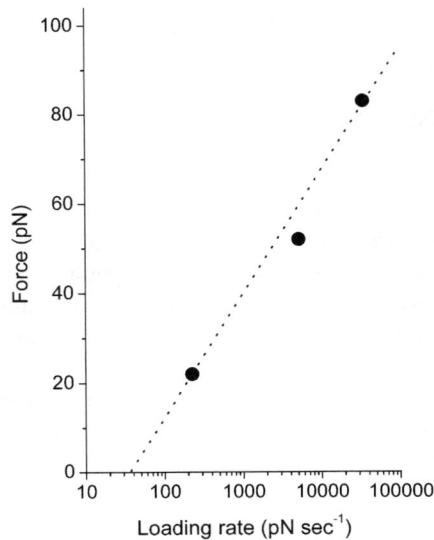

Figure 4. A dynamic force spectrum of data obtained using polymer linkers can be obtained by calculating or approximating the stiffness of the system at the mode forces measured.

where v_β characterizes the polymer. It is not simple to fit Eqn. 20 to a set of measured forces, however, due to its transcendental nature; and the equation also doesn't approximate well the forces measured with soft cantilevers. With the above data the equation fits best with $f_\beta = 11\,\text{pN}$ and off-rate $k_{off} = 3\,\text{sec}^{-1}$.

Histograms

Dynamic force spectroscopy usually requires the determination of the most probable rupture force f^* at a given loading rate r_f, and therefore requires finding the maximum of the distribution of rupture forces. With each force being a sample of a continuous distribution, the obvious way to find the maximum is to take many measurements and create a histogram, placing each measurement in a bin, or *class* of finite width. But what bin size should be used?

In 1926 Sturges [42] considered a histogram made of binomial coefficients

$$_nC_k = \frac{n!}{(n-k)!\,k!}, \tag{21}$$

where k is the number of combinations that can be selected from a set of n items. The histogram made of these coefficients has $n + 1$ bins, so the ith bin of a binomial histogram of j bins contains $_{j-1}C_i$ items. This histogram approaches that of a normal distribution for large j. The total sample size N is the sum of all the bins, as

$$N = \sum_{i=0}^{j-1} {}_{j-1}C_i. \tag{22}$$

With the binomial expansion [15]

$$(1+1)^{j-1} = \sum_{i=0}^{j-1} {}_{j-1}C_i, \tag{23}$$

therefore,

$$N = 2^{j-1}, \tag{24}$$

and the number of bins required to construct a histogram of N data normally distributed is

$$j = 1 + \log_2 N. \tag{25}$$

The bin width h_N is then just the data range divided by j. Sturges's rule is commonly used in statistical packages and graphing computer programs. As pointed out by Hyndman, though, the above reasoning is flawed, since any multiple of the coefficients could have been used; and whilst the resulting distribution would still be normal, the value of j changes.

Alternative methods for bin size selection have been proposed, including Scott's rule [43]. Scott showed that for a Gaussian distribution the optimal choice of bin size is

$$h_N = 3.49\sigma N^{-\frac{1}{3}}, \tag{26}$$

where σ is the estimation of the standard deviation. Scott showed that although Eqn. 26 is derived from an assumption of a Gaussian distribution, it is useful for a large variety of non-Gaussian densities. Interestingly for us the optimum value of h_N for mildly skewed distributions, like those of the unbonding probability $p(f)$ with negative skewness (Eqn. 7), are very similar to those estimated assuming a Gaussian distribution. (In theory one could derive the optimum value of h_N for the unbonding distribution by

$$h_N^* = \left\{ 6 \bigg/ \int_0^\infty P'(x)^2 \, dx \right\}^{\frac{1}{3}} N^{-\frac{1}{3}}, \tag{27}$$

but this can only be solved approximately, and to use it in practice would require prior knowledge of f). Scott's rule is a simple and convenient way to determine the bin size for constructing a histogram and with Sturges's rule gives an indication of the amount of force data required to adequately determine the modal rupture force at a given loading rate.

Experimental considerations have been used to choose h_N: specifically, measuring or estimating the measurement noise. As discussed, the largest source of noise is thermal fluctuation of the cantilever, giving

$$h_N = K \sqrt{\frac{k_B T}{K}}. \tag{28}$$

For AFM cantilevers ($K = 25 \, \mathrm{pN \, nm^{-1}}$), h_N equates to $10 \, \mathrm{pN}$. Sturges's rule suggests that over 500 measurements would be required to sample a simple distribution of forces ranging up to $100 \, \mathrm{pN}$ ($j = 10$); and Scott's rule suggests this would be sufficient to measure a bond of force scale ($f_\beta \sim 0.8\sigma$) of approximately $20 \, \mathrm{pN}$.

This experimental consideration suggested an alternative method of estimating the mode of the underlying distribution without resort to the histogram [44, 45]. Each measurement x can be considered as drawn from a Gaussian distribution of variance equal to that of the measurement error (cantilever noise) and centred on the measured value. This Gaussian distribution reflects the probability that the measured value is correct, given no other information. The cumulative probability distribution is then constructed as the sum of these Gaussians, as

$$\mathrm{CPD}(f) = \frac{1}{\sigma \sqrt{2\pi}} \sum_{i=1}^N e^{-\frac{1}{2}\left(\frac{f-x_i}{\sigma}\right)^2} \tag{29}$$

The distribution is created for values of f (usually $1 \, \mathrm{pN}$ apart) spanning the range of data, and the location of the maximum is noted. A distinct advantage of this method is that not only is the modal value estimated with arbitrary resolution of force, but also the curvature of the top of the distribution can be calculated numerically around the mode. Since the cumulative distribution is the convolution of the underlying rupture force distribution and the Gaussian measurement, it is possible to estimate the force scale by

$$f_\beta \approx \left\{ \left(-\frac{1}{\mathrm{cpd}\left(f^*\right)} \frac{\left(\mathrm{cpd}(f^* + 2\Delta f) - 2\mathrm{cpd}(f^*) + \mathrm{cpd}(f^* - 2\Delta f)\right)}{4\Delta f^2} \right)^{-1} - \sigma^2 \right\}^{\frac{1}{2}} \tag{30}$$

Table 1 illustrates the effectiveness of various methods of histogram and force analysis to pinpoint the true mode. Each of the methods was applied 50 times to synthetic data comprising 100 forces generated from Monte Carlo simulations of forced bond rupture with added noise ($f_\beta = 10 \, \mathrm{pN}$, $k_{off} = 1 \, \mathrm{sec^{-1}}$, $r_f = 10{,}000 \, \mathrm{pN \, sec^{-1}}$, $\sigma = 6.4 \, \mathrm{pN}$, $f^* = 69.1 \, \mathrm{pN}$). In these simple tests, Scott's method was the most accurate whilst CPD was the most precise. The CPD method could also make a stab at estimating the force scale at $11 \, \mathrm{pN}$.

Table 1 Analysis of various methods of estimating the mode of a force distribution comprising 100 forces. The true mode is 69.1 pN. The cumulative probability distribution (CPD) method also permits an estimation of the force scale (true value 10 pN).

Method	Mode average estimate (n = 50; pN)	SEM of mode (n = 50)	f_β average estimate (n = 50; pN)
Noise (h_N = 6.4 pN)	69.4 (+ 0.4%)	0.74	–
Sturges	69.5 (+ 0.6%)	0.62	–
Scott	68.8 (− 0.4%)	0.50	–
CPD	68.7 (− 0.6%)	0.34	11 (SEM = 0.67)

Loading Rate

With the use of many cantilevers it looks possible to measure f^* to within a few picoNewtons of the true value. The accuracy to which we can measure f therefore depends on the range of loading rate over which the data span, and, for k_{off}, on the orders of loading rate over which f_β is extrapolated. Practically current AFM cantilevers can be used with retract velocities ranging between 10 and 5000 nm sec^{-1} before thermal drift or hydrodynamics becomes significant. Where no polymer linker is used, the loading rate range can be expanded through the use of levers with differing stiffness; and in theory loading rates from 10^2 to 10^6 or 10^7 pN sec^{-1} can be achieved. As indicated previously, when linkers are used the stiffness of the system can be dominated by that of the polymer, and the loading rate range is restricted to that of the retract velocity. The spread in force scale is

$$\left\langle \sigma_{f_\beta} \right\rangle = \frac{2\left\langle \sigma_{f^*} \right\rangle}{\ln\left(r_f^{High}/r_f^{Low}\right)} \approx \frac{\left\langle \sigma_{f^*} \right\rangle}{\log_{10}\left(r_f^{High}/r_f^{Low}\right)}, \quad (31)$$

with $\log_{10}(r_f^{High}/r_f^{Low})$ the number of orders of loading rate used. A comparison with Eqn. 15 shows that the range of loading rates available to the AFM is a limiting factor.

The spread of extrapolated off-rates depends on the variance of the force scale and the lowest force value measured. The extrapolated rate can vary from

$$k_{off} < \frac{r_f^{Low}}{f_\beta + \left\langle \sigma_{f_\beta} \right\rangle} \exp\left[-\left(\frac{f_{Low}^* - \left\langle \sigma_{f^*} \right\rangle}{f_\beta + \left\langle \sigma_{f_\beta} \right\rangle}\right)\right], \quad (32)$$

and

$$k_{off} > \frac{r_f^{Low}}{f_\beta - \left\langle \sigma_{f_\beta} \right\rangle} \exp\left[-\left(\frac{f_{Low}^* + \left\langle \sigma_{f^*} \right\rangle}{f_\beta - \left\langle \sigma_{f_\beta} \right\rangle}\right)\right], \quad (33)$$

For accurate measurements of k_{off} one requires force scales measured over a large spread of rates and low forces from which to extrapolate. Low forces require low rate, hence polymer tethers and soft levers are required, thereby limiting the range of loading rates to around 2.5 orders of magnitude. The limitation of the AFM is the ability to achieve low loading

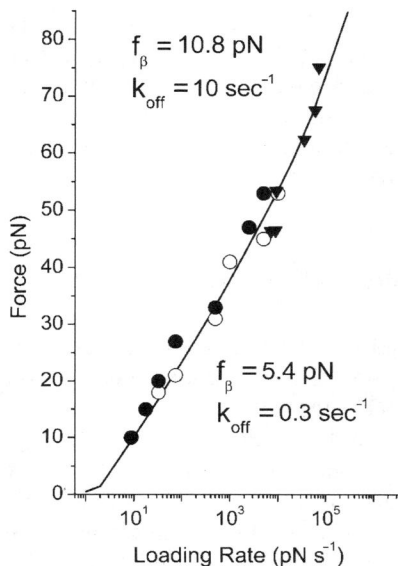

Figure 5. Dynamic force spectrum of the oligonucleotide system shown in Figure 1, taken over an extended range of loading rates with the aid of the biomembrane force probe (circles). Two transition states are seen; the outer transition state at 0.8 nm was missed by the AFM (triangles).

rates, with the necessary force sensitivity, and developments in lever technology, with usable spring-constants of less than 1 pN nm^{-1}, offer great promise. An example in which the AFM loading rates are too high was highlighted in our AFM studies of RNA dissociation [14]. We observed a single force scale for dissociation of a 12-base pair strand of RNA corresponding to a single transition state $x^{\ddagger} = 0.8$ nm. When a bulge was introduced along the strand, the force scale doubled, suggesting $x^{\ddagger} = 0.4$ nm, but still only one transition state was observed. We argued that here the AFM was only sensitive to the new transition state introduced with the bulge and couldn't achieve loading rates low enough to detect the original dissociation transition state at 0.8 nm. A few years later we revisited this study and applied our newly constructed biomembrane force probe to explore more of the bulge dissociation landscape. Pleasantly, both transition states were measured confirming our assumptions from the AFM experiment (Figure 5) [46].

By assuming that we can measure f* to within 1 pN through the use of many cantilevers (see above), and that we can measure a force of 20 pN (three times the thermal force noise of a 10 pN nm^{-1} lever), it is possible to estimate the spread of extrapolated off-rate and force scale possible with the AFM. For our example above ($f_{\beta} = 10$ pN, $k_{off} = 1$ sec^{-1}), the force scale is known within 2.5 pN (x_{β} is between 0.33 and 0.55 nm), and off-rate is extrapolated to be between 0.6 and 1.3 sec^{-1}.

Hydrodynamics

As discussed, fast retract rates are precluded due to the rapidly increasing effect of hydrodynamics [9, 19]. In free solution the effect of hydrodynamics around the cantilever can be visualized by the hysteresis between the approach and retract trace, and measurement of this can provide an estimate of the drag force acting on the lever [47]. The drag force on the lever in free solution is proportional to the velocity at which it is moved, and providing this force is larger than the force measured for the bond, is missing from the measurement. The bond experienced more force than that measured by the deflection. The drag force for standard sized AFM levers has been estimated to be approximately 2×10^{-3} pN sec nm^{-1}. However, the hydrodynamics of the lever near to the substrate do not reflect those in free solution [48–50], and an accurate assessment of the 'missing' force at high velocity AFM measurements has only just begun [51].

Application to Protein Unfolding

An alternative to measuring the unbinding of a ligand from its receptor is to study the forced unfolding of proteins [45, 52–54]. The most studied protein in this fashion is the 27th domain of the I-band of titin (I27)[55–58].

DFS studies of I27 undertaken with the AFM showed a single regime of force, suggesting the measurement of one unfolding transition state [56]. It is known, however, that I27 unfolds through an intermediate under force; this intermediate was first observed in modelling studies [59], confirmed by analysis of the force versus extension trace [60], and has been characterised by structural techniques [57]. As the intermediate is seen in AFM traces (as a plateau at around 100 pN), the unfolding event seen in AFM, which occurs at forces above 100 pN, is occurring from this intermediate. The intermediate is not seen in chemical denaturation studies. It is not expected, therefore, that the unfolding rate extrapolated from the DFS measurements should match the rate measured in chemical denaturation studies.

Analysis of protein unfolding force spectroscopy experiments is complicated, as most rely on the use of tandem protein domains stressed and unfolded in series [9, 16, 61, 62]. As demonstrated through the iminobiotin/strepatvidin experiments indicated previously, the effective unfolding (unbonding) rate is increased by a factor equal to the number of proteins (bonds) in the chain [28]. When pulling on a repeat of eight proteins, the kinetics of the first unfolding event is eight times that of the last. This is complicated further, as the number of proteins stressed by the AFM is not always known, since the AFM can attach to the chain anywhere along its length. The number of events seen in a trace is an indication of the minimum number of domains stressed, but one can only be sure of the number that have been stressed if all eight unfolding events are seen. Seven peaks in a trace, for instance, may mean that seven proteins were stressed or that the AFM detached before the eighth unfolded. A second complication in such experiments is the reduction of the loading rate with successive unfolding events. On unfolding, the protein adds to the length of the random polymer that is pulling on the remaining domains. Since most experiments are performed at constant velocity, this increase in contour length of the polymer l causes a reduction in the loading rate (Eqn. 19). Ideally, markers should be used to indicate the number of domains stressed by, for instance, making a heteropolymer with the protein of interest sandwiched between another domain that unfolds with a different length [63], making use of specific tags (His tags, cysteine, biotin, etc.) to attach the AFM to the end of the chain, and analyzing each peak (the 1st of 8, 2nd of 8, 3rd of 8, …, 1st of 7, 2nd of 7, … etc.) independently. This is very time consuming.

By studying the probability of picking up chains on one through to eight domains, and analysing the kinetics of each combination of event number and chain length, the DFS data of I27 indicated a force scale $f_\beta = 12.8$ pN ($x^\ddagger = 0.32 \pm 0.01$ nm) and unfolding rate (from the intermediate) $k_{off} = (2.9 \pm 1.1) \times 10^{-4}$ sec^{-1} [58]. The location of the unfolding transition state is again supported by molecular simulation (Figure 6)[64].

In 2003 Hummer and Szabo proposed a new method to analyze DFS results that is derived from a simple stochastic model of pulling on a molecular potential with a harmonic string [3]. Their model reduces to the phenomenological approach in a limit. The method

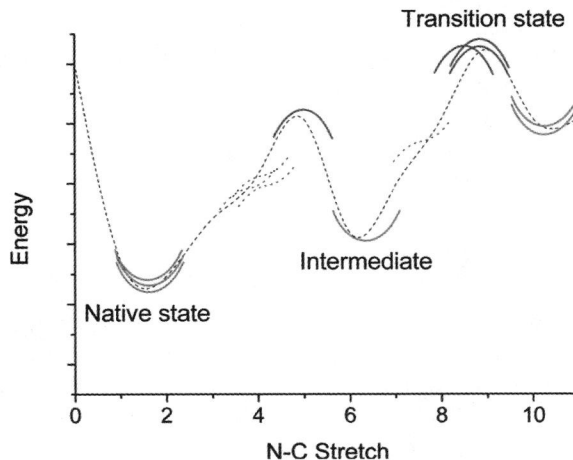

Figure 6. Molecular simulations of protein unfolding indicate transition state locations comparable to those measured in dynamic force spectroscopy [64]. Here, a simulation of the force-induced unfolding of I27 shows a transition state to unfolding 0.3 nm away from an intermediate. DFS of I27 produces a force scale $f_\beta = 12.8\,\mathrm{pN}$ ($x^\ddagger = 0.32$). (*See Color Plates*)

considers diffusional and forced motion across a landscape which is the sum of two potentials: one fixed for the bond, and one moving as the retracted AFM spring and molecular linker. Using a cusplike potential the model reduces the barrier height in time, not linearly by $tr_f x^\ddagger$ as in Eqn. 2, but by $0.5\kappa_s (vt)^2$ representing movement of a harmonic constraint. In summary, the mean force of rupture of a bond pulled with a velocity below that where friction dominates is

$$\bar{f} = k_B T \left\{ k_m x^\ddagger - \left[2\kappa \ln \left(\frac{k_{off}\, e^{\gamma + \kappa_m (x^\ddagger)^2 / 2}}{\kappa_s v x^\ddagger \left(\kappa_m / \kappa\right)^{3/2}} \right) \right]^{1/2} \right\} \tag{34}$$

with κ_m = stiffness of the bond, as ω_c in Eqn. 2. Here the bond potential is cusplike with a sharp transition state), κ_s = stiffness of the pulling system (AFM cantilever and linkers) scaled by $k_B T$, and $\kappa = \kappa_m + \kappa_s$. Application of Eqn. 34 for DFS relies on upwards curvature of the force data to fit κ_m. On the I27 data the unfolding rate was found to be $10^{-5} < k_{off} < 10^{-4}$ sec^{-1}, and force scale $f_\beta = 10\,\mathrm{pN}$ ($x^\ddagger = 0.42$). The difficulty in applying Eqn. 34 is the determination of three parameters (as opposed to two) from limited data over the short range of loading rates available to the AFM. Whilst Eqn. 34 will be of use in protein unfolding studies (although only through careful application to individual events, and not to data pooled from all events) where nonspecific events can often be discounted; being reliant on mean forces, its application is somewhat limited in ligand/receptor studies. Curvature in DFS data, as fitted by Eqn. 34, can arise through several processes: smooth transition states can shift towards to the bound state under force (akin to the Hammond effect in protein unfolding studies); the transition between two energy barriers is curved in the force spectrum; there may be an unaccounted for effect of polymers on the loading rate; etc. Advances in instrumentation and accuracy in force determination with the AFM are necessary before the full potential of Eqn. 34 can be realized.

The effect of increasing linker stiffness ks is to increase the probability that molecular rebinding can occur, and therefore affects the crossover from near equilibrium to the far equilibrium regime of the phenomenological model [9]. We found a crossover in force between these two regimes, above which the system can be adequately described by Eqn. 8; and below which rebinding is significant and the force plateaus at a level reflective of the equilibrium kinetics and stiffness of the molecular linkage

$$f_{\otimes} \sim \left[2 k_B T \kappa_s \ln \left(K_{eq} \right) \right]^{1/2}, \tag{35}$$

where K_{eq} is the equilibrium constant of the interaction. Here is little evidence that the crossover to equilibrium forces has been measured with AFM so far. Measurements without the use of linkers have mainly been in studies of biotin and its analogues with streptavidin; and at the lowest forces measured (circa 75 pN) using soft cantilevers (10 pN nm^{-1}), the system is far from equilibrium (K_{eq} would need to be over 1×10^{29}!). For long polymer linkages this crossover force is very low, given approximately by

$$f_{\otimes} \sim \frac{k_B T}{b} \left[\left(K_{eq} \right)^{b/l} - 1 \right]. \tag{36}$$

Final Remarks

DFS studies of many different types of molecular system, undertaken by the AFM, have revealed new insight into the physics, chemistry, and biology of molecular systems. Receptor ligand interactions have revealed multiple transition states in their dissociation energy landscapes that bestow functionality to regulate processes controlled by force [40, 65–67]. We have seen how protein constructs provide mechanical plasticity [58], gained insight into the shape of the protein (un)folding funnel [8, 39, 58], and seen the effect of mutations on stability and processing [14]. It is difficult to see how this information, gained from multiple transition states and the effects of force on dissociation kinetics, could have been enhanced by other studies. It is clear, though, that the potential of DFS can only be realized if a wide range of forces can be studied, and this means many orders of loading rate. In this regard, current AFM technology is far from ideal. Personally, our recent revelations of both multiple protein unfolding transition states and multiple kinetic barriers to oligonucleotide dissociation have stemmed from the application of other methodologies.

DFS has also provided a valuable benchmark for molecular simulation, where now we are able to calculate molecular trajectories across microseconds of time. We are still many years away from being able to simulate what is observable on the experimental timescale at the atomic level, and the data available from force spectroscopy is providing invaluable as a checkpoint, and to focus on methodological and algorithmic developments [24, 64, 68].

Current developments in AFM cantilever technology are more directed towards improvements in imaging. Levers are becoming smaller, which has the effect of moving much of their thermal noise above the sampling rate of the instrument. Whilst of significant advantage in imaging, this does not help the force spectroscopist much, as thermal noise will still be present (dependent upon the stiffness, not resonant frequency) but not be seen. Smaller levers will, however, reduce the hydrodynamic limitations and permit faster retract velocities. Great improvements are being made in the electronics that control the AFM, with the latest instruments capable of sampling at several MHz, both permitting the pinpointing of the point (and force) of rupture, and enabling the incorporation of new force protocols (such as constant ramp-of-force, as opposed to constant velocity).

To date, the expression first derived by Evans (Eqn. 8) has proved capable of explaining DFS, suggesting that all of the interactions studied are characterized by deep potentials and sharp barriers and have diffusion times orders of magnitude faster than the rate at which the landscape is perturbed by force. With improvements in both data analysis and experimentation, some of which have been summarized here, added information about the energy landscape and unbonding process may be gleaned from DFS through the application of models such as those of Hummer and Szabo.

References

1. Evans, E. and K. Ritchie, Dynamic strength of molecular adhesion bonds. Biophysical Journal, 1997. 72(4): pp. 1541–1555.
2. Merkel, R., et al., Energy landscapes of receptor-ligand bonds explored with dynamic force spectroscopy. Nature, 1999. 397(6714): pp. 50–53.
3. Hummer, G. and A. Szabo, *Kinetics from nonequilibrium single-molecule pulling experiments*. Biophysical Journal, 2003. **85**(1): pp. 5–15.
4. Hummer, G. and A. Szabo, *Free energy surfaces from single-molecule force spectroscopy*. Accounts of Chemical Research, 2005. **38**(7): pp. 504–513.
5. Dudko, O.K., G. Hummer, and A. Szabo, *Intrinsic rates and activation free energies from single-molecule pulling experiments*. Physical Review Letters, 2006. **96**(10).
6. Hyeon, C.B. and D. Thirumalai, *Can energy landscape roughness of proteins and RNA be measured by using mechanical unfolding experiments?* Proceedings of the National Academy of Sciences of the United States of America, 2003. **100**(18): pp. 10249–10253.
7. Nevo, R., et al., *Direct measurement of protein energy landscape roughness*. Embo Reports, 2005. **6**(5): pp. 482–486.
8. Schlierf, M. and M. Rief, *Temperature softening of a protein in single-molecule experiments*. Journal of Molecular Biology, 2005. **354**(2): pp. 497–503.
9. Evans, E. and P.M. Williams, eds. *Dynamic Force Spectroscopy: I. Single Bonds*. Les Houches Ecole de Physique LVVX. Physics of Bio-molecules and Cells, ed. H. Flyvbjerg. 2002, Springer.
10. Evans, E., *Probing the relation between force - Lifetime - and chemistry in single molecular bonds*. Annual Review of Biophysics and Biomolecular Structure, 2001. **30**: pp. 105–128.
11. Evans, E., et al., *Chemically distinct transition states govern rapid dissociation of single L-selectin bonds under force*. Proceedings of the National Academy of Sciences of the United States of America, 2001. **98**(7): pp. 3784–3789.
12. Green, N.M., *Avidin*. Advances in Protein Chemistry, 1975. **29**: p. 85–133.
13. Green, N.M., *Avidin and Streptavidin*. Methods in Enzymology, 1990. **184**: pp. 51–67.
14. Green, N.H., et al., *Single-molecule investigations of RNA dissociation*. Biophysical Journal, 2004. **86**(6): p. 3811–3821.
15. Abramowitz, M. and I.A. Stegun, eds. *Hanbook of Mathematical Functions*. 1972, Dover.
16. Williams, P.M., *Analytical descriptions of dynamic force spectroscopy: behaviour of multiple connections*. Analytica Chimica Acta, 2003. **479**(1): pp. 107–115.
17. Gergely, C., et al., *Unbinding process of adsorbed proteins under external stress studied by atomic force microscopy spectroscopy*. Proceedings of the National Academy of Sciences of the United States of America, 2000. **97**(20): pp. 10802–10807.
18. Evans, E., et al., *Mechanical switching and coupling between two dissociation pathways in a P-selectin adhesion bond*. Proceedings of the National Academy of Sciences of the United States of America, 2004. **101**(31): pp. 11281–11286.
19. Williams, P.M., et al., *On the dynamic behaviour of the forced dissociation of ligand-receptor pairs*. Journal of the Chemical Society-Perkin Transactions 2, 2000(1): pp. 5–8.
20. Lo, Y.S., Y.J. Zhu, and T.P. Beebe, *Loading-rate dependence of individual ligand-receptor bond-rupture forces studied by atomic force microscopy*. Langmuir, 2001. **17**(12): pp. 3741–3748.
21. Grubmuller, H., B. Heymann, and P. Tavan, *Ligand binding: Molecular mechanics calculation of the streptavidin biotin rupture force*. Science, 1996. **271**(5251): pp. 997–999.
22. Moore, A., et al., *Enthalpic approach to the analysis of the scanning force ligand rupture experiment*. Journal of the Chemical Society-Perkin Transactions 2, 1998(2): pp. 253–258.

23. Moore, A., et al., *Analyzing the origins of receptor-ligand adhesion forces measured by the scanning force microscope*. Journal of the Chemical Society-Perkin Transactions 2, 1999(3): pp. 419–423.

24. Galligan, E., et al., *Simulating the dynamic strength of molecular interactions*. Journal of Chemical Physics, 2001. **114**(7): pp. 3208–3214.

25. Williams, P.M., *Force Spectroscopy*, in *Scanning Probe Microscopies: Beyond Imaging*, P. Samori, Editor. 2006, WILEY-VCH. pp. 250–274.

26. Tawar, R.G., et al., *Complex bond architecture revealed through dynamic force spectroscopy*.

27. Ros, R., et al., *Antigen binding forces of individually addressed single-chain Fv antibody molecules*. Proceedings of the National Academy of Sciences of the United States of America, 1998. **95**(13): pp. 7402–7405.

28. Patel, A.B., et al., *Influence of architecture on the kinetic stability of molecular assemblies*. Journal of the American Chemical Society, 2004. **126**(5): pp. 1318–1319.

29. Sulchek, T.A., et al., *Dynamic force spectroscopy of parallel individual Mucin1-antibody bonds*. Proceedings of the National Academy of Sciences of the United States of America, 2005. **102**(46): pp. 16638–16643.

30. Sulchek, T., R.W. Friddle, and A. Noy, *Strength of multiple parallel biological bonds*. Biophysical Journal, 2006. **90**(12): pp. 4686–4691.

31. Cleveland, J.P., et al., *A Nondestructive Method for Determining the Spring Constant of Cantilevers for Scanning Force Microscopy*. Review of Scientific Instruments, 1993. **64**(2): pp. 403–405.

32. Walters, D.A., et al., *Short cantilevers for atomic force microscopy*. Review of Scientific Instruments, 1996. **67**(10): pp. 3583–3590.

33. Heim, L.O., M. Kappl, and H.J. Butt, *Tilt of atomic force microscope cantilevers: Effect on spring constant and adhesion measurements*. Langmuir, 2004. **20**(7): pp. 2760–2764.

34. Proksch, R., et al., *Finite optical spot size and position corrections in thermal spring constant calibration*. Nanotechnology, 2004. **15**(9): pp. 1344–1350.

35. Bonaccurso, E. and H.J. Butt, *Microdrops on atomic force microscope cantilevers: Evaporation of water and spring constant calibration*. Journal of Physical Chemistry B, 2005. **109**(1): pp. 253–263.

36. Clifford, C.A. and M.P. Seah, *The determination of atomic force microscope cantilever spring constants via dimensional methods for nanomechanical analysis*. Nanotechnology, 2005. **16**(9): pp. 1666–1680.

37. Stiernstedt, J., M.W. Rutland, and P. Attard, *A novel technique for the in situ calibration and measurement of friction with the atomic force microscope*. Review of Scientific Instruments, 2005. **76**(8).

38. Higgins, M.J., et al., *Noninvasive determination of optical lever sensitivity in atomic force microscopy*. Review of Scientific Instruments, 2006. **77**(1).

39. Ng, S.P., et al., *Mechanical unfolding of TNfn3: The unfolding pathway of a fnIII domain probed by protein engineering, AFM and MD simulation*. Journal of Molecular Biology, 2005. **350**(4): pp. 776–789.

40. Wojcikiewicz, E.P., et al., *Force spectroscopy of LFA-1 and its ligands, ICAM-1 and ICAM-2*. Biomacromolecules, 2006. **7**(11): pp. 3188–3195.

41. Evans, E. and K. Ritchie, *Strength of a weak bond connecting flexible polymer chains*. Biophysical Journal, 1999. **76**(5): pp. 2439–2447.

42. Sturges, H., *The choice of a class-interval*. Journal of the American Statistical Association, 1926. **21**: pp. 65–66.

43. Scott, D., *On optimal and data-based histograms*. Biometrika, 1979. **66**(3): pp. 605–610.

44. Baumgartner, W., et al., *Cadherin interaction probed by atomic force microscopy*. Proceedings of the National Academy of Sciences of the United States of America, 2000. **97**(8): pp. 4005–4010.

45. Clarke, J. and P.M. Williams, *Unfolding Induced by Mechanical Force*, in *Protein Folding Handbook*, J. Buchner and T. Kiefhaber, Editors. 2005, WILEY-VCH. pp. 1111–1142.

46. Wahab, O., et al., *Multiple transition states measured in RNA strand dissociation*.

47. Tees, D.F.J., R.E. Waugh, and D.A. Hammer, *A microcantilever device to assess the effect of force on the lifetime of selectin-carbohydrate bonds*. Biophysical Journal, 2001. **80**(2): pp. 668–682.

48. Clarke, R.J., et al., *The drag on a microcantilever oscillating near a wall*. Journal of Fluid Mechanics, 2005. **545**: pp. 397–426.

49. Clarke, R.J., et al., *Stochastic elastohydrodynamics of a microcantilever oscillating near a wall*. Physical Review Letters, 2006. **96**(5).

50. Clarke, R.J., et al., *Three-dimensional flow due to a microcantilever oscillating near a wall: an unsteady slender-body analysis*. Proceedings of the Royal Society a-Mathematical Physical and Engineering Sciences, 2006. **462**(2067): pp. 913–933.

51. Janovjak, H.J., J. Struckmeier, and D.J. Muller, *Hydrodynamic effects in fast AFM single-molecule force measurements*. European Biophysics Journal with Biophysics Letters, 2005. **34**(1): pp. 91–96.

52. Best, R.B. and J. Clarke, *What can atomic force microscopy tell us about protein folding?* Chemical Communications, 2002(3): pp. 183–192.

53. Best, R.B., et al., *Force mode atomic force microscopy as a tool for protein folding studies*. Analytica Chimica Acta, 2003. **479**(1): pp. 87–105.

54. Clarke, J. and G. Schreiber, *Folding and binding - new technologies and new perspectives - Editorial overview.* Current Opinion in Structural Biology, 2003. **13**(1): pp. 71–74.

55. Carrion-Vazquez, M., et al., *AFM and chemical unfolding of a single protein follow the same pathway.* Biophysical Journal, 1999. **76**(1): pp. A173-A173.

56. Carrion-Vazquez, M., et al., *Mechanical and chemical unfolding of a single protein: A comparison.* Proceedings of the National Academy of Sciences of the United States of America, 1999. **96**(7): pp. 3694–3699.

57. Fowler, S.B., et al., *Mechanical unfolding of a titin Ig domain: Structure of unfolding intermediate revealed by combining AFM, molecular dynamics simulations, NMR and protein engineering.* Journal of Molecular Biology, 2002. **322**(4): pp. 841–849.

58. Williams, P.M., et al., *Hidden complexity in the mechanical properties of titin.* Nature, 2003. **422**(6930): pp. 446–449.

59. Lu, H., et al., *Unfolding of titin immunoglobulin domains by steered molecular dynamics simulation.* Biophysical Journal, 1998. **75**(2): pp. 662–671.

60. Marszalek, P.E., et al., *Mechanical unfolding intermediates in titin modules.* Nature, 1999. **402**(6757): pp. 100–103.

61. Williams, P.M. and E. Evans, eds. *Dynamic Force Spectroscopy: II. Multiple Bonds.* Les Houches Ecole de Physique LVVX. Physics of Bio-molecules and Cells, ed. H. Flyvbjerg. 2002, Springer.

62. Zinober, R.C., et al., *Mechanically unfolding proteins: The effect of unfolding history and the supramolecular scaffold.* Protein Science, 2002. **11**(12): pp. 2759–2765.

63. Best, R.B., et al., *Can non-mechanical proteins withstand force? Stretching barnase by atomic force microscopy and molecular dynamics simulation.* Biophysical Journal, 2001. **81**(4): pp. 2344–2356.

64. Toofanny, R.D., P.M. Williams, and R. Elber, *Long time-scale simulations of protein unfolding under force using the stochastic difference equation in length algorithm.* Biophysical Journal, 2005. **88**(1): pp. 185A-185A.

65. Zhang, X.H., D.F. Bogorin, and V.T. Moy, *Molecular basis of the dynamic strength of the sialyl Lewis X-selectin interaction.* Chemphyschem, 2004. **5**(2): pp. 175–182.

66. Zhang, X.H., et al., *Molecular basis for the dynamic strength of the integrin alpha(4)beta(1)/VCAM-1 interaction.* Biophysical Journal, 2004. **87**(5): pp. 3470–3478.

67. Bayas, M.V., et al., *Lifetime measurements reveal kinetic differences between homophilic cadherin bonds.* Biophysical Journal, 2006. **90**(4): pp. 1385–1395.

68. Toofanny, R.D. and P.M. Williams, *Simulations of multi-directional forced unfolding of titin I27.* Journal of Molecular Graphics & Modelling, 2006. **24**(5): pp. 396–403.

6

Simulation in Force Spectroscopy

David L. Patrick

1 Introduction

Simulation has played an important role in the study of molecular-scale forces since the 1970s, almost from the time such forces could first be measured experimentally. Over the past three decades, as experimental probes have grown in sophistication and sensitivity, the scope and accuracy of computer modeling have developed in step. Today simulation and experiment are so closely linked in the field of force spectroscopy that it seems hardly possible to consider one without the other.

Although the influence of simulation is certainly growing, with some notable exceptions the frontier has to this point been driven largely by experiment. In comparison to pencil-and-paper theory, which has been extremely successful in helping to define unifying concepts and analytical frameworks, the principle contribution of computation has been to fill in missing details, and to clarify and help explain experimental findings. This role is beginning to change, however, as steady advances in methodology and computer hardware expand the versatility and reliability of modeling approaches, and improving software makes it increasingly feasible for scientists without backgrounds in simulation to adopt these techniques in their own research.[1] Modeling is becoming both increasingly predictive, and increasingly accessible.

Another emerging role for simulation in the field of force spectroscopy is to investigate questions experiments cannot address, such as the study of phenomena occurring on very short time and length scales, atomic-scale structure or energy flow, and the distribution of forces among simultaneously interacting contacts. A simple example occurs in the study of force-separation curves. In an experiment, mechanical instability in the transducer may prevent portions of the force spectrum from being measured, or limit sensitivity by requiring a stiffer spring constant. In a simulation however, where forces can be applied in any arbitrary way, or parts held fixed at arbitrary positions, one can study scenarios which may not be achievable by experiment, even in principle. As we will see throughout, this powerful advantage has been used in a number of studies to shed light on phenomena that might have otherwise gone unrecognized, or worse, been misinterpreted.

This chapter aims to provide an introduction to the role of simulation in force spectroscopy. I survey the most important computational techniques, including a discussion of their strengths and weaknesses. Particular attention is given to the difference in timescales probed by experiments and simulation, and the consequences this has for comparing results from the two approaches. The chapter concludes with a discussion of some selected case studies, which while far from comprehensive, is intended to give some indication of the breadth of problems to which simulation can be fruitfully applied, while illustrating the range of strategies that can be employed.

2 Simulation Methods

The theoretician interested in studying force spectroscopy has at his disposal a wide selection of methods. The optimum choice is influenced primarily by the time, length, and energy scales characterizing the problem; the importance of chemical reactivity; and whether the problem involves forces varying in time, in space, or both. In force spectroscopy, the key interactions usually occur at molecular scales, chemical reactions (i.e., covalent bond breaking and formation) are usually avoided, and the forces of interest vary in both space and time. Due to these considerations, the two most widely-used approaches are molecular dynamics (MD) and Monte Carlo (MC) simulation, with forces computed using empirical interaction potentials.

MD simulations model the motion of interacting particles through time, beginning from a set of initial coordinates and velocities, and treating the natural evolution of the system due to internal and external forces. The sequence of configurations generated in this way is intended to mimic what one would observe viewing the system through an idealized microscope, with the added benefit that the particle positions, velocities, forces, and energies are known at every moment. MD adjusts the particles' positions in a series of discrete timesteps, each about 0.1 τ in duration, where τ is a timescale characterizing the fastest natural motion of the system. In an atomistic model τ is the period of a vibrating chemical bond, about 10 femtoseconds. Temperature and/or pressure may also be controlled during the simulation through thermostat and barostat algorithms that occasionally rescale the atomic velocities and positions.

MD simulation is the most common method for studying the temporal evolution of nanoscale systems and can reach time scales of up to several hundred nanoseconds for systems with a few thousand particles using classical equations of motion (i.e., Newton's Laws) and empirical (non-quantum mechanical) interaction potentials. Particularly for the novice modeler, MD simulation is an attractive choice for getting started relatively easily. Reference 2 lists several comprehensive reviews of MD simulation.

MC is the second most important modeling technique used to study force spectroscopy. Like MD simulation, MC generates a series of configurations for interacting particles, and both methods produce ensembles with physically realistic samplings of phase space, i.e., with Boltzmann-weighted distributions of energies. In their conventional implementations, they are also equally computationally efficient for the determination of many equilibrium properties. However, unlike MD, which deterministically generates a trajectory of states through time, MC produces trajectories according to a stochastic algorithm, with no simple temporal connection between successive states. MC algorithms also differ from MD in that they are based entirely on a calculation of the system's energy; dynamical quantities such as velocity are not meaningful. This has advantages in certain situations, such as when it is difficult to determine the force law from the intermolecular potential (for example, with hard sphere potentials), or when particle mass is irrelevant to the problem (for example, in simulations of the Ising Model). Detailed introductions to the MC method may be found in Reference 3.

MD and MC have been in development for about 50 years,[4] and today come in many different specialized forms optimized for studying particular kinds of systems or for measuring particular observables. In addition to more or less routine structural, energetic, and kinetic properties, they can also be used to model thermodynamics, phase transitions and the statistics of infrequent events, such as barrier crossings.

In addition to MD and MC simulation, energy minimization has also sometimes been used in modeling force spectroscopy. Normally it is the potential energy that is minimized, equivalent to computing a system's nominal ground state at zero temperature, where the entropic contribution vanishes. This is not the same as an equilibrium MD or MC simulation, or an experimental measurement performed under equilibrium conditions, where it is the free energy that is minimized. However, because energy minimization calculations are less time consuming than equilibrium simulations, they can be useful for modeling certain types of rigid or structurally constrained systems, where the entropic contribution to the force is either small or constant.

3 Forces and Energies

Definition of the forces (or energies) between interacting atoms and molecules is at the heart of every simulation. Their description, more than any other single factor, determines the accuracy of the results. Ideally one would use quantum mechanics to compute the forces. However, force spectroscopy simulations tend to involve many hundreds or thousands of particles, which is too many to treat quantum mechanically. Empirical potentials are therefore usually used instead, with quantum mechanical effects included implicitly in the parameterization. These potentials compromise accuracy for computational efficiency and speed; whereas the cost of quantum mechanical calculations scales at least as N^3, where N is the number of atoms, for empirical potentials the burden is usually less than N^2. Highly refined potentials exist for nearly all classes of materials, and for treating interactions between dissimilar materials.[5] Some widely used examples are listed in Table 1.

The empirical potentials used most often to model force spectroscopy are of two different kinds. For treating molecules, spring-and-ball type molecular mechanics (MM) force fields offer a good balance between accuracy and speed of computation. MM force fields are developed by fitting a collection of adjustable parameters, such as spring constants, bond lengths, and atomic charges, to a dataset created from experimental measurements or high level quantum mechanical calculations. A simple MM forcefield is shown in Figure 1. The total potential energy is computed as a sum over all interaction and intramolecular strain terms.

MM performs most reliably when applied to model systems under conditions similar to those used to create the parameterization dataset, and therefore one finds a multitude of choices available in the literature for a single material, each tailored to best describe a certain property or set of properties.[15, 16]

A second class of widely-used potential functions is designed to allow alterations in the chemical bonding of atoms during the course of a simulation.[17] The embedded atom method for metals,[9] and Tersoff potential[11] for covalent materials are two examples. Unlike

Table 1. Some empirical energy potentials.

Bio/Organic compounds	AMBER[6], CHARMM[7], CVFF[8]
Metals and alloys	The embedded atom model[9]
Covalent materials (e.g., Si)	Stillinger-Weber potential[10], Tersoff potential[11]
Simple surfaces	The Steele potential[12]
Water	TIP4P[13], SPC[14]

$$U = \sum_{bonds} U_{bond} + \sum_{angles} U_{angle} + \sum_{dihedrals} U_{torsion} + \sum_{atom\_pairs} U_{LJ} + \sum_{atom\_pairs} U_{ee} + \sum_{atoms} U_{wall}$$

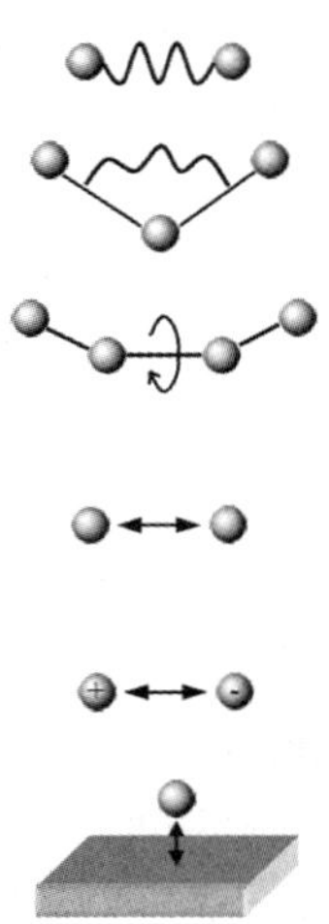

Bond stretching

$$U_{bond} = \frac{1}{2} k_b (r - r_0)^2$$

Angle bending

$$U_{angle} = \frac{1}{2} k_a (\theta - \theta_0)^2$$

Bond torsion

$$U_{torsion} = \sum_{i=0}^{n} a_i \cos^i (\phi)$$

Van der Waals interactions

$$U_{LJ} = 4\varepsilon \left[\left(\frac{\sigma}{r} \right)^{12} - \left(\frac{\sigma}{r} \right)^{6} \right]$$

Electrostatic interactions

$$U_{ee} = \frac{q_i q_j}{4\varepsilon_o \varepsilon_m r}$$

Smooth wall potential

$$U_{wall} = \frac{C_{12}}{(z - z_0)^{12}} - \frac{C_3}{(z - z_0)^3}$$

Figure 1. Molecular mechanics. A typical MM force field has energy terms describing intramolecular distortions and intermolecular interactions, and terms to account for influences exerted by the outside world. The overall energy U is recomputed at every step in a simulation using the atomic postions, bond lengths and angles, etc., and summed over all terms to give the total energy. Modern MM force fields are more sophisticated than this simple example, with improved accuracy achieved through the use of anharmonic potentials, 3-body interactions, and by including non-additive cross coupling terms.

spring-and-ball models, where bonds are defined at the outset, this second class of models computes the potential energy in terms of both the local geometry and the electron density contributed by surrounding atoms, updating the energy as the arrangement changes with each dynamical step. To illustrate, Equation 1 shows the form of the so-called "glue model".[18] The Hamiltonian of a collection of atoms is computed as the sum of two terms:

$$E_{tot} = \frac{1}{2} \sum_{ij} V(r_{ij}) + \sum_{i} F(\bar{\rho}_i) \tag{1}$$

where $V(r_{ij})$ is a pair potential which depends on the separation between atoms i and j, and reflects the shape and chemical bond-forming properties of their valence orbitals, and $F(\bar{\rho}_i)$ is the "embedding energy" which depends on the local electron density contributed by all other atoms. Although based on a quantum mechanical picture of electron sharing and covalent bonding, both terms are empirical functions, containing fitted parameters. Multi-body potentials like the glue model are especially useful for modeling phenomena such as atomic flow and fracture in solids, alloy formation, and the atomic structure of clusters and surfaces.

    Empirical force fields often do quite well predicting small-scale structural features such as bond angles and lengths, vacancy energies in solids, and in comparing the relative energies of compounds with similar bonding arrangements, such as conformational isomers. They do less well in treating highly strained systems, and situations where the chemical bonding is unusual in some way. State-of-the-art force fields for biomolecules are also growing increasingly accurate in their predictions of biomolecular secondary and tertiary structure, especially when augmented by a rule-based algorithm.[19]

In modeling force microscopy, we are often interested in determining quasi-macroscopic attributes, such as elastic constants and interfacial energies, which arise from the collective behavior of many interacting particles. These can be more difficult for force field developers to include in their parameterizations, so care must be taken to ensure that properties of interest are faithfully reproduced. As a rule of thumb, agreement between modeling and experiment for collective properties is best for simple, single-component systems, and grows less reliable as complexity increases. Thus although empirical force fields are available that reproduce mechanical properties like bulk moduli to within a few percent for materials such as pure metals, silicon, and diamond, for relatively complex systems such as proteins, supramolecular complexes, carbon nanotubes, or self-assembled monolayers, simulation and experiment can easily differ by 25% to 100%.[20]

In some circumstances, quantum mechanical effects are not sufficiently well accounted for by their implicit incorporation into classical force fields. Covalent bond breaking or formation, highly strained systems, and charge transfer reactions are typical cases. Although significant progress has been made in extending the scope of quantum mechanical calculations to include ever larger systems, most especially through Carr and Parrinello's integrated approach combining density functional theory and MD,[21] ab initio MC and MD methods still remain limited to the treatment of at most a few hundred atoms. So-called hybrid methods, which combine classical and quantum mechanical treatments by dividing the system into regions modeled with different levels of theory, are extending the reach of quantum mechanics to systems containing many more particles,[22] but so far they have received limited application in modeling force spectroscopy.

More generally, in recent years there has been a growing trend toward treating complex, large-scale systems through hierarchical schemes in which the parameterization of increasingly coarse levels of description is derived from lower level, more detailed models.[23] For example, the mesoscopic length-scale and multi-second time-scale dynamics of lipid membranes can be modeled based on a spring-and-ball network, which is parameterized from atomistic molecular mechanics simulations on nanometer-sized domains over nanosecond timescales, using force fields based on single-molecule ab initio calculations. In the future one can envision the ultimate seamless integration of the hierarchical approach, which may someday allow the nonexpert to reliably study systems of arbitrary complexity without having to be concerned with the underlying technical details.

4 The Simulation Timescale

In force spectroscopy experiments the interacting components are moved several orders of magnitude more slowly than it is feasible to simulate using an atomistic model (see Figure 2). The most common force spectroscopy techniques, including the surface forces apparatus,[24] atomic force microscope (AFM),[25] optical tweezers,[26] and biomembrane force probe,[27] measure forces over seconds to microseconds, whereas simulations rarely last longer than a few tens of nanoseconds. The simulation timescale is therefore much shorter—and the speeds at which interacting components are moved and forces changed are far larger—than the experiments one would like to compare them to.

This timescale difference has several important consequences for the interpretation of simulation data and for the way simulations relate to experiments. One of the most serious is to bias the kinetic pathways followed in simulations toward fast processes. Phenomena that

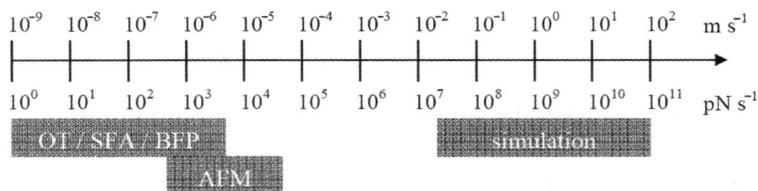

Figure 2. MD simulations model probing velocities many orders of magnitude faster than experiments. Optical tweezers (OT), surface forces apparatus (SFA), bomembrance force probe (BFP), atomic force microscopy (AFM).

occur too slowly, like large-scale conformational changes in a protein, or grain boundary diffusion in a solid, may have insufficient time to proceed in a rapid simulation. The system is forced to find an alternative route through momentum and conformational space, sometimes at the expense of avoiding pathways that would have a lower free energy on longer timescales.[28, 29] The qualitative picture of atomic motion one observes in a relatively short simulation may therefore not coincide with the trajectory in a relatively long experiment, even if the model used to compute forces and energies is exact.

A second consequence of short simulation times arises because the formation and rupture of molecular contacts involves transitioning over a sequence of energy barriers, which implies a certain timescale for thermally-activated crossing. At times longer than the thermal timescale, these transitions occur spontaneously and the bonds have no strength, but on shorter timescales an external probing force is needed to increase the transition rate.[30] This probing force deforms the free energy landscape, changing the both the energetics and kinetics of the system, and biasing motion along a single probing coordinate. In most cases, applying a force shortens bond lifetime by reducing barrier heights along the relaxation pathway, but the opposite can also be true if "jamming" occurs or if molecules deform so that they bind more tightly.[31]

Despite these differences, comparison of simulations to experiments remains an important test of the former, and can provide useful insights into the latter. With these caveats in mind, we therefore summarize some of the strategies that have been taken to attempt to scale the forces measured at simulation timescales to those at experimental timescales.

Perhaps the most obvious approach is to perform a series of simulations over increasingly longer times, then attempt to extrapolate the trend to the experimental regime. Simple models using nonequilibrium statistical mechanics to describe motion across a free energy landscape with barriers lead to the prediction that the maximal force required to overcome the highest barrier (the "unbinding force", which is often what is measured in an experiment) should usually scale as $F \sim \ln(v)$, where v is the velocity at which the interacting components are brought together or separated. This relationship has indeed been observed in many (but certainly not all) simulations and experiments, and potentially applies to a range of circumstances, including cases involving reversible and irreversible bonds, soft and stiff transducers, and whether one bond or many bonds are being probed.[32] The logarithmic scaling applies only in a certain velocity window, however; the timescale must be shorter than the natural (thermal) barrier crossing rate, but not so large that the free energy landscape is distorted to the point that the barriers are masked completely.

Grubmüller, Evans, and others have considered the effects of velocity over timescales relevant to MD simulation, identifying two qualitatively different regimes that appear to have the strongest contribution: a *friction regime* for fast pulling velocities, and an *activated regime*, where pulling is slow enough for thermal barrier crossing to become important.[33, 34] At MD timescales and longer, a good approximation for the overall force is then: $F(v) = F_{friction} + F_{act}$.

In the friction regime, the force required to move the system across its free energy landscape is approximately linear in the velocity $F_{friction} \approx \gamma v$, where γ is the Stoke's friction coefficient. In the activated regime, transition state theory gives $F_{act} \approx \dfrac{k_B T}{L} \ln\left(\dfrac{v}{k_o \Delta L}\right)$, where k_B is Boltzmann's constant, T is temperature, L is the "unbinding length", or width of the barrier separating metastable states along the probing coordinate, ΔL is the spread in L, reflecting alternative transition pathways, and k_o is the spontaneous barrier crossing rate, which is the rate that would be observed in the absence of an external probing force. Combining these terms gives

$$F(v) \approx \gamma v + \frac{k_B T}{L} \ln\left(\frac{v}{k_o \Delta L}\right) \qquad (2)$$

which allows extrapolation of the forces measured at short timescales to longer timescales, so long as some of the simulations take place in both the friction and thermal regimes. With the exception of the spontaneous barrier crossing rate k_o, which must be determined experimentally, all other terms in this expression can be estimated by performing a set of simulations over a range of loading velocities.

Grubmüller and co-workers used this formalism in a study of the unbinding of the AN02–DNP-hapten antibody complex, shown in Figure 3, to extrapolate MD forces to those which would be measured on experimental time scales.[34, 35] The small hapten ligand was pulled from the antibody binding pocket using a harmonic spring whose tethering point was moved at a range of different speeds, from 10–$0.1\,\text{ms}^{-1}$. The individual force profiles all shared a similar global shape, with a large barrier early in the unbinding pathway that determines the unbinding force, and a number of smaller, somewhat more variable barriers farther along the pulling coordinate. Plotting the maximal force against pulling velocity, the authors fit Eq. 2 to determine L, ΔL, and the friction coefficient γ, using a spontaneous dissociation constant k_o estimated from experimental data on two structurally similar complexes. The extrapolation expected for pulling velocities typical of an experiment ($\sim 10^{-6}\,\text{ms}^{-1}$) is shown with a diamond symbol.

In this case the authors argued it was possible to extrapolate from MD to experimental timescales because, although most simulations occurred in the friction regime (as evidenced by the linear relationship between F and v for much of the data, shown in the inset) the slower pulling velocities extended into the activated regime as well. It should be recognized, however, that in most cases extrapolation of MD forces to experimental timescales is hazardous. Besides the problems already mentioned having to do with biased kinetics, many simulations take place at velocities where the force is dominated by friction effects, masking the individual activation barrier(s) probed at the slower experimental timescales.[36, 37] A detailed understanding of the energy landscape along the coordinate is required to ensure a proper interpretation of the force, since the existence of multiple barriers or multiple transition pathways can greatly complicate the picture.[38]

To further illustrate this point, consider the stretching and unfolding of the modular protein polyubiquitin, which was both simulated and measured experimentally by Fernandez and co-workers.[39] Polyubiquitin is a chain of ubiquitin proteins found as several structural isomers, linked either through their N-C termini, or between their C terminus and one of four different residues occurring in various places in the monomer. The type of linkage affects both the biological function and mechanical properties of the polymer. Fernandez and co-workers performed MD stretching simulations of polyubiquitin with linkages between

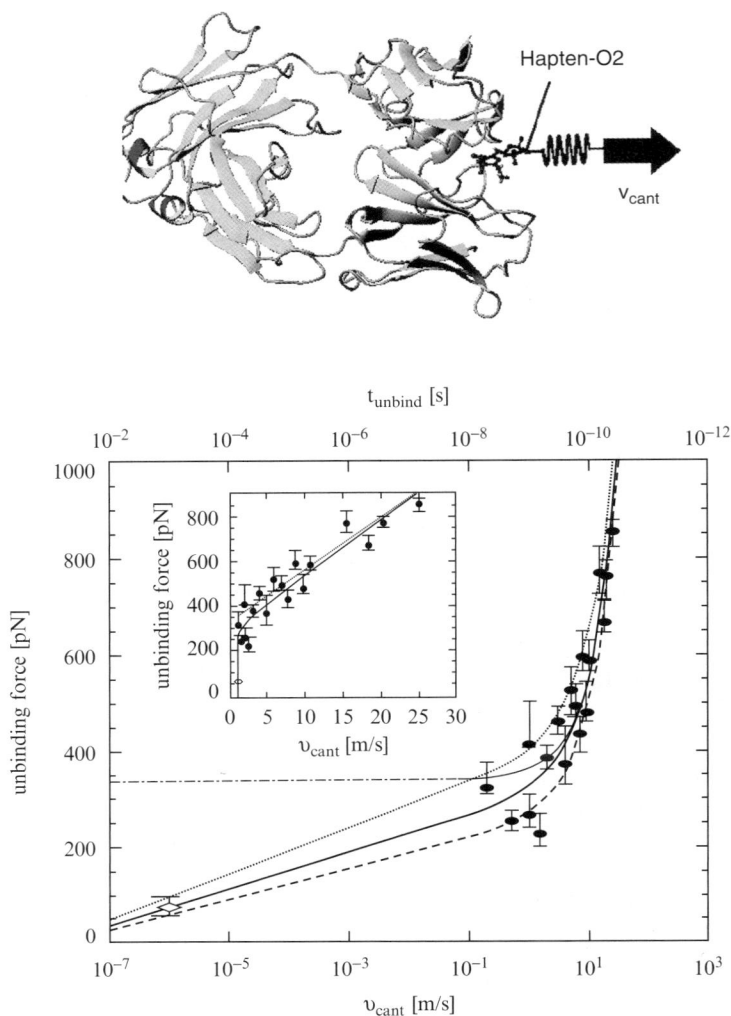

Figure 3. Grubmüller and co-workers performed MD simulations of the unbinding of a DNP-hapten ligand from the monoclonal antibody AN02 F_{ab} at different pulling velocities to extrapolate the unbinding force to experimental timescales. (top) A model of the proteinligand complex. (bottom) Computed unbinding forces (filled circles) and predicted experimental unbinding force (diamond) as a function of pulling velocity. The lines show the best fit on Eq. 2 to the computed unbinding forces for two different values of the spontaneous rupture rate k_o. The inset shows the same data on a linear velocity scale, indicating that most of the simulaations occurred in the friction regime. Reprinted with permission from Reference 34. Copyright 1999 Elsevier. (*See Color Plates*)

different domains and compared them to AFM measurements. Because the stretching velocity used in the simulations was so much larger than the experimental measurements ($100\,\text{ms}^{-1}$ vs. $\sim 10^{-6}\,\text{ms}^{-1}$), no direct comparison of the forces was possible; the simulated rupture force was about 20 times larger than that observed in the experiments. Thus although such simulations produce a wealth of information about protein structure, dynamics, and unfolding pathways at high strain rates, at a detailed level the experimental and simulated force curves may bear only qualitative resemblance.

　　We have been discussing force spectroscopy simulations conducted in a manner similar to most experiments—with the load applied using a soft transducer whose position is smoothly varied in time. This method is now sometimes referred to as "steered MD", because an external force "steers" the dynamics along a selected pathway. In a simulation, however, one has much greater freedom to choose how the load is varied, including the use of mechanisms that would be difficult to realize in an experiment. For example, the probing force can be held constant or indeed varied in any arbitrary way, providing complementary information to the soft transducer approach by assisting in delineating barriers along the probing coordinate.[40]

One widely used method is to increment the probing force (or alter the positions of interacting components) in discrete steps, then while holding it fixed, simulate for a period until equilibrium has been achieved.[†] An example from the author's own work is presented in Figure 4.[78] The force curve shows the interaction between a chemical force microscope stylus and a smooth wall. Unlike the previous examples, where the load was smoothly increased, in this case the force was computed at discrete positions by taking points from a simulation of continuous motion, fixing the tip height, then simulating for several nanoseconds until steady-state conditions were obtained. Except for the penultimate point near the height where contact formation/rupture occured, the force curve showed no hysteresis, an indication that the interaction was reversible.

Although obviously different from the quasi-continuous motion in an actual experiment, this approach has the advantage of allowing portions of the force spectrum to be studied

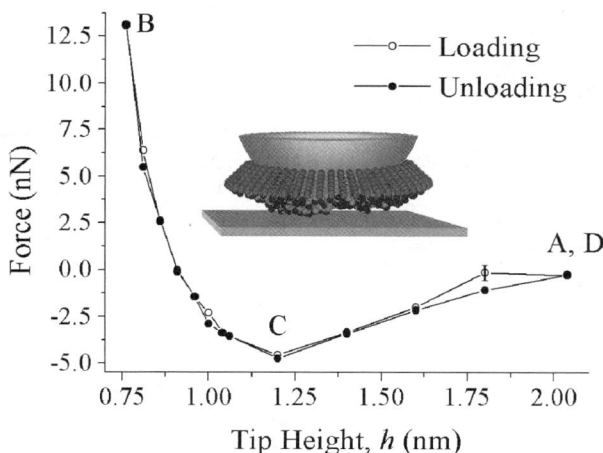

Figure 4. In a simulation one has the luxury of being able to move the interacting components in ways that would be difficult to realize in an experiment. This example presents results from a series of MD simulations modeling the interaction between a chemical force microscope stylus and a smooth wall during both loading and unloading. The figure illustrates a widely-used strategy for approximating infinitely slow motion in simulations of force spectroscopy—to hold the interacting components rigidly fixed at discrete separations and simulate until steady-state conditions have been achieved. Reprinted with permission from Reference 78. Copyright 2003 American Chemical Society. (*See Color Plates*)

[†] The problem of achieving equilibrium in a simulation (and of recognizing when you have) is not an easy one. This is because for the complex models typical of force spectroscopy, there is no single, definitive test that a simulated system has achieved equilibrium. Instead, one gathers evidence for (or against) equilibrium by examining time-dependent fluctuations, and when these have reached what appears to be a steady state, an equilibrium condition is assumed. Equilibrium simulations are especially challenging when a system has deep local minima on its free energy landscape, as is often the case in the measurement of force-distance curves, for example. Of course this issue affects experimental measurements as well, but it serves to highlight the importance of taking a cautious approach to the interpretation of results from kinetic modeling.

under near equilibrium conditions, and is the method that comes closest to approximating the limiting case of infinitely slow motion.

Force spectroscopy experiments often involve systems containing components whose characteristic motions occur over a wide range of time scales, which presents a different challenge to modeling. In conventional MD simulation, the timestep must be small enough to smoothly describe the highest frequency motion, which can make modeling of slower phenomena—like the response of a relatively massive AFM cantilever—prohibitively expensive.

One approach to this problem is to employ a temporally hybrid method in which the equations of motion for different components are integrated using different size timesteps.[41] For instance, the characteristic timescales associated with intra- and intermolecular motions differ by an order of magnitude or so, allowing one to use different sized timesteps to integrate vibrational and molecular center-of-mass motion, in analogy to the Born-Oppenheimer approximation for separating nuclear and electronic motion.

Jiang and co-workers have used this scheme in a series of studies examining adhesive and frictional forces between an AFM stylus and surfaces covered by self-assembled monolayers.[42] In the case shown in Figure 5, the stylus was modeled attached to a harmonic spring whose tethering point was gradually moved to simulate lateral or vertical scanning.[43] The mass of the stylus and the cantilever spring constant were chosen to be representative of an actual experiment, which led to a free resonance frequency of $\sim 10^5$ s^{-1}. The molecular components on the other hand were described pseudo-atomistically, and their highest frequency motion was $\sim 10^{14}$ s^{-1}.

Because the characteristic frequencies of motion of the stylus and molecules were so different, their vibrational energy modes were weakly coupled, and hence their short timescale motions could be approximated as occurring independently. To capture both of them, the authors performed simulations in a series of alternating stages with the equations of motion integrated using different size timesteps. In one stage the molecules were held frozen as the

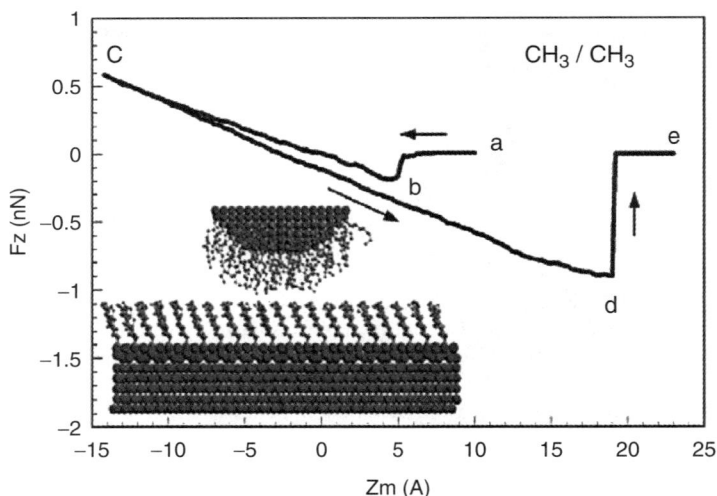

Figure 5. Another challenge related to the short timescales of MD simulation concerns how to model systems with both slow and fast motions. A CFM stylus interacting with a molecular film is one such case. The figure illustrates one solution to this challenge, based on a hybrid simulation technique. The force curve includes mesoscale characteristics, such as snap-to and -from contact (points b and d), as well as chafacteristics from molecular-scale interactions, such as hysteresis. Reprinted with permission from Reference 43. Copyright 2002 American Chemical Society.

tip was equilibrated using a timestep of $0.25\,\mu s$, while in the next stage the film was equilibrated with a timestep of 3×10^{-15} s while the stylus was held fixed. This allowed both the atomic-scale motion of the film and the nanometer-scale motion of the stylus to occur in a single simulation, even though the timescales governing the two types of motion differed by about nine orders of magnitude. The resulting force curve displays the main characteristics of an experimental one, including jump to/from contact.

In concluding this section, we note that, while these and other approaches take us some distance toward simulating the timescales covered in an experiment, each involves certain approximations that complicate direct comparison of the two. One naturally wonders therefore if the timescale problem in MD simulation will eventually be solved as computers continue to increase in speed and in the number of parallel processors. An argument that it will not was made by Chan and Dill in 1993.[44] They showed that even for a very large parallel computer, where the task of calculating each particle's trajectory is assigned to a separate processor, the overall rate of computation will remain limited by the speed at which processors can communicate. Using a reasonable estimate for this "communication bottleneck", a future parallel computer with an unlimited number of infinitely fast processors working for a full year has been estimated to have a limiting capacity of simulating less than one second of real time.[45]

A more fruitful approach is therefore likely to come from focusing on alternative algorithms. As is widely recognized, often the problem presented by force spectroscopy is not that the dynamics are slow per se, but rather that the dynamics involve passing through states that occur infrequently. Long simulation times are required in conventional algorithms because states are generated according to their Boltzmann weighted probabilities, and the highly energetic barrier crossing states so important in force spectroscopy occur rarely. Most simulation time is therefore spent exploring regions of phase space that are not directly pertinent to the problem. Alternative sampling schemes, using different algorithms to preferentially locate and probe such states may eventually provide the solution.[46]

5   Case Studies

With this short introduction to the most prevalent modeling techniques, we turn now to a discussion of several case studies. Our coverage is far from comprehensive, but hopefully provides some flavor of both the diversity of problems modeling can help to address, as well as the broad range of strategies that can be employed.

5.1   Biological Forces

In recent years simulation has had the largest and most visible impact in the study of biological forces. Motivated both by a better understanding of fundamental biophysics, and because stretching of some biopolymers is believed to regulate their physiological function *in vivo*,[47] recent experiments show that when biomolecules are subjected to a mechanical load, they rarely behave like simple entropic springs. Instead, what is often observed is that at critical loadings, they undergo abrupt structural transitions as subunits unfold, noncovalent interactions are disturbed, or a ligand slips from its binding pocket, producing rich and complex force spectra.[48] Even within a single family of biopolymers, a wide range of elastic behaviors can be observed.[49, 50]

In the mid 1990s, molecular modeling began to be used to investigate these forces and structural transitions.[51, 52, 53, 54] Grubmüller and co-workers reported the first detailed simulations of ligand-receptor forces in a landmark paper modeling the unbinding of biotin and streptavidin in 1997,[36] work inspired by Gaub and co-workers' experimental study of the same system a few years earlier.[55] Streptavidin is a tetrameric complex, able to accommodate up to four biotin ligands with high specificity and one of the strongest binding constants known to biology. Although the crystal structure of the protein complex was known, and the thermodynamics of binding well characterized by solution-phase measurements, as with many other ligand-receptor interactions, a detailed, atomic-scale explanation for the high specificity and large binding strength was largely lacking.

Grubmüller's group used MD simulation to treat one monomeric streptavidin unit, tethered to a fixed point in space, interacting with a biotin molecule attached to a flexible spring. The model is shown in Figure 6. As the spring was gradually stretched, the ligand withdrew

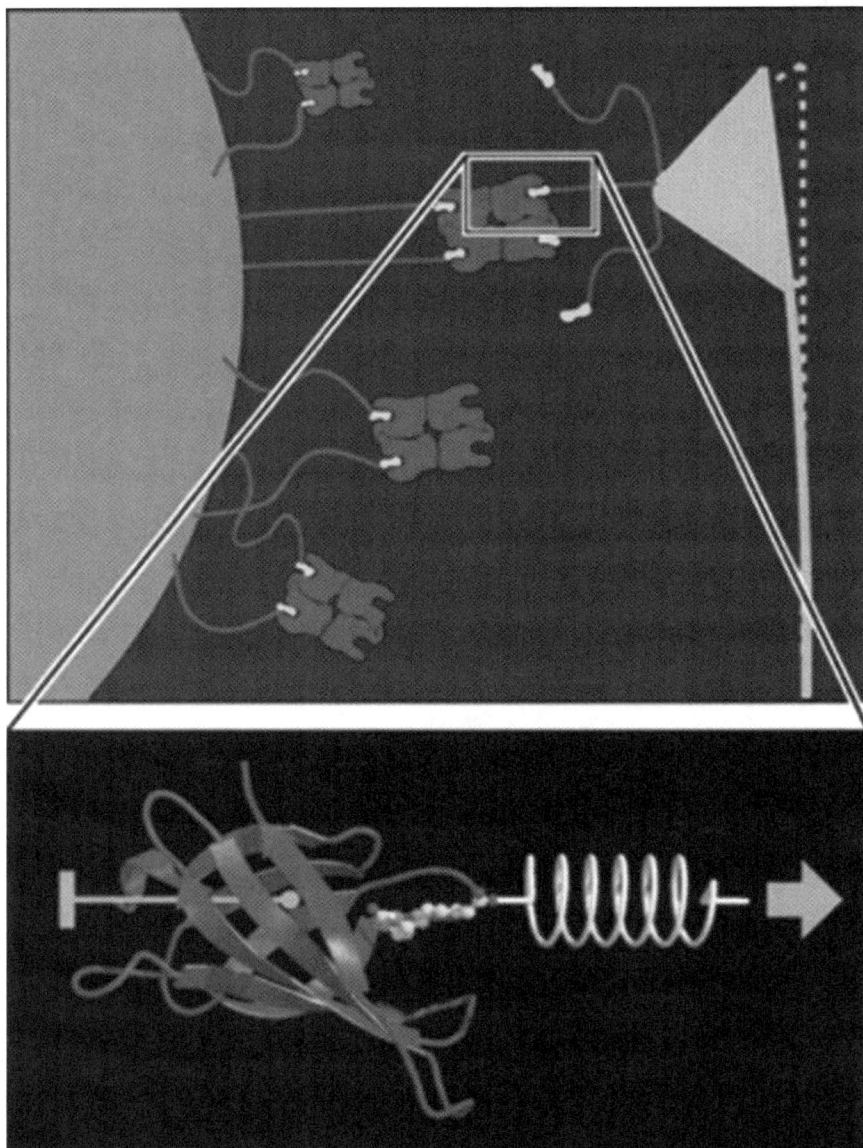

Figure 6. Simulations modeling the pulling force of an AFM tip to induce unbinding of a biotin ligand from a streptavidin receptor. (top) a cartoon of the experimental measurement. (bottom) The system modeled by simulation. Reprinted with permission from Reference [36]. Copyright 1996 AAAS. (*See Color Plates*)

from its binding pocket, and a series of snapshots showing atomic positions, interaction energies and forces, and the motion of interacting residues was recorded. These revealed that the biotin molecule is held within the binding pocket by a web of interactions, principally involving van der Waals contacts and hydrogen bonds. When the external load provided by the spring exceeded the strength of these bonds, the molecule began to move. Interestingly, rupture did not occur all at once; rather, the ligand detached from its binding pocket in a stepwise manner involving the sequential formation and disruption of a series of interactions along the unbinding pathway. A detailed analysis of these interactions showed which residues were involved, and at what stages of rupture. They also provided an explanation for the experimental observation that the effective length of the ligand-receptor bond was about 0.5 nm, much larger than a single hydrogen bond.[36, 55]

The Grubmüller study was significant both for the detailed insights it provided into the biotin-streptavidin system, and for the potent demonstration it gave of the value of modeling as a tool to complement force spectroscopy experiments. Numerous other ligand-receptor interactions have since been modeled, usually using MD simulation. These range from binding to HIV-1 reverse transcriptase,[56] avidin,[57] and retinoic acid receptor,[58] to studies pulling single proteins[59] and phospholipid molecules[60] from bilayer membranes.

5.2 Interfacial Forces

The study of interfacial forces has been the subject of theoretical investigation for over a hundred years, beginning with the pioneering work of Hertz on contacting solids in the late nineteenth century, and continuing almost without interruption to the present. Many interesting and challenging questions arise at the molecular scale, where modeling has contributed some seminal insights. In this section we review two representative examples: liquid forces, and contact mechanics.

Some of the earliest simulations of force spectroscopy concerned the study of forces in liquids. Solvation forces have been investigated using computer simulation from the mid 1970s,[61] since around the same time they were first observed using the surface forces apparatus.[62] More recently, AFM experiments probing the chemical and spatial properties of fluid forces have been performed with the tip immersed in a liquid.[63] While the mesoscopic forces between objects in liquids are generally well explained in terms of classical continuum theories,[64] atomistic simulation has proven extremely useful in explaining the role of the liquid on a molecular scale.

We take as an example the work of Lynden-Bell and co-workers, who were the first to model solvation forces relevant to the geometry of AFM, i.e., between a sharp stylus and flat surface.[65, 66] In their early studies, the tip was treated as a large sphere, immersed in a Lennard-Jones fluid of smaller spheres adjacent to a smooth surface. Later work investigated more realistic, all-atom models.[67] MD simulations were used to study fluid structure, dynamics, and interaction forces, with the step-wise method of tip motion described above used to approximate thermodynamic equilibrium.

As had been found in prior simulations of fluids confined between flat surfaces,[61, 68] the AFM geometry also produced a series of oscillations in the force-separation curve whose amplitude decreased with distance. The main differences were that the force oscillations decayed faster, and that additional fluid structure developed around the perimeter at the bottom of the stylus, in the crevice between the stylus and the surface. The authors also compared their results to predictions from a formulation of the Ornstein-Zernike relations for an isotropic ternary colloidal mixture, a continuum model for describing fluid structure in simple liquids.[69] As shown in Figure 7, except at very small tip heights (less than about two fluid particle diameters), the continuum model and atomistic simulations agreed well. One interesting conclusion from this body of work is that the solvation force can be comparable

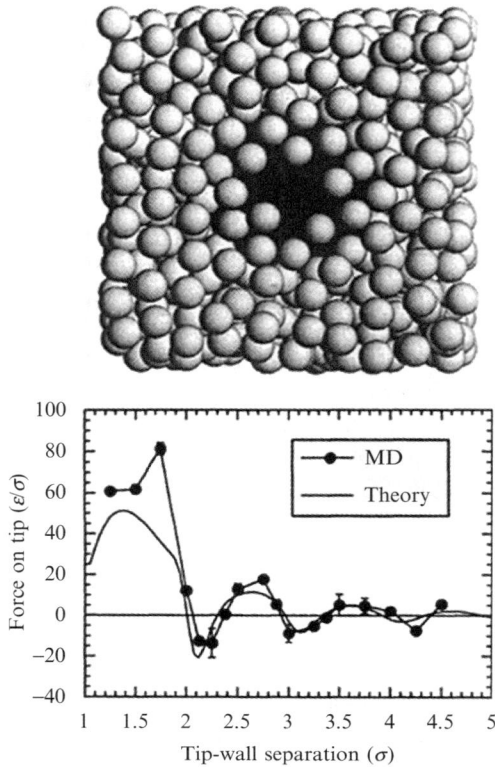

Figure 7. Comparing force-distance curves predicted by continuum theory and MD simulation. The simulation modeled a spherical tip of diameter 5σ immersed in a fluid of spherical particles of diameter σ interacting with a smooth, planar wall. Reprinted with permission from Reference 66. Copyright 1994 American Physical Society.

to the direct stylus-sample interaction measured in some noncontact force spectroscopy and imaging experiments (up to several hundred pN). The fluid forces are spread out over a relatively large contact area, effectively "cushioning" the tip-surface interaction, and are very sensitive to the detailed structure of the end of the stylus, as well as the arrangement of atoms on the sample surface. The simulations thus help shape our understanding of the way force spectroscopy and imaging measurements in liquids should be interpreted.

In addition to simple liquids, force spectroscopy simulations of many other types of interfacial materials have also been performed, including grafted molecular films,[70] polymers,[71] metals,[72] ionic crystals,[73] and organic fluids.[74] These simulations have brought to light a range of fascinating nanoscale rheological and tribological phenomena which together with experiment are rapidly revolutionizing our understanding of adhesion, friction, and lubrication.[75]

Investigation of the atomic-scale deformation of contacting solids has emerged as another rich and fruitful application of force spectroscopy. Part of the reason is foundational: the interpretation of most interfacial force measurements rests upon an understanding of the contact mechanics of the interacting components. For example, in order to extract binding energies for molecular pairings from a chemical force microscopy experiment, one must know the number of interacting molecules, and hence the contact area. The analysis of force curves to calculate elastic properties, interfacial energies, and coefficients of friction are all similarly grounded. Simulation has played an important role bridging a gap left unaddressed by experiments in this area, where many of the critical atomic-scale details are obscured because the forces are averaged over multiple interacting molecules or atoms.

To illustrate, we take as an example the calculation of the contact area between an AFM tip and a flat surface. Among experimentalists, two formalisms in particular have proven especially popular for analyzing force data of this sort. These are the Johnson-Kendall-Roberts (JKR),[76] and Derjaguin-Muller-Toporov (DMT) models.[77] Originally developed to explain meso- and macroscopic deformations of elastic bodies, both models provide closed-form expressions relating important parameters that cannot be measured (such as the contact area) to experimental observables like the pull-off force and tip radius of curvature.

Although widely used, application of these models at the nanoscale presents certain problems. For example, the JKR model is appropriate for describing contacts with high adhesion energy, large radius of curvature, and small elastic modulus, whereas the DMT model is appropriate for contacts involving weak adhesion, small radius of curvature, and large elastic modulus. It turns out, however, that most AFM experiments probe an intermediate regime, making the choice of the most appropriate model somewhat unclear. Moreover, while the validity of both models is well established at larger size scales, the applicability of continuum treatments to describe atomic-scale phenomena has some clear limitations. In fact the JKR and DMT models are widely used in the interpretation of molecular force spectroscopy experiments not because they have been proven accurate for this application, but simply because they are convenient.

To shed further light on these issues, the author and his students have used MD simulation to investigate the accuracy of these and some related models in describing the contact mechanics of a chemical force microscope stylus interacting with a flat surface.[78] The stylus was modeled atomistically and the state of the system recorded at a series of tip heights to directly measure the contact area, indentation, distribution of forces, and other features. A complete loading-unloading sequence was simulated using the step-wise method, approximating the case of infinitely slow tip motion (Figure 3). The simulations therefore provided more information about the model than can be obtained from an experiment, allowing the JKR and DMT models to be directly tested.

While the JKR model provided the best overall description of the simulation data, it is apparent from Figure 8 that the results revealed a number of discrepancies. These concern both the quality of the fit to the data, and the way in which these models are normally applied to interpret experimental results. For example, one common strategy is to independently measure the stylus radius of curvature, then use the pull-off force to calculate the energy of adhesion. The relevant equations are:

$$F_{pull-off} = -2\pi WR \ (DMT \ theory) \tag{3}$$

$$F_{pull-off} = -\frac{5}{6}\pi WR \ (JKR \ theory, \ fixed \ grips) \tag{4}$$

Here $F_{pull-off}$ is the maximum adhesive force, R is the tip radius of curvature, and W is the energy of adhesion. This was the approach taken for example in Reference 79, where the DMT theory was used to help understand measurements performed with the interfacial force microscope.

In the case of the simulation results however, it is clear that inconsistencies between the numerical data and continuum theories prevent this approach from providing a meaningful analysis. Beginning with the DMT theory, one predicts that the separation force—which is also the point of greatest attraction on the force curve—should occur just as the contact

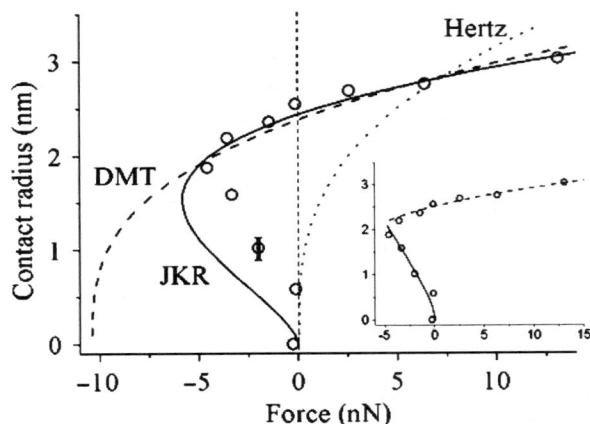

Figure 8. Simulation has been used to test models of contact mechanics, uncovering potential inconsistencies not directly observable by experiment. Comparison of simulated and fitted force—contact radius curves for a chemical force microscope stylus interacting with a flat surface reveals significant deviations. Simulation data are shown as open circles with the estimated uncertainty in the contact radius shown for one representative point. The best fits of the JKR and DMT and Hertz theories are shown. The inset (which shares units with axes in the main figure) shows the results of dividing the simulation data into two subsets, before and after the point of maximum adhesion, and fitting each subset separately in the JKR limit. The two subsets were fit using a single, fixed value for the work of adhesion while the reduced modulus was allowed to vary separately. The quality of the resulting fits demonstrates the effect of changing compliance during unloading. Reprinted with permission from Reference 78. Copyright 2003 American Chemical Society.

radius reaches zero. As can be seen in Figure 8, however, these two forces are very different: $F_{min} = -4.56\,nN \neq F_{separation} \sim 0\,nN$. In fact, the points of separation and of greatest attractive force are separated by more than $0.6\,nm$. In an actual experiment, where the contact radius would not be known, this discrepancy would go undetected. One would normally assume (perhaps incorrectly) that the separation force corresponded to the most attractive point on the force curve, and use this point for the calculation of W.

Applying the JKR theory in an analogous way to compute W presents a different difficulty. If the JKR model is assumed to hold, the point of separation should be identifiable as the position where the force (and contact radius) jumps discontinuously to zero. However, within the resolution of the discretized data, it appears that the force decreases smoothly to zero, i.e., that $F_{separation} \sim 0\,nN$. This result would only be possible within the JKR framework if the adhesion energy were infinitely large.

Some of the discrepancies revealed by the simulations would be overlooked in an actual experiment, where unlike the simulations, contact area is not separately known, possibly producing a misleading or incorrect interpretation of experimental results.

5.3 Liquid Crystal Forces

We conclude with a discussion of the recently discovered and apparently unique class of forces that occurs between particles immersed in liquid crystals (LCs).[80] Modeling these forces presents special challenges because long-range, micron-scale order in LCs arises from short-range, molecular-scale interactions. Boundary conditions must be handled carefully, and in a molecular simulation, a large number of particles is required to encompass the relevant length scales. Several different approaches have been developed to address these problems, making LCs an instructive case study illustrating strategies for designing computer models that incorporate just enough of the essential physics to make them relevant, while remaining computationally tractable.

In addition to the kinds of interactions that occur between particles in an isotropic solvent (e.g., electrostatic, van der Waals, steric forces, etc.), colloidal particles in LCs interact with one another through several mechanisms characteristic to organized media.[81] The strongest and most important of these is associated with distortions in the director field arising from the competitive effects of orientational anchoring on the particle surface, and bulk curvature elasticity, which prefers a uniform director configuration. The result is an anisotropic interaction that can be either repulsive or attractive depending on particle size, separation, shape, and orientation relative to the bulk director.[82]

Several different experimental schemes have been used to probe these forces, usually involving magnetic or electric fields.[83] Figure 9, from the work of Weitz et al., shows one example.[84] Here a magnetic field applied to a suspension of small ferrofluid droplets in a nematic LC was used to balance the attractive LC forces with repulsive magnetic forces, and in so doing to measure the strength of the potential.

Somewhat like the situation discussed above for force spectroscopy measurements in conventional, isotropic fluids, in a LC the forces between colloidal particles arise from the structure of the fluid. However, unlike an isotropic fluid, in a LC the forces are long-range, extending over a size scale up to many hundreds or thousands of times larger than a single fluid molecule. A fully atomistic description of the LC molecules would therefore involve so many particles that the computational burden would be unrealistically large. Much work has been devoted to developing models for LCs that reduce the computational

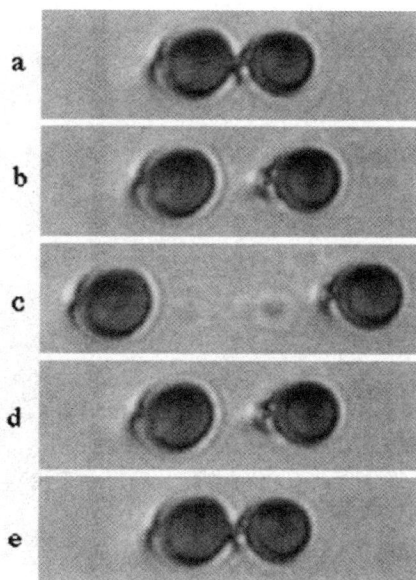

Figure 9. Colloidal particles in a LC solvent interact with one another through long-range distortions in the director field. The figure shows application of a magnetic field to induce a repulsive interaction which drives the droplets apart in (B) and (C). When the field is switched off, the elastic attractive interaction pulls the droplets together, in (D) and (E). Modeling such interactions poses special challenges. Reprinted with permission from Reference 84. Copyright 1997 American Physical Society.

cost, while still retaining enough of the essential physics to make them useful for studying real phenomena.

One approach, which might be described as "pseudo atomistic" is to model the LC as composed of anisotropic particles interacting through simplified force laws such as the Gay-Berne potential, which is a generalized version of the Lennard-Jones potential for non-bonded interactions between spherical particles:[85]

$$U_{GB} = 4\varepsilon\left(\vec{u}_i,\vec{u}_j,\vec{r}\right)\left[\left\{\frac{\sigma_s}{r-\sigma\left(\vec{u}_i,\vec{u}_j,\vec{r}\right)+\sigma_s}\right\}^{12} - \left\{\frac{\sigma_s}{r-\sigma\left(\vec{u}_i,\vec{u}_j,\vec{r}\right)+\sigma_s}\right\}^{6}\right]$$

Where $\vec{u}_i$ and $\vec{u}_i$ are unit vectors defining the orientation of particles i and j, $\vec{r}$ is the vector connecting the centers of the two particles, the orientation dependent size parameter σ is defined as:

$$\sigma\left(\vec{u}_i,\vec{u}_j,\vec{r}\right) = \sigma_s\left\{1-\frac{\chi}{2}\left[\frac{\left(\vec{u}_i\cdot\vec{r}+\vec{u}_j\cdot\vec{r}\right)^2}{1+\chi\left(\vec{u}_i\cdot\vec{u}_j\right)} + \frac{\left(\vec{u}_i\cdot\vec{r}-\vec{u}_j\cdot\vec{r}\right)^2}{1-\chi\left(\vec{u}_i\cdot\vec{u}_j\right)}\right]\right\}^{1/2}$$

χ is a shape anisotropy parameter ($\kappa = \sigma_e / \sigma_s$ = particle length / particle breadth):

$$\chi = \frac{\kappa^2-1}{\kappa^2+1}$$

and $\varepsilon(\vec{u}_j, \vec{u}_j, \vec{r})$ is an orientation-dependent energy well depth.

Figure 10. A snapshot from an MD simulation of 8000 Gay-Berne particles solvating two colloidal rods. Shading emphasizes particle orientation. MD simulations like these are used to model interaction forces in LCs. Reprinted with permission from Reference 86. Copyright 2003 American Physical Society. (*See Color Plates*)

The Gay-Berne potential is emblematic of a commonly used trick in the modeling of large systems, namely to combine together groups of atoms into larger entities, having in certain respects the average properties of their constituents, and thereby reducing the total number of particles involved in the calculation. Similar approaches are frequently encountered in the modeling of polymers, membranes, and other complex systems.

Figure 10 shows a snapshot from a simulation where this approach was used to study the interaction between two rod-shaped colloids in a nematic LC.[86] Simulation cells were employed containing up to 64,000 Gay-Berne particles, which despite lacking any atomistic detail, managed to capture much of the behavior of bulk LC fluids.

To study the interaction forces between the rods, boundary conditions at the edges of the simulation cell were established to favor a parallel molecular orientation (with the director roughly vertical in the figure), while at the surface of the colloids a radial orientation is favored. This results in a distortion of the director field, the energy of which depends on the placement and orientation of the large rods.

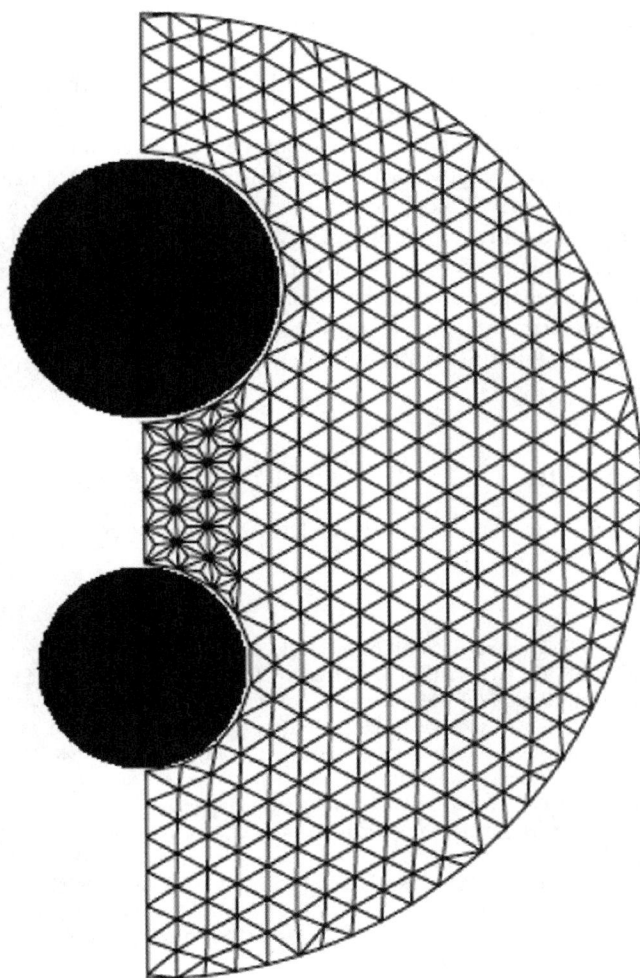

Figure 11. Lattice models on an adaptive mesh provide a different way of studying LC forces, and can be used to model much larger length scales. This figure shows the triangulation used to compute forces between two colloidal rods in a nematic LC solvent. Reprinted with permission from Reference 87.

Although this approach reduces the number of particles by a factor of 25 or more compared to an all-atom model, and also allows a larger timestep to be used in the MD integration algorithm, the total system size is still modest. In the case of Figure 10 for example, the cell measured $10 \times 50 \times 50$ molecules in size, and the colloids only three molecules in diameter. Unfortunately, much of the interesting phenomena observed in experimental systems occurs over greater distances and involves much larger particles.

The way around this limitation is to use an even more coarse-grained description of the LC. One approach dispenses with molecules altogether, replacing them by a network, or mesh, in which each cell contains information about the average molecular orientation at that point. An example from the work of Stark et al. is presented in Figure 11.[87] The overall free energy is minimized using MC simulation or an iterative minimization scheme based on nearest-neighbor interactions in a lattice spin model. To improve computational efficiency, an adaptive mesh can be used wherein space is partitioned into smaller volumes in those regions where the director changes sharply.

Conclusions

As the examples discussed in this chapter illustrate, computer simulation provides a diverse and powerful set of tools for studying molecular scale forces. The simulation timescale is normally much shorter than the experimental one, a fact which, while complicating direct comparisons between the two, enhances their complementarity in many situations. Steady improvements in methodology and hardware are extending the reach of modeling to include ever larger and more complex systems and increasing the ability to model those systems with greater levels of realism and accuracy. While for many years force spectroscopy simulations have been used primarily as an explanatory technique, filling in details left unaddressed by experiments, increasingly they are being used in a predictive way, suggesting new avenues of experimental research and contributing more broadly to development of the field.

Acknowledgements

The author thanks the Camille and Henry Dreyfus Foundation for support of this work.

References

1. See, for example, Phillip Walsh, Andrey Omeltchenko, Kalia, R. K.; Nakano, A.; Vashishta, P.; Saini, S., *Appl. Phys. Lett.* **2003**, *82*, 118; (b) Abraham, F. F.; Schneider, D.; Land, B.; Lifka, D.; Skovira, J.; Gerner, J.; Rosenkrantz, M., *J. Mech. Phys. Solids* **1997**, *45*, 1461; (c) Omeltchenko, A.; Bachlechner, M. E.; Nakano, A.; Kalia, R. K.; Vashishta, P.; Ebbsjo, I.; Madhukar, A.; Messina, P., *Phys. Rev. Lett.* **2000**, *84*, 318; (d) Bachlechner, M. E.; Omeltchenko, A.; Nakano, A.; Kalia, R. K.; Vashishta, P., *Appl. Phys. Lett.* **1998**, *72*, 1969.
2. (a) Understanding Molecular Simulation From Algorithms to Applications, Frenel, D.; Smit, B., Academic Press, San Diego, **2002**; (b) Molecular Dynamics Simulation Elementary Methods, Haile, J. M., John Wiley and Sons, New York, **1992**; (c) Computer Simulation of Liquids, Allen, M. P.; Tildesley, D. J., Oxford university Press, Oxford, **1987.**
3. (a) Simulation and the Monte Carlo Method, Rubinstein, R. Y., John Wiley and Sons, New York, **1981**; (b) Monte Carlo Concepts Algorithms and Applications, Fishman, G. S., Springer Verlag, **1996**.
4. (a) Fermi, E.; Pasta, J.; Ulam, S., *Los Alamos Report LA 1940*, **1955**, "Studies in nonlinear problems"; (b) Alder, B. J.; Wainwright, T. E., *J. Chem. Phys.* **1957**, *27*, 1208.
5. (a) Molecular Mechanics Energy Functions, Murrell, J. N.; Carter, S.; Farantos, S. C.; Huxley, P. (eds.) John Wiley and Sons, New York, 1984; (b) Erkoc, S., *Phys. Rep.* **1997**, *278*, 79; (b) Engler, E. M.; Andose, J. D.; Schleyer, P. v. R., *J. Am. Chem. Soc.* **1973**, *95*, 8005; (c) Hall, D.; Pavitt, N., *J. Comp. Chem.* **1984**, *5*, 441; (e) Halgren, T. A.; Nachbar, R. B., *J. Comput. Chem.*, **1996**, *17*, 587.

6. Weiner, S. J.; Kollman, P. A,; Case, D. A,; Singh, U. C.; Ghio, C.; Alagona, G.; Profeta, S., Jr.; Weiner, P. *J. Am. Chem. Soc.* **1984**, *106*, 765.

7. Brooks, B. R.; Bruccoleri, R. E.; Olafson, B. D.; States, D. J.; Swaminathan, S.; Karplus, M. *J. Comput. Chem.* **1983**, *4*, 187.

8. Halgren, T. A., *J. Comput. Chem.*, **1996**, *17*, 490; (b) *ibid* **1996**, *17*, 520; (c) *ibid* **1996**, *17*, 553; (d) *ibid*, **1996**, *17*, 616.

9. (a) Finnis, M. W.; Sinclair, J. E., *Philos. Mag. A*, **1984**, *50*, 45; (b) Sutton, A. P.; Chen, J., *Philos. Mag. Lett.*, **1990**, *61*, 139; (c) Daw, M.S.; Baskes, M. I., *Phys. Rev. B* **1984**, *29*, 6443.

10. Stillinger, F. H.; Weber, T.A., *Phys. Rev. B*, **1985**, *31*, 5262.

11. (a) Tersoff, J., *Phys. Rev. B*, **1988**, *38*, 9902; (b) Tersoff, J., *J. Phys. Rev. B*, **1988**, *38*, 9902.

12. (a) Steele, W. A., *Surf. Sci.* **1973**, *36*, 317; (b) Steele, W. A.; Vernov, A. V.; Tildesley, D. *J. Carbon* **1987**, *25*, 7.

13. (a) Jorgensen, W. L., *J. Chem. Phys.* **1982**, *77*, 4156; (b) Jorgensen, W. L.; Chandresekhar, J.; Madura, J. D.; Impey, R. W.; Klein, M. L., *J. Chem. Phys.*, **1983**, *79*, 926.

14. Berendsen, H.J.C; Postma, J.P.M.; van Gunsteren, W.F.; Hermans, J., *Intermolecular Forces*, Pullman, B., (ed.), Reidel, Dordrecht, Holland **1981**, 331.

15. (a) Engler, E. M.; Andose, J. D.; Schleyer, P. v. R., *J. Am. Chem. Soc.* **1973**, *95*, 8005; (b) Gundertofte, K.; Palm, J.; Pettersson, I.; Stamvik, A., *J. Comp. Chem.* **1991**, *12*, 200; (c) Gundertofte, K.; Liljefors, T.; Norrby, P-O; Pattesson, I., *J. Comp. Chem.* **1996**, *17*, 429

16. Carlsson, A. E., *Solid State Phys.*, **1990**, *43*, 1.

17. Erkoc, S., *Phys. Rep.* **1997**, *278*, 79.

18. Ercolessi, F.; Tosatti, E.; Parrinello, M., *Phys. Rev. Lett.*, **1986**, *57*, 719.

19. See, for example, Computational methods for Protein Folding, Friesner, R. A.; Prigogine, I. (Eds.), John Wiley and Sons, New York, **2002**.

20. See, for example, the review of experimental and computational measurements of elasticity in carbon nanotubes reported by: Gupta, S.; Dharamvir, K.; Jindal, V. K., *Phys. Rev. B* **2005**, *72*, 165428.

21. Car, R.; Parrinello, M., *Phys. Rev. Lett.* **1985**, *55*, 2471.

22. See, for example, Sherwood, P., *Modern Methods and Algorithms of Quantum Chemistry*, Proc. 2nd Ed., Grotendorst, J. (Ed.), John von Neumann Inst. For Computing, Julich, **2000**.

23. Uhlherr, A.; Theodorou, D.N., *Curr. Opin. Solid State and Mat. Sci.* **1998**, *3*, 544.

24. (a) Tabor, D.; Winterton, R. H. S. *Proc. R. Soc. London A* **1969**, *312*, 435; (b) Israelachvili, J. N.; Tabor, D. *Proc. R. Soc. London A* **1972**, *331*, 19.

25. Binnig, G.; Quate, C. F.; Gerber, G., *Phys. Rev. Lett.* **1986**, *56*, 930.

26. A. Ashkin, Dziedzic, J. M., *Science* **1987**, *235*, 1517.

27. Evans, E.; Ritchie, K.; Merkel, R., *Biophys. J.* **1995**, *68*, 2580.

28. (a) Chen, Y. L.; Helm, C. A.; Israelachvili, J. N., *J. Phys. Chem.* **1991**, *95*, 10736. (b) Joyce, S. R.; Michalske, R. A.; Crooks, R. M., *Phys. Rev. Lett.* **1992**, *68*, 2790.

29. Blackman, G. S.; Mate, C. M.; Philpott, M. R., *Phys. Rev. Lett.* **1990**, *65*, 2270.

30. Bell, G. I., *Science* **1978**, *200*, 618.

31. Marshall, B. T.; Long, M.; Piper, J. W.; Yago, T.; McEver, R. P.; Zhu, C., *Nature* **2003**, *423*, 190.

32. See, for example, Seifert, U., *Phys. Rev. Lett.* **2000**, *84*, 2750.

33. Evans, E.; Ritchie, K., *Biophys. J.* **1997**, *72*, 1541.

34. Heymann, B.; Grubmüller, H., *Chem. Phys. Lett.* **1999**, *303*, 1.

35. Heymann, B.; Grubmüller, H., *Biophys. J.* **2001**, *81*, 1295.

36. Grubmüller, H.; Heymann, B.; Tavan, P., *Science* **1996**, *271*, 997.

37. Izrailev, S.; Stepaniants, S.; Balsera, M.; Oono, Y.; Schulten, K., *Biophys. J.*, **1997**, *72*, 1568.

38. (a) Bartolo, D.; Derenyi, I.; Ajdari, A., *Phys. Rev. E.*, **2002**, *65*, 051910; (b) Derenyi, I.; Bartolo, D.; Ajdari, A., *Biophys. J.* **2004**, *86*, 1263.

39. Carrion-Vazquez, M.; Li, H.; Lu, H.; Marszalek, P. E.; Oberhauser, A. F.; Fernandez, J. M., *Nature Struct. Biol.* **2003**, *10*, 738.

40. (a) Paci, E.; Karplus, M., *J. Mol. Biol.*, **1999**, *288*, 441; (b) Paci, E.; Caflisch, A.; Pluckthun, A.; Karplus, M., *J. Mol. Biol.* **2001**, *314*, 589.

41. Tuckerman, M. E.; Berne, B. J.; Martyna, G. J., *J. Chem. Phys.*, **1992**, *97*, 1990.

42. (A) Leng, Y. S.; Jiang, S. *J. Chem. Phys.* **2000**, *113*, 8800; (b) Leng, Y. S.; Jiang, S. *Phys. Rev. B* **2001**, *63*, 193406; (c) Leng, Y. S.; Jiang, S. *Tribol. Lett.* **2001**, *11*, 111; (d) Zhang, L.; Leng, Y.; Jiang, S., *Langmuir* **2003**, *19*, 9742.

43. Leng, Y.; Jiang, S., *J. Am. Chem. Soc.* **2002**, *124*, 11764.

44. Chan, H. S.; Dill, K. A., *Physics Today*, **1993**, *46*, 24.

45. Beyond The Molecular Frontier Challenges For Chemistry And Chemical Engineering, Committee on Challenges for the Chemical Sciences in the 21st Century, US National Academies Press, 2003.

46. See, for example, Ruiz-Montero, (a) Bolhuis, P. G.; Dellago, C.; Chandler, D., *Faraday Discuss.*, **1998**, *110*, 421; (b) M. J.; Frenkel, D.; Brey, J. J., *Mol. Phys.* **1997**, *90*, 925.

47. Erickson, H.P. *Proc. Natl. Acad. Sci. USA*, **1994**, *91*, 10114.

48. (a) Smith, S.B.; Cui, Y.; Bustamante, C., *Science* **1996** *271*, 795; (b) Rief, M.; Oesterhelt, F.; Heymann, B.; Gaub, H.E., *Science* **1997**, *275*, 1295.

49. Marszalek P.E.; Li, H.; Oberhauser, A. F.; Fernandez, J. M., *Proc. Natl. Acad. Sci USA*, **2002**, *99*, 4278.

50. Marszalek, P. E.; Li, H.; Fernandez, J. M., *Nature Biotech.* **2001**, *19*, 258.

51. Konrad, M. W.; Bolonick, J. I., *J. Am. Chem. Soc.* **1996**, *118*, 10989.

52. Kosikov, K. M.; Gorin, A., Zhurkin, V. B.; Olson, W. K., *J. Mol. Biol.* **1999**, *289*, 1301.

53. Labrun, A.; Lavery, R., *Nucleic Acids Res.* **1996**, *42*, 383.

54. Cizeau, P.; Viovy, J.-L., *Biopolymers*, **1997**, *24*, 2260.

55. Florin, E.-L.; Moy, V. T.; Gaub, H. E., *Science* **1996**, *271*, 997.

56. Shen, L.; Shen, J.; Luo, X.; Cheng, F.; Xu, Y.; Chen, K.; Arnold, E.; Ding, J.; Jiang, H., *Biophys. J.*, **2003**, *84*, 3547.

57. Izrailev, S.; Stepaniants, S.; Balsera, M.; Oono, Y.; Schulten, K., *Biophys. J.*, **1997**, *72*, 1568.

58. Kosztin, D.; Izrailev, S.; Schulten, K., *Biophys. J.*, **1999**, *76*, 188.

59. (a) Müller, D.J.; Baumeister, W.; Engel, A., *Proc. Natl. Acad. Sci. USA* **1999**, *96*, 13170; (b) Engel, A.; Müller, D.*J. Nature Struct. Biol.* **2000**, *7*, 715.

60. Marrink, S.-J.; Berger, O.; Tieleman, P.; Jahnig, F., *Biophys. J.*, **1998**, *74*, 931.

61. (a) Rowley, L. A.; Nicholson, D.; Parsonage. N. G., *Mol. Phys.* **1976**, *31*, 365; (b) Abraham, F. A., *J. Chem. Phys.* **1978**, *68*, 3713.

62. Israelachvili, J. N.; Adams, G. E., *J. Chem. Phys. Faraday Trans. I*, **1978**, *74*, 975.

63. (a) O'Shea, S. J.; Welland, M. E.; Rayment, T., *Appl. Phys. Lett.* **1992**, *60*, 2356; (b) O'Shea, S. J.; Welland, M. E.; Pethica, J. B., *Chem. Phys. Lett.* **1994**, *223*, 336; (c) Han, W.; Lindsay, S. M., *Appl. Phys. Lett.* **1998**, *72*, 1656.

64. Hansen, J. P.; McDonald, I. R., *Theory of Simple Liquids*, Academic Press, London (**1976**).

65. Gelb, L. D.; Lynden-Bell, R. M., *Chem. Phys. Lett.*, **1993**, *211*, 328.

66. Gelb, L. D.; Lynden-Bell, R. M., *Phys. Rev. B* **1994**, *49*, 2058.

67. Patrick, D. L.; Lynden-Bell, R. M., *Surf. Sci.* **1997**, *380*, 224.

68. Magda, J; Tirrell, M.; Davis, H. T.; *J. Chem. Phys.* **1985**, *83*, 1888; (b) Douglas, L. J.; Lupkowski, M.; Dodd, T. L.; van Swol, F., *Langmuir*, **1993**, *9*, 1442.

69. Henderson, D.; Plischke, M., *J. Chem. Phys.* **1992**, *97*, 7822.

70. (a) Grest, G. S., *Phys. Rev. Lett.* **1996**, *76*, 4979; (b) Glosli, J. N.; McClelland, G. M., *Phys. Rev. Lett.* **1993**, *70*, 1960; (c) Bonner, T.; Baratoff, A., *Surf. Sci.* **1997**, *377–379*, 1082.

71. Sumpter, B. G.; Getino, C.; Noid, D. W.; Wunderlich, B., *Makromol. Chem., Theory Simul.* **1993**, *2*, 55.

72. (a) Katagiri, M.; Patrick, D. L.; Lynden-Bell, R. M., *Surf. Sci.*, **1999**, *431*, 260; (b) Landman, U.; Luedtke, W. D.; Nitzan, A., *Surf. Sci.* **1989**, *210*, L177; (c) Landman, U.; Luedtke, Gao, J., *Langmuir*, **1996**, *12*, 4514.

73. Shluger, A. L.; Wilson, R. M.; Williams, R. T., *Phys. Rev. B.*, **1994**, *49*, 4915.

74. (a) Gao, J.; Luedtke, W.D.; Landman, U., *J. Chem. Phys.* **1997**, *106*, 4309; (b) Gao, J.; Luedtke, W.D.; Landman, U., *J. Phys. Chem. B* **1997**, *101*, 4013; (c) Gao, J.; Luedtke, W.D. Landman, U., *Phys. Rev. Lett.*, **1997**, *79*, 705.

75. For reviews see: (a) Bhushan, B.; Israelachvili, J.N.; Landman, U., *Nature*, **1995**, *374*, 607; (b) Robbins, M.O.; Muser, M.H. in: *Handbook of Modern Tribology*, ed. B. Bhushan (CRC Press, Boca Raton, **2000**); (c) Landman, U.; Luedtke, W. D.; Ringer, W. M., *Wear*, **1992**, *153*, 3.

76. Johnson, K. L.; Kendall, K.; Roberts, A. D., *Proc. R. Soc. London, A* **1971**, *324*, 301.

77. Derjaguin, B. V.; Muller, V. M.; Toporov, P., *J. Colloid Interf. Sci.* **1975**, *53*, 314.

78. Patrick, D. L.; Flanagan IV, J.; Lynden-Bell, R. M., *J. Am. Chem. Soc.* **2003**, *125*, 6762.

79. Thomas, R. C.; Houston, J. E.; Crooks, R. M.; Kim, T.; Michalske, T. A., *J. Am. Chem. Soc.* **1995**, *117*, 3830.

80. Poulin, P.; Raghunathan, V. A.; Richetti, P.; Roux, D., *J. Phys. II* **1994**, *4*, 1557.

81. Poulin, P.; Stark, H.; Lubensky, T. C.; Weitz, D. A., *Science*, **1997**, *275*, 1770.

82. (a) Stark, H., *Phys. Rep.* **2001**, *351*, 387; (b) Stark, H.; Ventzki, D.; Reichert, M., *J. Phys.: Condens. Matter* **2003**, *15*, S191.

83. Loudet, J. C.; Poulin, P., *Phys. Rev. Lett.* **2001**, *87*, 165503.

84. Poulin, P.; Cabuil, V.; Weitz, D. A., *Phys. Rev. Lett.* **1997**, *79*, 4862.

85. Gay, J. G.; Berne, B. J., *J. Chem. Phys.* **1981**, *74*, 3316.

86. Andrienko, D.; Tasinkevych, M.; Patricio, P.; Allen, M. P.; Telo da Gama, M. M., *Phys. Rev. E*, **2002**, *68*, 051702.

87. Stark, H.; Stelzer, J.; Bernhard, R., Eur. Phys. J. B **1998**, *10*, 515.

Tip Functionalization: Applications to Chemical Force Spectroscopy

Craig D. Blanchette, Albert Loui, and Timothy V. Ratto

Introduction

Adhesion events in chemistry, biology, and material science occur through intermolecular interactions between distinct chemical functionalities. Chemical force spectroscopy has recently emerged as a powerful and versatile tool for the quantitative characterization of these molecular forces. Based on atomic force microscopy (AFM), chemical force spectroscopy relies on the modification of probe tips as a means of introducing chemical specificity into force measurements. In practice, a chemically modified tip is brought into contact with a substrate of specific functionality and as the tip is withdrawn, the interaction force between the functionalized tip and substrate is measured.

Chemical force spectroscopy has been used to study a rich diversity of chemical and biological interactions, including antibody-antigen[1-3] and lectin-carbohydrate pairs,[4-7] adhesive properties of polypropylene films,[8] interactions between gold and self-assembled monolayers,[9] adhesion forces between organic molecules and calcium oxalate crystals related to kidney stone formation,[10] and adhesive forces between complementary DNA strands.[11] Access to such a broad array of interactions depends critically on the successful surface modification of AFM tips. Over the past decade, several chemical strategies have been developed for tip functionalization, including silanization,[3-5, 7, 12-15] esterification,[2, 16-20] metal deposition (particularly gold for additional thiol chemistry),[21-28] self-assembled monolayers,[11, 29-31] and polymer cross-linking.[2, 4, 6, 7, 15, 19, 20, 32, 33]

This chapter serves as a basic introduction to these tip functionalization schemes. The material is presented at a basic level such that one with no experience in surface functionalization can learn the fundamentals of the different chemical strategies that have been developed thus far. The chapter is organized into four sections. Each section begins with a brief description and an overview of the type of chemistry that is the focus of the section, followed by subsections which discuss specific types of chemical methods. Within each subsection we present a specific experimental protocol that has been used in the literature. Tip functionalization often involves sequentially adding functionality in several steps, and the four sections are structured according to the sequence of chemical modifications that need to occur to reach a desired functionality. In other words, the earlier sections deal with the chemical reactions occurring closer to the tip surface and the later sections deal with the chemical reactions occurring further from the tip (Figure 1). The first section discusses efficient methods to clean AFM tips prior to functionalization. The second section reviews

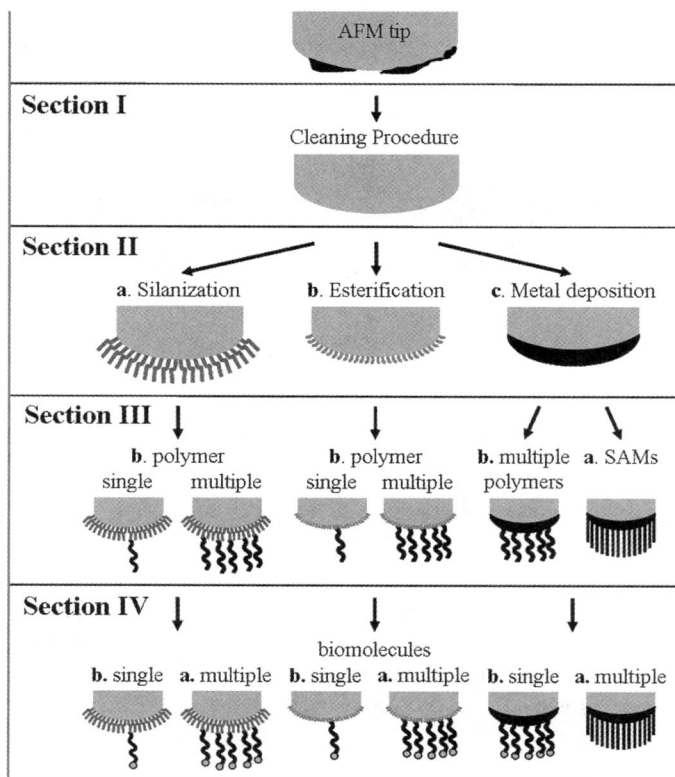

Figure 1. Outline of the structure of the book chapter.

the methods of adding functional groups (metals, organic functional groups, etc.) directly to cleaned AFM tips (silanization in Section IIa, esterification in Section IIb, and metal deposition in Section IIc). In the third section we review different methods that have been developed to add additional chemical entities to the first layer of tip functionality (polymers in Section IIIa and self-assembling monolayers [SAMs] in Section IIIb). In the fourth section we briefly discuss the chemical strategies for attaching biological molecules to the second layer of functionalization (polymers and SAMs, Section IVa). In this section we also discuss one of the milestones of CFM—single molecule force spectroscopy—again with a focus on the addition of biomolecules such as proteins (Section IVb). In this subsection (IVb) we primarily discuss the aspects of force spectroscopy as it applies to biological systems. The methods presented are a combination of the different chemical strategies discussed in the first three sections as they apply to single-molecule attachment. It is our hope that this chapter can be used as a guide for researchers who are attempting tip functionalization with an application to CFM experiments for the first time.

I Tip Cleaning

Commercially available AFM cantilevers are made from silicon nitride (Si_3N_4), which forms a silanol (SiOH) surface layer under ambient atmospheric conditions. For the methods that will be described in the subsequent sections, the hydroxyl terminated surface serves as a chemically-active interface between the substrate and the functionalizing species of interest. It is therefore important that new cantilevers be subjected to an aggressive cleaning procedure in order to remove any contaminants which might interfere with the formation of a uniform silanol layer.

A spectroscopic study done by Lo et al. revealed that silicone oils are pervasive contaminants on commercial AFM cantilevers, directly derived from the poly(dimethylsiloxane) (PDMS) commonly comprising shipping and packaging materials.[34] These authors determined that the most effective way of completely removing these low-molecular weight polysiloxanes is by immersion in a mixture of concentrated sulfuric acid and hydrogen peroxide.[34]

The method developed in this work, and subsequently employed by many research groups, involves an initial rinse of a cantilever chip with ethanol and subsequent drying under nitrogen at a very low rate of flow (to prevent structural damage to the cantilevers). Then the chip is immersed in a standard piranha solution, H_2SO_4/H_2O_2, 70:30 (v/v), for 30 minutes. As in the handling of any strongly oxidizing acids, personal protective equipment should be worn at all times, including gloves (preferably butyl rubber), goggles, and a face shield for protection; in addition, all work should be performed in a ventilated chemical hood. Given the small size of cantilever chips and the extremely caustic nature of the etchant, sparing use of piranha solution is strongly recommended (e.g., a few milliliters or less). The chips should then be thoroughly rinsed with deionized water, followed again by an ethanol rinse and dried under nitrogen. To help drive off any residual moisture, the cantilever chips should finally be baked in an oven or on a hotplate at 100–150 °C for approximately 30 to 60 minutes.

Once the cantilever chips have been cleaned, storage in a similarly cleaned glass petri dish is necessary to prevent recontamination. Storage in polymer-based containers may result in a redeposition of polysiloxanes as a result of silicone-based additives in the plastic.[34] Long-term maintenance of clean chip batches requires storage in a dry nitrogen atmosphere free of volatile organic contaminants.

Small quantities of gold contamination may occasionally occur on the tip side of an optical AFM cantilever due to the surface diffusion of gold from the reflex coating, which may adversely affect both functionalization and force measurements.[34] While gold is readily removed by immersion in a solution of aqua regia (HCl/HNO$_3$), this treatment will indiscriminately remove both the contaminant and reflex coating. Considering the current average cost per AFM probe ($20–30), the most expedient solution to undesirable gold contamination is simply to employ cantilevers with aluminum reflex coatings.

II Direct Functionalization of Tip Surfaces

In this section we present methods of attaching functional organic groups or metal layers (with a specific emphasis on gold) directly onto the silicon nitride tips through the Si-OH layer. For many applications this step may be the only one required. If one is simply interested in measuring the covalent interactions between organic compounds or metal surfaces, this initial functionalization will suffice; however many applications require adding additional molecules to the AFM tip. In several of these applications this first step is necessary to attach a highly reactive group onto the AFM tip for subsequent attachment of the molecules of interest. For example, a gold coating permits thiol chemistry, and silane chemistries can be used for the attachment of a variety of different molecules to oxide layers. Therefore, we will present the first step of functionalization in general terms and in some select cases discuss more specifically the parameters one must be mindful of for subsequent tip functionalization.

IIa Silanization

Silanization is the process by which silicon-containing molecules are attached to hydroxylated silicon surfaces through the formation of Si-O-Si bonds. These organosilane reagents are typically represented by the chemical formula $RSi(OR')_3$, where R is often an aminoalkyl chain and R' is a potentially surface reactive functionality. Some substituents (e.g., alkoxyl) can react with inorganic materials such as glass and silica; others, including vinyl, acryl, and thiol functional groups, can readily bond with various organic species. The organosilanes can therefore act as linkers between the silicon nitride tip surface and the functional molecules of interest; for this reason, these molecules are sometimes referred to as silane coupling agents.

Figure 2a illustrates the silanization of a silicon nitride tip surface with (3-aminopropyl) triethoxysilane (APTES), an organosilane commonly used in tip functionalization.[4, 7, 15, 29] In this case, the three alkoxyl groups form Si-O-Si linkages to the substrate, while the amine group may be used for subsequent functionalization (e.g., attachment of a polymer spacer). Such secondary modifications will be discussed in more detail in Sections III and IV. While this method can be used directly to form a self-assembled monolayer (SAM) on an AFM tip,[29] it is not commonly employed for this purpose; instead, SAMs are generally formed on gold-coated tips through alkanethiol chemistry (Section IIIa).

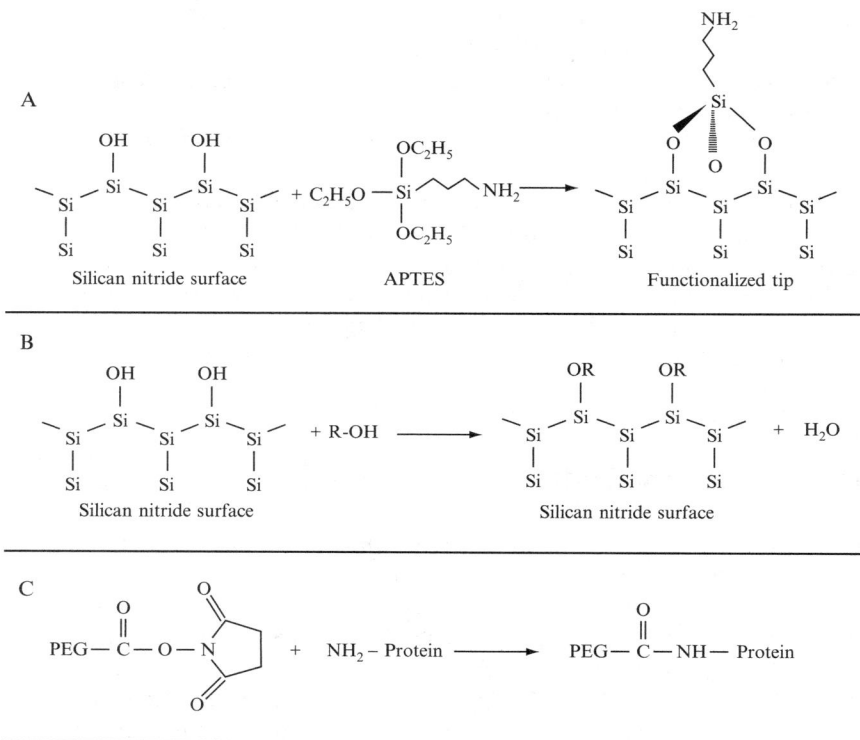

Figure 2. Chemical reactions involved in tip functionalization. (A) Silanization reaction of 3-aminopropyl-triethoxysilane (APTES) with surface silanols. (B) General esterification reaction with surface silanols. (C) Reaction between the highly reactive N-hydroxysuccinimide (NHS) attached to one end of a PEG polymer and a free amine group (lysine) of a protein, this is used for attaching proteins to tips functionalized with PEG polymers containing NHS.

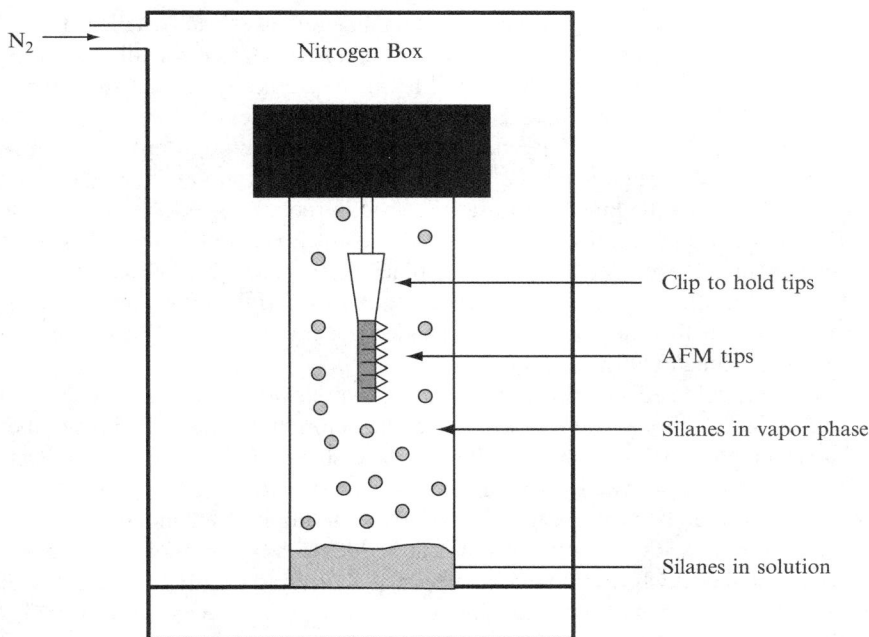

Figure 3. Diagram of experimental set up for tip functionalization from the vapor. Tips are clamped in a sealed vial containing ~ 400 mL of silane in solution. AFM tips are to be clamped such that the tips do not contact the silane solution. Silane will evaporate and silanization of the AFM tips will occur in the vapor phase. This entire process should take place under nitrogen or in a nitrogen box to prevent water from polymerizing the silanes on the AFM tip. (*See Color Plates*)

To achieve a uniform monolayer of organosilanes, tip silanization should occur in the vapor phase. A cartoon diagram of this simple process is provided in Figure 3; cleaned AFM cantilever chips are mounted in a closed container under dry nitrogen above a solution containing the silanes of interest. A moisture-free reaction vessel is required, since silanes readily polymerize in the presence of water. The solution may contain more than one type of organosilane; for example, a mixture of two silanes, one containing a reactive group (such as an amine) and the other a nonreactive group (such as methyl), may be desirable as a means of limiting the number of reactions studied per force curve. This optimization becomes critically important when performing single-molecule studies (Section IV).

To complete the tip silanization process, the cantilever chips must be completely dried in an oven at ~ 110 °C for 15 minutes and then placed back into a dry nitrogen atmosphere. Since the organosilane surface layer is hydrolytically unstable, it is highly recommended that tips be used within a few hours of preparation.[32] Silanization is the most commonly used method of tip functionalization, and has been successfully used in the study of a wide range of systems.[3–5, 7, 12–15]

IIb Esterification

Esterification describes a condensation reaction in which two chemicals (typically an alcohol and an acid) form an ester as the reaction product. The esterification of silanol with alcohols is well known, and has been found to lead to the covalent attachment of organic molecules to a silanol covered surface.[35] This process is illustrated in Figure 2b. By a judicious selection of alcohols, the resulting organic functionalization layer can be tailored to suit the experimental needs. In most cases, the direct esterification of surface silanol groups requires heating above 400 K over the course of several hours. Therefore, the clean substrate is usually first treated with tetrachlorosilane followed by diethylamine, or directly with $Si(NEt_2)_4$ (synthesized from $SiCl_4$ and NEt_2H);[35] the resulting surface amino groups can then readily react with the selected organic alcohols. This intermediate activation of the silanol-covered surface greatly increases the kinetics of esterification at room temperature. As discussed in the silanization section (Section 2A) the vast methods of tip functionalization schemes are directed to form amine-terminated functionalization.[2, 16–20, 36] Therefore we have focused on a specific type of esterification reaction that results in amine-terminated tip functionalization; but keep in mind that this method can be expanded generally to include different types of organic compound functionality.

For amine-terminated functionality the esterification reaction is conducted in a ethanolamine chloride solution. As discussed above, the chloride functions to activate the reaction. Upon cleaning the AFM tip through the piranha etching method described in Section 1, surface bound water is removed by drying the tips in an oven for ~ 1 hour at ~ 110–150 °C. Tips are then immediately esterfied in a 55% (wt/vol) solution of ethanolamine chloride in dimethyl sulfoxide (DMSO) for ~ 10–15 hours at 100 °C with 0.3 nm molecular sieve beads. This process is achieved while applying an aspirator vaccum to trap H_2O in a $CaCl_2$ tower.[2] After the reaction has been completed, tips are subsequently cleaned with DMSO. More recently this protocol had been modified to allow the reaction to occur at room temperature. In this procedure initially 30%(mol/mol) of 2-aminoethanol-Cl are melted in dry DSMO at 100 °C in the presence of 0.3 nm molecular sieve beads. After the solution has cooled to room temperature the tips are added and incubated for 15 hours. Following this incubation the AFM tips are extensively washed in bare DMSO and then dried under a flow of nitrogen. The amine-functionalized tips must then be stored in a dessicator until further use.[20]

The process discussed above can readily be applied to any reaction containing alcohols with different functionality. But it is worth noting that one may need to use a modified version due to the sensitivity of the functionality contained on the reacting alcohol. In an attempt to functionalize AFM tips with other types of alcohols aside from ethanolamine, one will need to tailor the conditions of the reaction (temperature, concentration, etc.) such that the functional group of the alcohol is not altered. As discussed in Section IIa, one may want to use a ratio of two different alcohols (one containing a reactive group such as an amine and the other a nonreactive group such as methyl) to limit the number of reactive groups on the tip. This becomes more relevant when dealing with single-molecule studies, which will be discussed in more detail in Section IV.

IIc   Metal Deposition/Coating

Generally this method needs to be conducted by someone with experimental expertise in the process of depositing or coating surfaces or substrates with metals. Therefore the purpose of this section will be to present the background information on metal deposition as it applies to depositing metal coatings on AFM cantilevers. We will not go into the specific experimental details (such as temperature, deposition time, pressure, etc.) of this process, because these parameters will vary drastically for different metals and different deposition rates. In addition, AFM tips with a wide variety of metal coatings are commercially available including aluminum, gold, nickel, chromium, and platinum.

There have been a wide variety of different methods developed for processes that involve metal deposition. Reviewing these methods is beyond the scope of this chapter, therefore we will focus on vapor deposition, one of the most common methods for depositing metals on microstructures. It is worth mentioning that when this technique is applied to tip functionalization, generally a whole wafer of silicon AFM chips is coated at once; therefore, in the rest of this section when we discuss metal deposition on an AFM tip it will be in relation to a whole silicon wafer of AFM tips. Vapor deposition, in general, refers to processes in which vapor state materials are condensed onto a surface to form a solid material. These processes are used to form coatings to alter the mechanical, electrical, thermal, optical, corrosion resistance, wear, and reactive properties of the substrates. Vapor deposition processes generally take place within a vacuum chamber. There are two categories of vapor deposition processes, physical vapor deposition (PVD) and chemical vapor deposition (CVD). In PVD processes, the source material is heated, resulting in evaporation of the metal followed by condensation on the substrate. In CVD processes, the gases in the coating chamber are heated to drive the deposition reaction (i.e., the metal deposition process occurs through a chemical reaction). The vapor process generally used to coat AFM tips with a metal layer is PVD. PVD methods are dry vacuum deposition methods in which the coating is deposited over the entire object simultaneously, rather than in localized areas. There are two general PVD methods, sputtering and evaporation; both methods have been used for metal coating or deposition on AFM tips.

In evaporation the substrate is placed inside a vacuum chamber, in which a block of the material to be deposited is also located (i.e., the source of metal for deposition). The source material is then heated to the point where it begins to evaporate. A low pressure environment is required so that the molecules evaporate freely in the chamber and subsequently condense on all surfaces. This principle is the same for all evaporation technologies; only the method used to heat the source material differs. There are two popular evaporation technologies: e-beam evaporation and resistive evaporation, each referring to the heating method. In e-beam evaporation, an electron beam is aimed at the source material causing local heating and evaporation. In this method an electron beam evaporator fires a high-energy beam from an electron gun to boil a small spot of source material. As a result, heating is not uniform, so lower vapor pressure materials can be deposited. Generally the beam is bent through an angle of 270° in order to ensure that the gun filament is not directly exposed to the evaporating material. This method results in deposition rates that range from 1–10 nm per second. A cartoon diagram of an electron beam evaporator is shown in Figure 4A. With resistive or thermal evaporation, a "boat", made of a high melting point material and containing the source material, is heated electrically with a high current to make the material evaporate. This must be done in a high vacuum to allow the vapor to reach the substrate without reacting with or scattering against other gas-phase atoms in the chamber, and to reduce the presence of impurities from the residual gas in the vacuum chamber. Many materials are restrictive in terms of what evaporation method can be used (e.g., aluminum is quite difficult to evaporate using resistive heating). This limitation is typically related to the phase transition properties of that material. A cartoon diagram of a resistive thermal evaporator is shown in Figure 4C. Both of these methods, electron beam evaporation[22] and resistive evaporation,[21, 23, 26] have been applied in the metal deposition of the coating of AFM cantilevers. Generally tips are gold-coated in order to reduce the possibility of damage due to corrosion and/or in order to ease the use of using thiol chemistry to add functionality. When gold is used, generally an initial 1–5 nm thick layer of chromium[28, 37] or titanium[29, 38] is used to increase the adhesion of the gold to the AFM tip.

In addition sputtering is also often used for metal deposition. Sputtering utilizes a mechanism by which atoms are dislodged from the surface of a material as a result of collision with high-energy particles. Thus in this technique atoms or molecules are ejected from a target material by high-energy particle bombardment such that the ejected atoms or molecules can condense as a thin film on a substrate. This method in several cases has proven more useful in forming uniform thin metallic films. The process can be described as a sequence

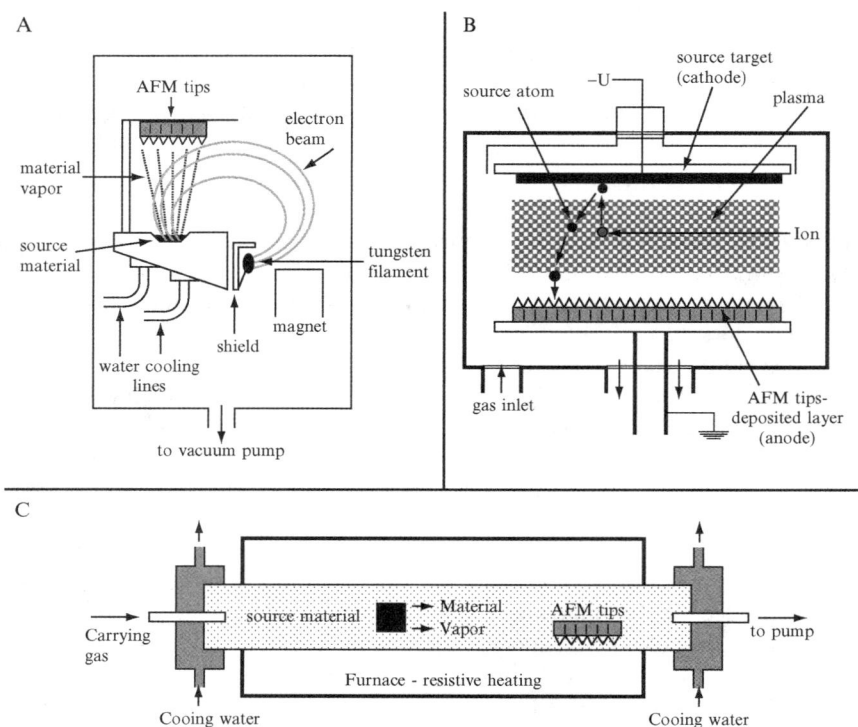

Figure 4. Methods to coat AFM cantilevers with metals. (A) Schematic of electron beam evaporation. (B) Schematic of sputtering. (C) Schematic of resistive evaporation. (*See Color Plates*)

of four steps: 1) ions are generated and directed at a target material; 2) the ions sputter atoms from the target; 3) the sputtered atoms get transported to the substrate through a region of reduced pressure; and 4) the sputtered atoms condense on the substrate, forming a thin film. This method has been applied for metallic coatings of AFM tips.[39–41] In this work AFM tips were gold-coated using 2 X 300 coating times at an argon pressure of 0.0001 mbar and an electrical current of 45 mA, which resulted in a globular gold coating with an average diameter of 120 nm and thickness of 60 nm.[39, 40] A cartoon diagram of the sputtering mechanism is shown in Figure 4B.

In the variety of different methods used to coat tips, the parameters that are most regulated are temperature and pressure. These two parameters dictate the deposition rate and the uniformity of the deposited metal layer. Therefore when conducting this type of experiment it is important to understand how those parameters will affect the deposited metal coating.

III Additional Layers of Functionality

Many applications of chemical force microscopy (and force spectroscopy in general) require additional layers of functionality to be coupled to the AFM tip. These include self-assembled monolayers (SAMs) onto gold-coated tips,[11, 29–31, 42, 43] polymers (such as DNA and polyethylene glycol (PEG)) onto gold-coated and amine-functionalized AFM tips,[2, 4, 6, 7, 15, 19, 20, 32, 33, 44] and organic functional groups and bio-molecules onto chemically active polymer tethers.[4, 7, 15, 27] The later functionality is applicable to single-molecule force spectroscopy in relation to biological systems, which will be discussed in more detail in Section IV. It may seem redundant to dedicate two separate sections to SAMs and polymer functionalization, since SAMs are essentially a monolayer of alkane polymers, which may or may not contain a terminal functional group. However, they differ in that SAMs self-assemble into monolayer structures through specific processes that are based on interactions between the surface and the polymer molecules (van der Waals forces). Polymers do not typically self-assemble into monolayers, and thus warrant a separate discussion.

Tip functionality containing self-assembled monolayers has been used for a variety of measurements, including characterizing adhesive and frictional interactions between substrates and SAMs terminated with -NH$_2$, -COOH, -OH, and -CH$_3$ functional groups as a function of solution pH and ionic strength,[29] to determine the contact potentials,[30] the effect of wetting properties,[31] and the properties of chiral molecules.[42] This work has provided a wealth of information on how the chemistry of a surface affects its physical properties. On the other hand, the addition of polymers generally serve two primary purposes, to provide a spacer for the subsequent addition of biomolecules and functional groups[4, 7, 15, 20, 32, 36, 45] (such as polyethylene glycol [PEG]) or to study the stretching mechanics or intrastrand binding strengths[11, 28, 46–50] (such as DNA). Thus for these applications the initial functionalization (silanization, esterification, and metal coating) is necessary to facilitate in the subsequent addition of either SAMs or polymers in general.

IIIa   Self-Assembled Monolayers (SAMs)

The most common method of functionalizing AFM tips with SAMs is done through thiol chemistry on gold-coated tips, where the tips are coated through the methods described in Section IIc.[11, 29, 30, 41, 42, 51] It is worth noting that silanization (Section IIa) has also been used to functionalize tips with SAMs, although it is not a common method.[29] Because of their importance in tip functionalization, we will briefly discuss the chemical properties of thiol molecules. A thiol is a compound that contains a functional group which is composed of a sulfur atom and a hydrogen atom (-SH). The thiol group is the sulfur analog of the hydroxyl group (-OH). Therefore the chemistry of thiols is closely related to the chemistry of alcohols. One feature of thiols that distinguish them from alcohols is the extremely high affinity sulfur displays for gold (20–35 kcal/mol). Therefore when alkanes that are functionlized on one end with a thiol group (alkylthiols) come in contact with a gold substrate, the sulfur end of the alkane readily adsorbs to the surface. The unique combination of alkane polymers functionalized with terminal thiol groups results in the self-assembly of these molecules into ordered monolayers. Due to the wide applications of this system, a great deal of work has been done to examine the processes by which alkyl thiols self-assemble onto gold surfaces. It is generally believed that alkyl thiols initially bind to the gold surfaces in a horizontal orientation, in which the alkyl tails lie flat on the surface. This initial interaction is driven by the binding strength between thiols and gold (~ 20–35 kcal/mole) and is modeled as a Langmuir binding isotherm. As these binding events continue, the density of the horizontally oriented alkyls will begin to interact laterally (mainly through van der Waals forces). At increased densities these interactions will cause the alkyl chains to lift off the substrate but still remain tethered through the thiol-gold bond. At some point during this absorption process there exists a mixture of horizontally oriented alkyl chains and islands of upright alkyl chains tilted ~ 30° to the normal. When this stage is reached, binding kinetics become much more complicated and can no longer be modeled with a simple Langmuir binding isotherm. Eventually these domains of upright alkane chains merge and cover the bulk of the substrate. This process is similar to a two-dimensional crystallization process. Figure 5 contains a cartoon representation of the process. The initial stage in the SAM formation generally occurs on the order of minutes at

Figure 5. Formation of SAMs on gold coated AFM tips. Initially, the thiolyated alkyls bind to the surface via the gold-thiol interaction in a horizontal orientation. As the density increases the polymers will begin to interact, causing the polymers to lift off of the substrate to form islands. As the density of islands increases, they merge to eventually form a uniform monolayer where all the polymer tails have lifted off of the surface and are oriented ~ 30° from the normal. (*See Color Plates*)

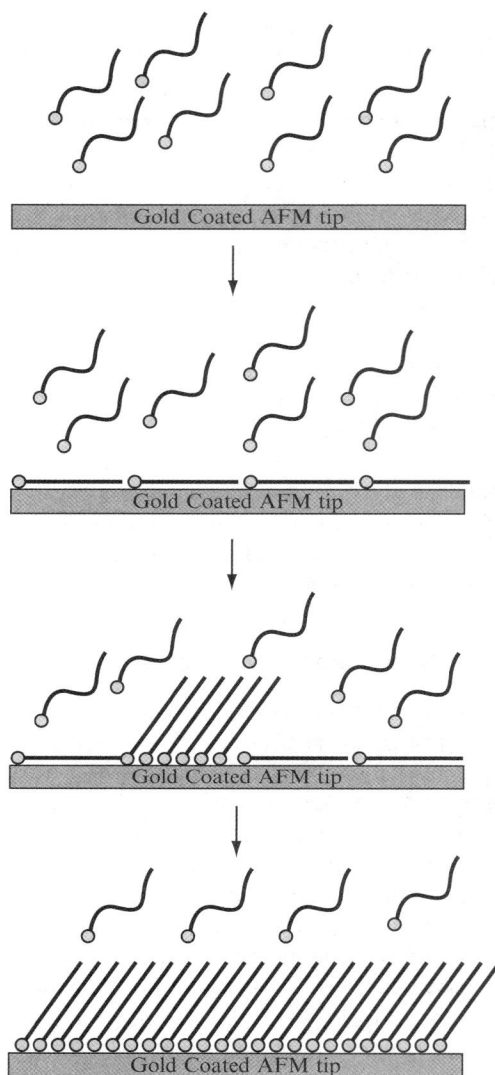

normal alkyl thiol concentrations (0.1–10 mM), but the ordering of the assembly may take days or longer, depending on the system.

One aspect of essential importance for the method of SAM functionalization is to clean the gold-coated tips right before functionalization. One method of cleaning that has proven effective involves initially oxidizing the tips in an ozone cleaner for ~30 minutes, followed by immersion of the tips in pure hot ethanol at ~65 °C for ~30 minutes. This process of cleaning results in more oxidation than reported in the literature and has been shown to be effective at removing surface contaminants.[51] One drawback of extensive oxidation is that higher temperatures of ethanol may be required to assure the complete reduction of the heavily oxidized gold surfaces.

Once the tips have been properly cleaned, formation of the SAMs is completed by incubating the tips in a solution containing the appropriate alkyl thiols. Generally the reaction of the alkyl thiols with the gold-coated tips is performed in ethanol solutions. As discussed above, the concentration of alkyl thiols used will affect the time for ordered SAM formation, but typically the concentration of alkyl thiols in ethanol used ranges of 1–4 mM. Despite this general range the alkyl thiol concentration should still be treated as an experimental parameter, which may need to be adjusted for optimal coverage and time for formation. As an example of this process, Fujihira et. al. formed SAM functionalized AFM tips of 1-decanethiol ($CH_3(CH_2)_9SH$) and 11-mercaptoundecanoic acid ($HOOC(CH_2)_{10}SH$) by immersing the cleaned gold tips into a 1 mM 1-decanethiol ($CH_3(CH_2)_9SH$) and 11-mercaptoundecanoic acid ($HOOC(CH_2)_{10}SH$) ethanol solution for 24 h at room temperature. This was followed by thoroughly rinsing the tips in pure ethanol and then drying under nitrogen.[51] Again, this is only one example; it will be necessary to determine both the concentration of alkylthiols and incubation time for the particular system one is interested in studying.

IIIb Polymer Attachment

One of the primary differences between the functionality that will be discussed in this section and that of SAMs is that these polymers lack the ability to self-assemble into structured monolayers (e.g., PEG and DNA). The primary reason for the attachment of polymers with these types of physical properties is to study the mechanical stretching properties of the polymer to act as a tether for further attachment of low numbers of molecules (such as biomolecules). Therefore polymer attachment of this form requires prior functionalization as discussed in Section II (gold deposition for thiol chemistry, silanization or etherification for a reactive group, or the addition of functional SAMs). The two most predominant polymers attached through these methods are PEG and DNA. Therefore we will primarily focus the discussion of this section on these two polymers, but the techniques can be applied to other types as well.

The attachment of polymers through thiol-gold absorption depends on one end of the polymer containing a thiol group to adsorb to the gold-coated tip. As discussed above, the polymer will attach to the tip via the strong thiol-gold interaction (~ 20–35 kcal/mole). Although this method of polymer functionalization is relatively simple, it is generally used for substrate functionalization as opposed to tip functionalization. The reason is that there is only limited control over the number of polymers that attach to the tip, since any thiol group will adsorb to gold; this will be discussed in more detail in Section IV. Despite this drawback it has been used for DNA attachment.[28, 52] In this work the authors examine the effect of salt concentration and the contact force on the unbinding of DNA. The method of tip attachment involved initially coating the tips with gold. Immediately following this the tips were placed in a solution composed of 100 μM 6-mercapto-1-hexanol (97%) (MCH) and 1.0 μm thiolated single-stranded DNA. The tips were incubated for 20 h in this solution prior to use in force measurements. As can be seen, this procedure follows closely the methods described for the attachment of SAMs, except for thiolated molecule concentration and incubation time. As mentioned in Section IIIa, these parameters will need to be determined experimentally for any thiol-gold functionalization to ensure efficient polymer adsorption. It is worth noting that for experiments involving the characterization of the mechanical properties of DNA under an applied force, DNA strands can be anchored to the substrate through the above chemistry. However, adding DNA to a tip does not always require chemical functionality, as DNA can be physisorbed to an AFM tip through strong nonspecific electrostatic interactions between the tip and DNA.[50, 53, 54] Needless to say, for studies involving more complex DNA interactions—such as hybridization—a specific linkage will most likely be required.

Polymers may also be attached to AFM tips through chemical reactions between conjugated polymers (i.e., those containing a reactive group) and tips already containing reactive functionality (through silanization or esterification). This method has been extensively employed to attach polymers.[4, 7, 15, 20] As mentioned above, a commonly used polymer is PEG and it is typically attached to tips functionalized with a reactive amine. This has been historically the chosen method because the method of adding amine functionality to the tips through esterification (Section IIb) and silanization (Section IIa) is very straightforward. For subsequent polymer attachment the polymers are typically conjugated with a highly reactive ester such as N-hydroxysuccinimide (NHS), which reacts with tip surface amines to form a stable amide covalent linkage (Figure 2c). One of the most common NHS esters for this type of reaction is succinimidyl propionate (SPA). In general the polymers are conjugated with functionality on both ends. One end is then used to react with the aminated tip surface, and the other end contains either the functional group of interest or another highly reactive group for further conjugation with biological molecules (Section IV). A typical example of a protocol used to functionalize AFM tips with PEG using the above reaction scheme is to initially add amine functionality through silanization (APTES, Section IIa) or esterification (ethanolamine chloride, Section IIb). The tips are then immersed in chloroform containing 10–25 mg/ml of SPA-PEG for thirty minutes, followed by extensive rinsing in chloroform and then drying under a slow flow of nitrogen.[4, 7, 15] Although the amine-ester reaction is the most common method for the addition of polymers to AFM tips (versus the thiol-gold reaction), there are several different chemical combinations that could achieve the same goal. Since the purpose of this chapter is to present the general method of tip functionalization, we will leave it up to the reader to explore the variety of different chemical combinations that can be applied to attach polymers to tip surfaces containing a reactive group.

IVa Attachment of Additional Molecules (Proteins) to Adsorbed Polymers and SAMs

Several applications require further functionalization after attachment of polymers and SAMs. Some of the techniques of functionalizing AFM tips discussed thus far require fairly harsh conditions (chloroform and ethanol solvents, high temperatures, low or high pH, etc.). Thus this additional chemistry step is necessary if the molecule of interest is sensitive to these factors. In particular this is relevant to the attachment of biological molecules such as proteins; therefore the focus of this section will be on methods to attach proteins to polymers.

At this point it may not be apparent why the addition of the polymer is necessary, since it is conceivable that the proteins could be directly attached to the first layer of tip functionality (i.e., thiol chemistry on gold-coated surfaces or amines attached through silanization and esterification). For these studies the primary purpose of the polymer is to act as a spacer separating the tip from the substrate. At close separations a wide range of nonspecific forces will dominate the tip-sample interactions. These nonspecific forces include: van der Waals forces, electrostatic forces, repulsive forces, as well as structural, depletion, and hydrophobic forces. Therefore when conducting these experiments it is nearly impossible to distinguish these nonspecific interactions from the specific interaction of interest. By employing a polymer spacer the specific interaction is moved further away from the surface and one can readily distinguish the interaction of interest from the nonspecific events experimentally (Figure 6). This issue becomes much more important when conducting single-molecule experiments, which will be the focus of Section IVb.

The chemistry involved in attaching proteins to polymer tethers is largely dependent on the types of surface amino acids that can react covalently with the polymers. Generally these reactions are directed toward lysine and cysteine. Lysine contains a terminal amine, which can readily be attached to polymers conjugated with the reactive esters discussed in the previous section (NHS, SPA). Cysteine, on the other hand, contains a terminal thiol.

Color Plates

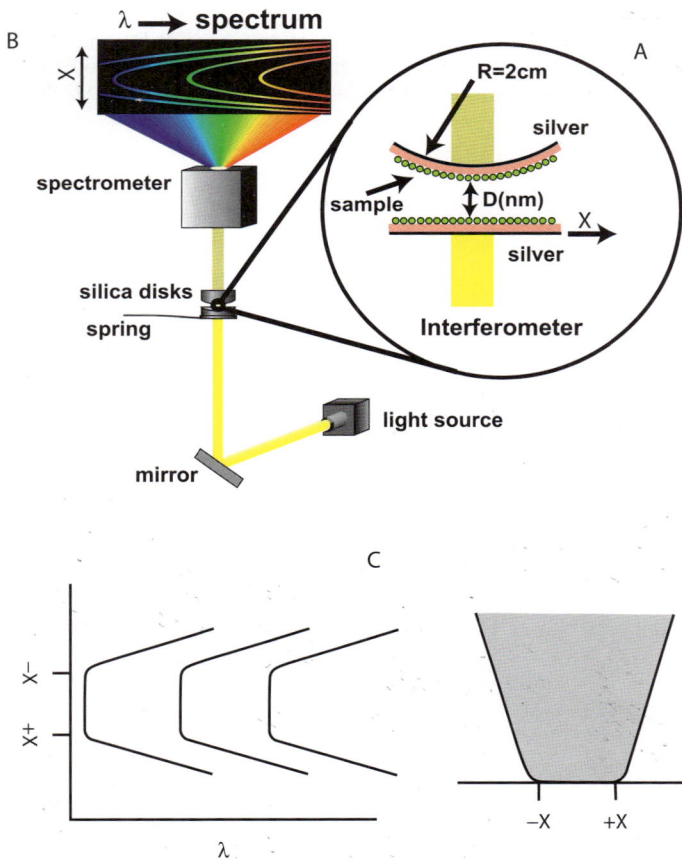

Figure 1-1. The Surface Force Apparatus. (A) Samples in the SFA are supported on two hemicylindrical lenses oriented at right angles to each other. The equivalent geometry is a sphere interacting with a flat plate. (B) The samples with the reflecting silver mirrors form the resonant cavity of a Fabry-Perot interferometer. White light passed through the samples generates a series of interference fringes. The curvature of the fringes corresponds to the curvature of the contact region between the samples. (C) Example of the distortion in the interference fringes resulting from surface deformation (flattening). The substrate deformation (right) is reflected in the shapes of the fringes (left). The distance from −X to +X is the diameter of the contact area (2x), and is measured directly from the interference fringes.

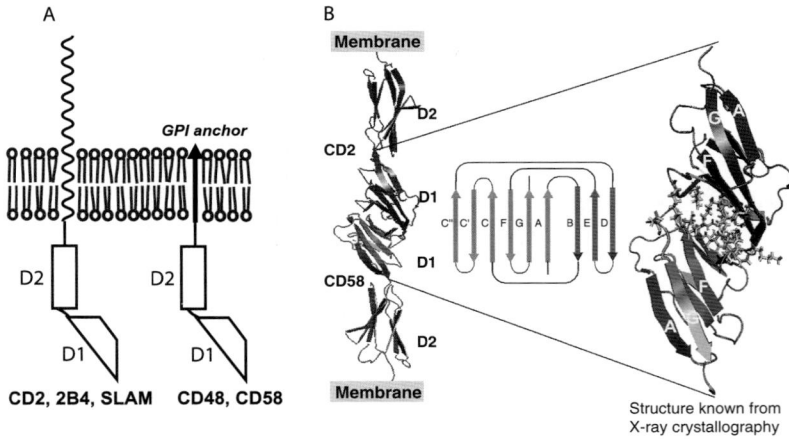

Figure 1-5. (A) General structure of the CD2 protein family. The proteins are anchored to the membrane via hydrophobic tails. The extracellular region consists of two domains D1 and D2. This figure shows two different hydrophobic anchors observed in this family. (B) Proposed head-to-head binding alignment between CD2 and its ligands. The structure of the complex between the outer D1 domains is shown on the right side.

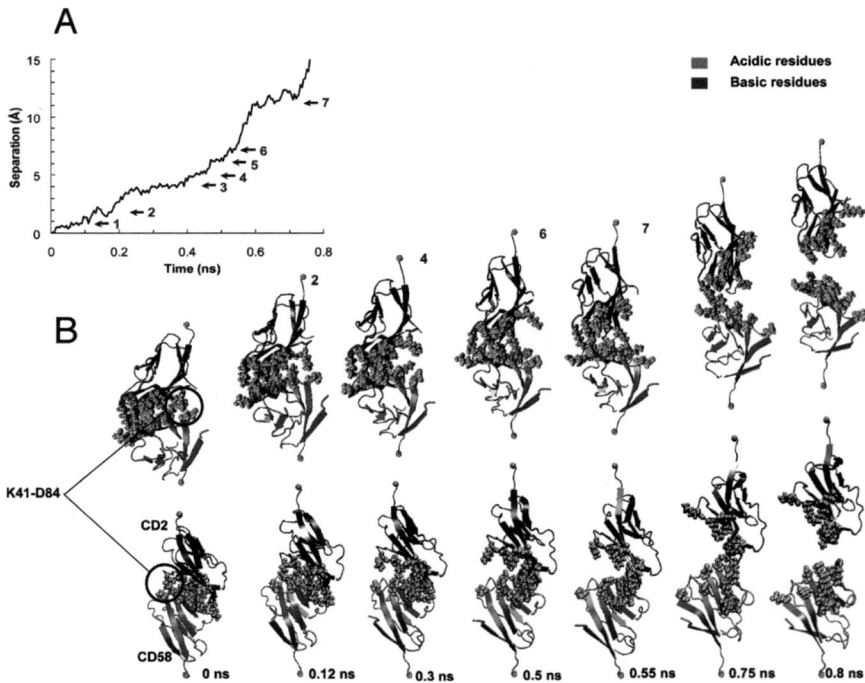

Figure 1-7. (A) Simulated separation vs. time trajectory as CD2 and CD58 are pulled apart. (B) Snapshots of the complex when discrete salt bridges rupture during protein detachment. The corresponding positions on the separation-time trajectory are indicated in (A).

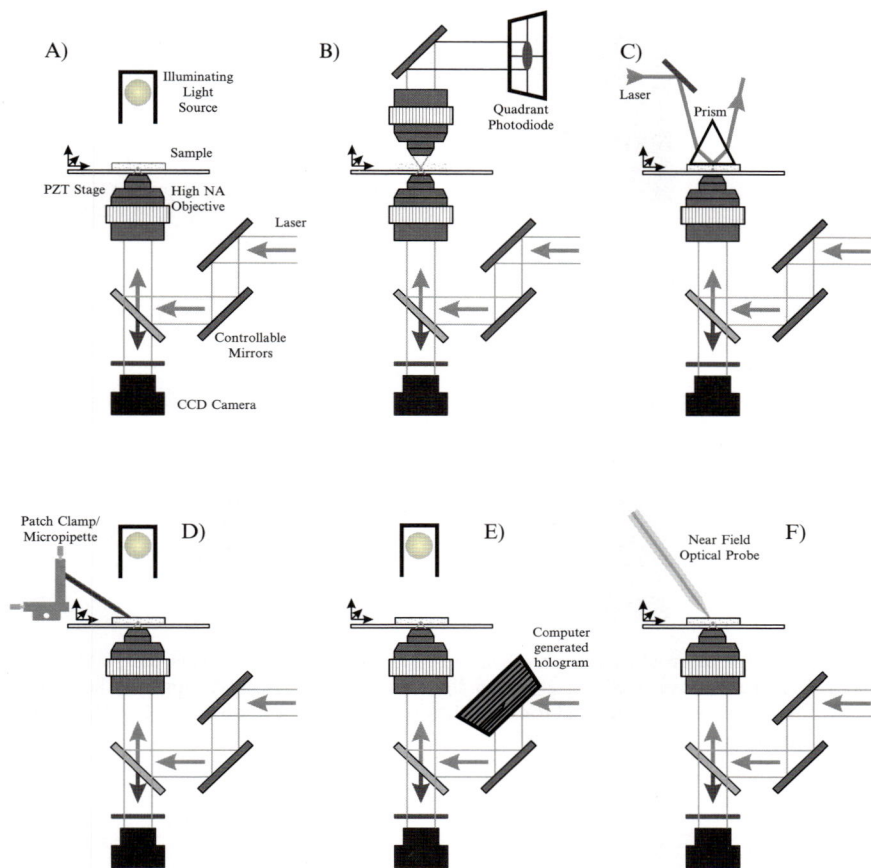

Figure 2-5. Common extensions to optical traps: (A) steerable mirrors to move trap position; (B) imaging of trapped particle onto a quadrant photodiode; (C) evanescent field excitation of trapped particles near a surface; (D) incorporation of a micropipette or patch-clamp; (E) dynamic trap configurations with a spatial light modulator; (F) excitation and imaging with a near-field optical probe.

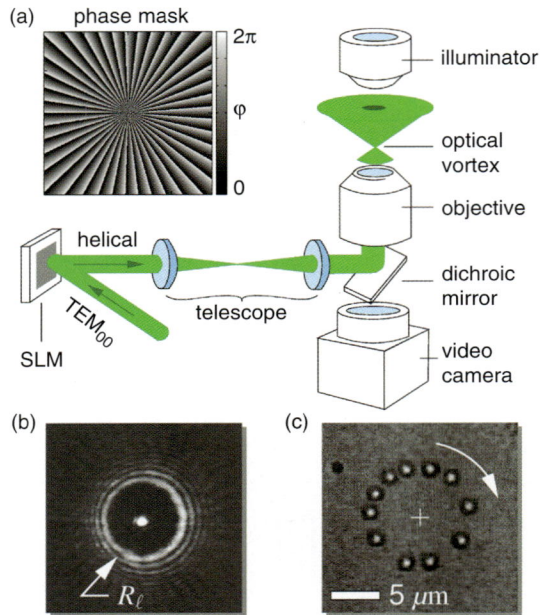

Figure 2-7. (A) Generation of a Laguerre-Gaussian mode from a Hermite-Gaussian TEM_{00} beam using a computer generated phase mask on a spatial light modulator, (B) producing a radially symmetric intensity profile, (C) which can be used for the controlled rotation of trapped objects [Reprinted Figure with permission of Ref. 309 by the American Physical Society].

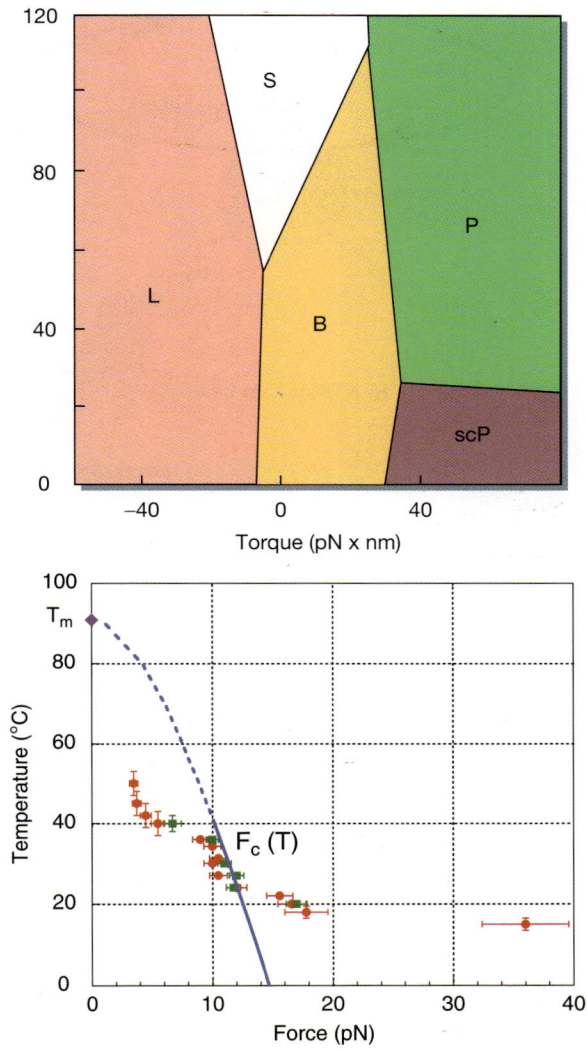

Figure 2-14. (A) Force-supercoiling phase diagram for dsDNA identifying five phases of double-stranded DNA and (B) Force-temperature phase diagram for unzipping dsDNA, where the blue line represents a simple thermodynamic model [(A) reprinted from Ref. 152 by permission of Macmillan Publishers Ltd; (B) reprinted Figure, with permission, from Ref. 160 by the American Physical Society].

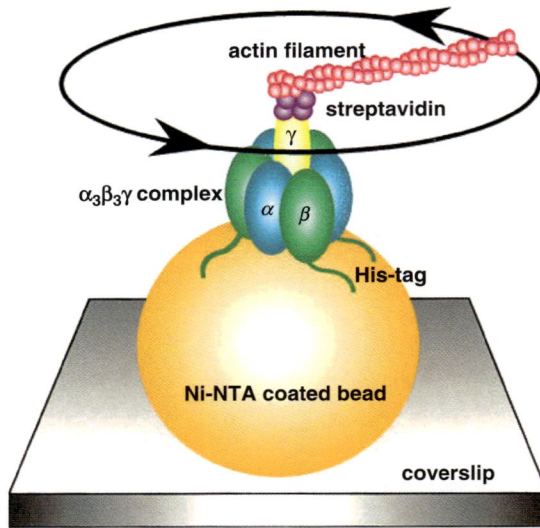

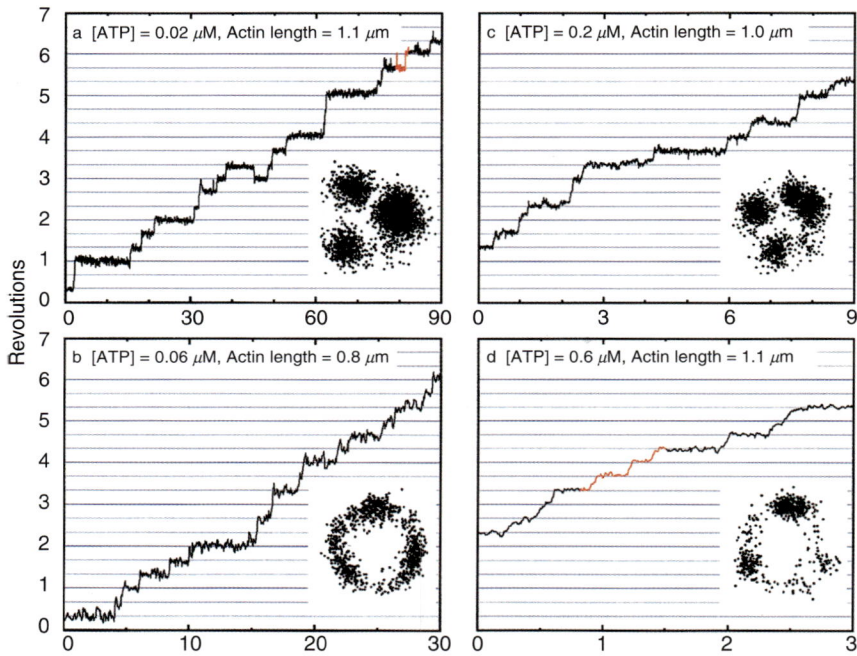

Figure 2-20. (A) Experimental setup used to measure the rotation of the F1 ATPase subunit. (B) Discrete rotations of 120° can be seen in the time course of the rotating filament [reprinted from Ref. 229 by permission of Elsevier].

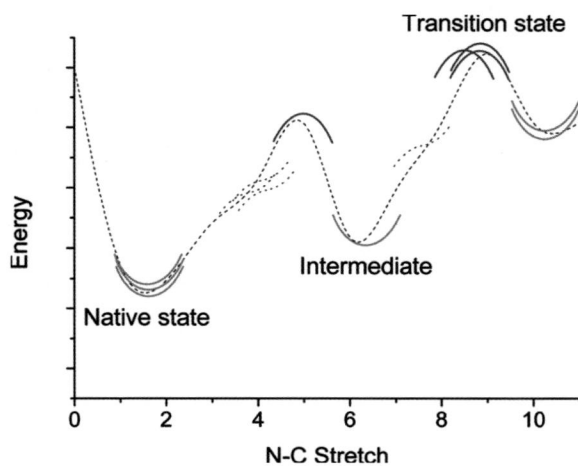

Figure 5-6. Molecular simulations of protein unfolding indicate transition state locations comparable to those measured in dynamic force spectroscopy [64]. Here, a simulation of the force-induced unfolding of I27 shows a transition state to unfolding 0.3 nm away from an intermediate. DFS of I27 produces a force scale $f_\beta = 12.8\,\text{pN}$ ($x^\ddagger = 0.32$).

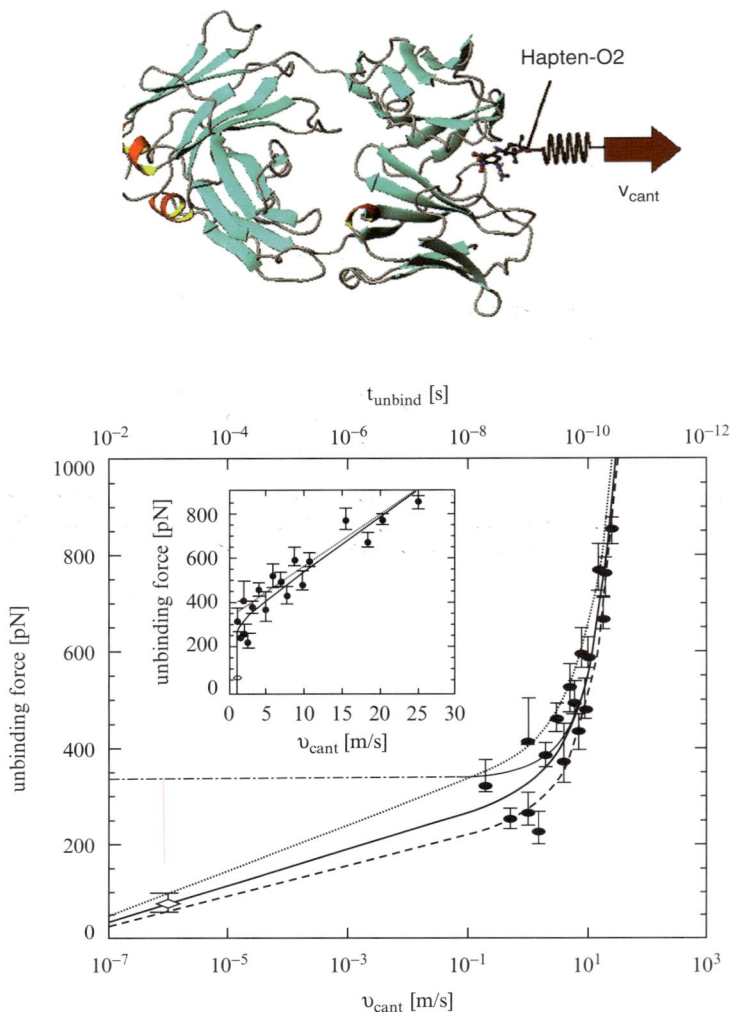

Figure 6-3. Grubmüller and co-workers performed MD simulations of the unbinding of a DNP-hapten ligand from the monoclonal antibody AN02 F$_{ab}$ at different pulling velocities to extrapolate the unbinding force to experimental timescales. (top) A model of the proteinligand complex. (bottom) Computed unbinding forces (filled circles) and predicted experimental unbinding force (diamond) as a function of pulling velocity. The lines show the best fit on Eq. 2 to the computed unbinding forces for two different values of the spontaneous rupture rate k_o. The inset shows the same data on a linear velocity scale, indicating that most of the simulaations occurred in the friction regime. Reprinted with permission from Reference 34. Copyright 1999 Elsevier.

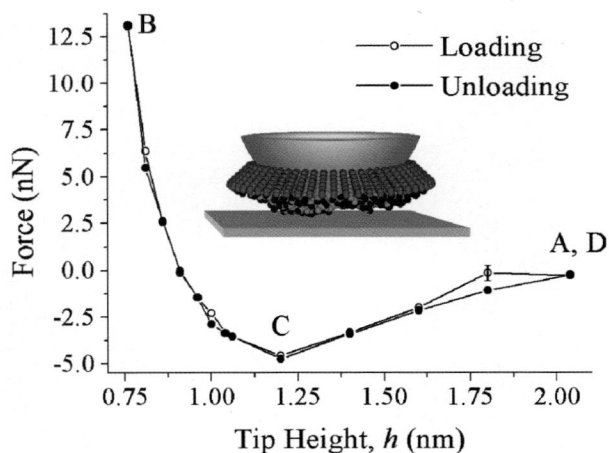

Figure 6-4. In a simulation one has the luxury of being able to move the interacting components in ways that would be difficult to realize in an experiment. This example presents results from a series of MD simulations modeling the interaction between a chemical force microscope stylus and a smooth wall during both loading and unloading. The figure illustrates a widely-used strategy for approximating infinitely slow motion in simulations of force spectroscopy—to hold the interacting components rigidly fixed at discrete separations and simulate until steady-state conditions have been achieved. Reprinted with permission from Reference 78. Copyright 2003 American Chemical Society.

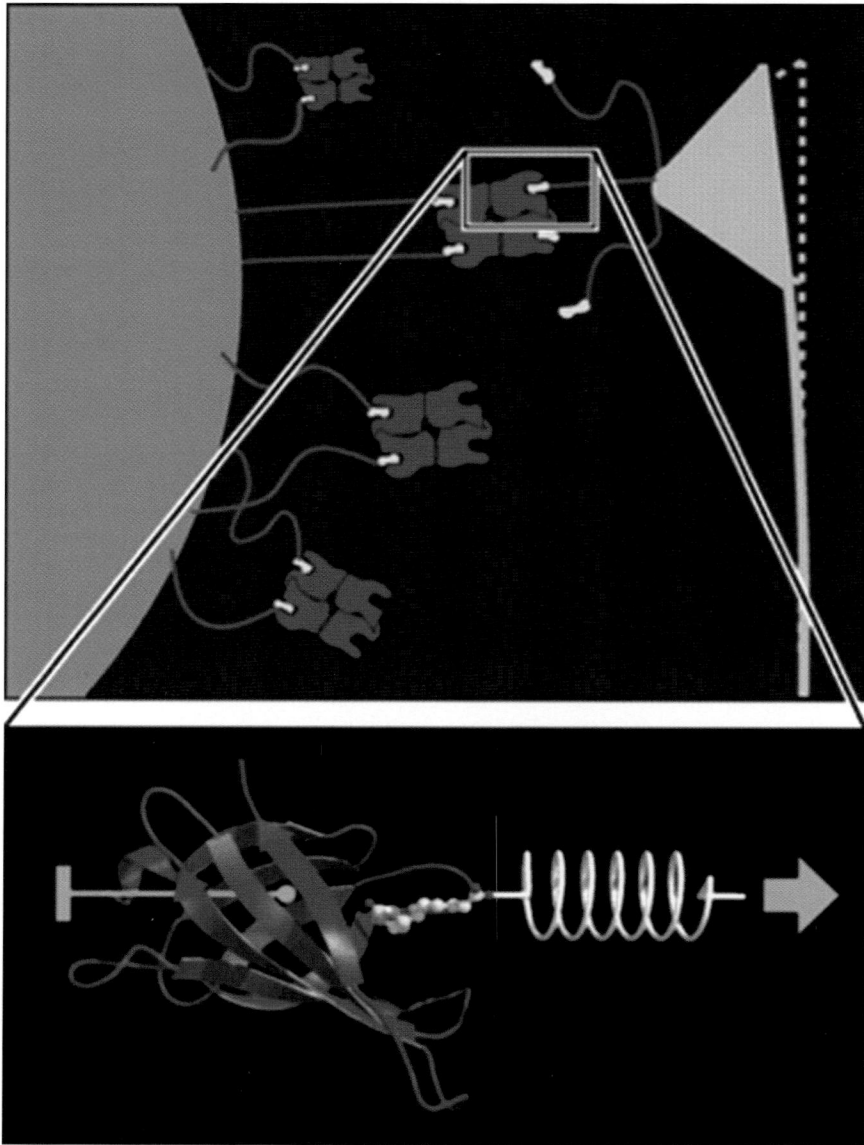

Figure 6-6. Simulations modeling the pulling force of an AFM tip to induce unbinding of a biotin ligand from a streptavidin receptor. (top) a cartoon of the experimental measurement. (bottom) The system modeled by simulation. Reprinted with permission from Reference [36]. Copyright 1996 AAAS.

Figure 6-10. A snapshot from an MD simulation of 8000 Gay-Berne particles solvating two colloidal rods. Shading emphasizes particle orientation. MD simulations like these are used to model interaction forces in LCs. Reprinted with permission from Reference 86. Copyright 2003 American Physical Society.

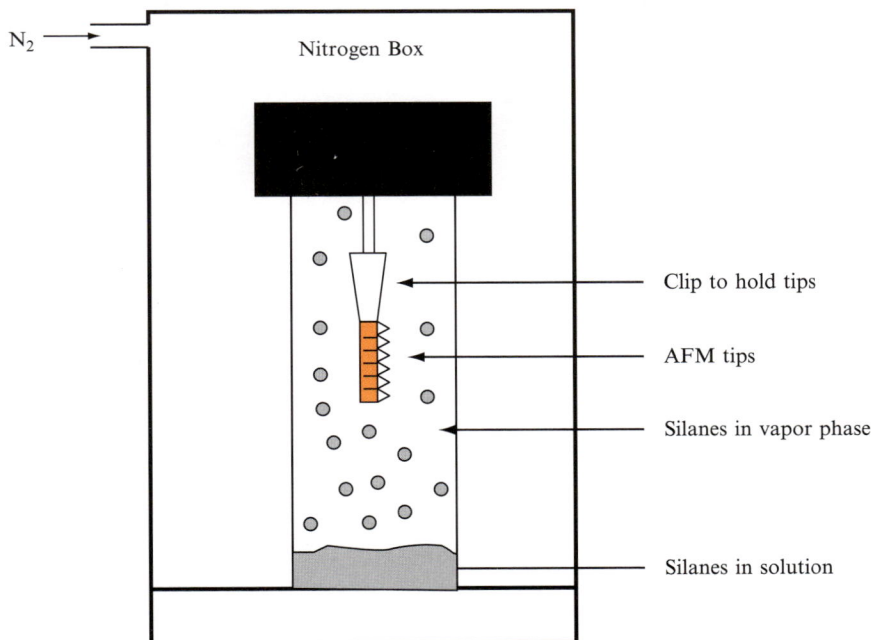

Figure 7-3. Diagram of experimental set up for tip functionalization from the vapor. Tips are clamped in a sealed vial containing ~ 400 mL of silane in solution. AFM tips are to be clamped such that the tips do not contact the silane solution. Silane will evaporate and silanization of the AFM tips will occur in the vapor phase. This entire process should take place under nitrogen or in a nitrogen box to prevent water from polymerizing the silanes on the AFM tip.

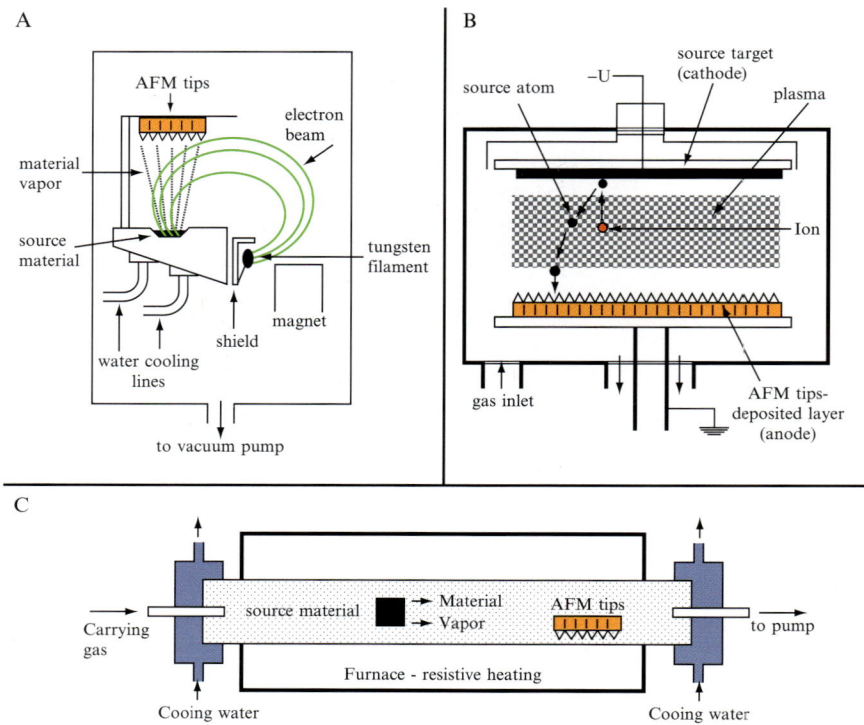

Figure 7-4. Methods to coat AFM cantilevers with metals. (A) Schematic of electron beam evaporation. (B) Schematic of sputtering. (C) Schematic of resistive evaporation.

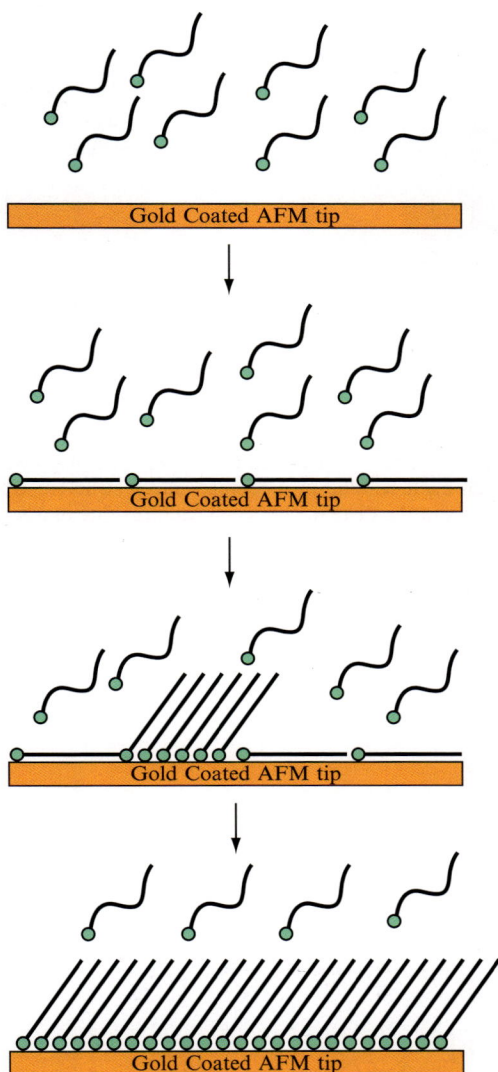

Figure 7-5. Formation of SAMs on gold coated AFM tips. Initially, the thiolyated alkyls bind to the surface via the gold-thiol interaction in a horizontal orientation. As the density increases the polymers will begin to interact, causing the polymers to lift off of the substrate to form islands. As the density of islands increases, they merge to eventually form a uniform monolayer where all the polymer tails have lifted off of the surface and are oriented ~ 30° from the normal.

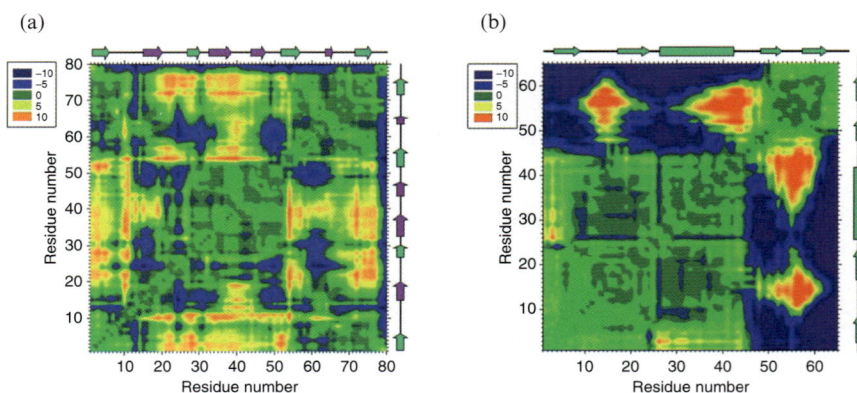

Figure 8-11. Difference distance maps for (A) E2lip3 when extended between the N-terminus and lys 41, and (B) protein L when extended by the N and C-termini. The maps were calculated by subtraction of the distance between each residue in a representative structure (from CVMD simulations) just after the transition state from the distance between each residue in a representative structure just before the transition state. An increase in the distance between pairs of amino acids is shown in blue, a reduction in red, and areas that remain constant in green. The position of secondary structural elements within each protein in the native state is shown top and right of each map. Inspection of each map suggests that upon extension, force is transmitted globally through E2lip3, extensively deforming the protein from its native state. protein L, however, shows a highly local response that results in the shearing of two mechanical sub-units past each other.

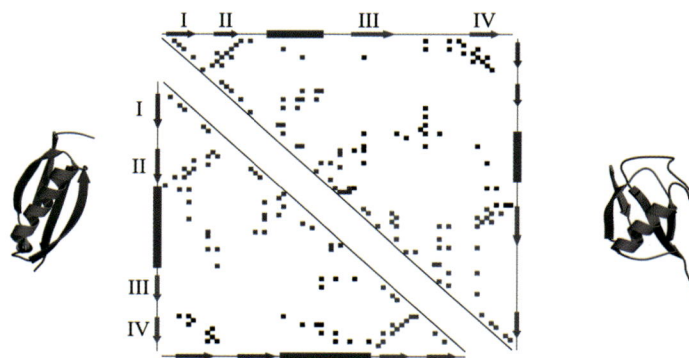

Figure 8-20. Contact map of protein L (bottom left) and ubiquitin (top right). Side chain contacts (the smallest distance between atoms of two residues < 5Å, calculated by CSU software[121]) made by pairs of amino acids within structural unit 1 (β-hairpin 1 and the helix) or within structural unit 2 (β-hairpin 2) are shown by green and red squares, respectively. Contacts made between these structural units are shown in black. β-strands (labelled I to IV) and α-helices were defined from the solution or X-ray structures using DSSP[108] and are shown as arrows and rectangles, respectively, alongside each contact map. The two structural units are coloured green (unit 1) and red (unit 2) in each protein and are also shown superimposed onto the three-dimensional structure of protein L (left) and ubiquitin (right)[54]. Used by permission.

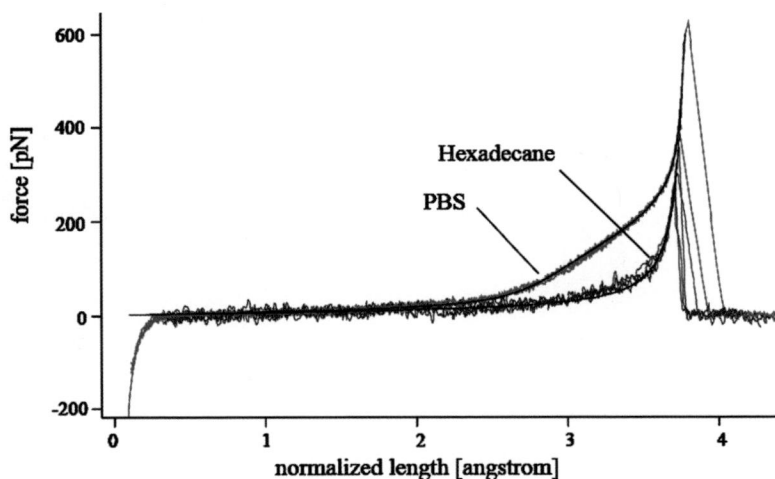

Figure 9-6. Force extension curves of PEG in a polar solvent (PBS) and a nonpolar solvent (hexadecane). The polar solvent stabilizes helical structures between PEG monomers that lead to distortion from the eFJC model. The freely-jointed chain with two-state conformation transition elements (TLS-eFJC) provides a more accurate fit (used with permission from Oesterhelt et al., Reference 26).

A.

B.

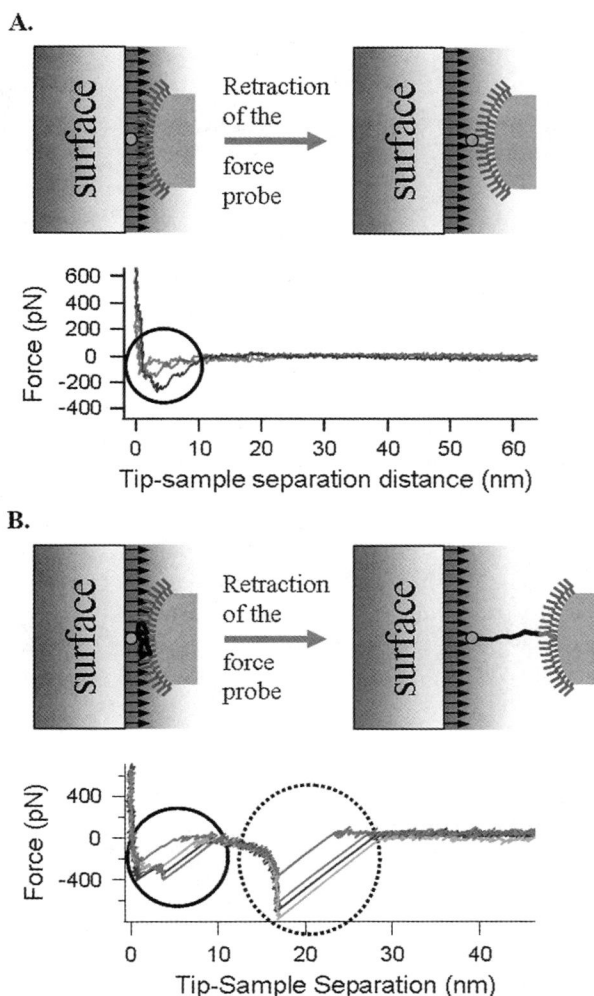

Figure 6. Effect of spacer for determining specific interactions. (A) When a spacer is not present, non-specific interactions occur at the same tip-sample separations as non-specific interactions (solid black circle), making it impossible to distinguish the interactions of interest. (B) When a polymer space is present the specific interactions (dashed black circle) occur further out from the surface than non-specific interactions (solid black circle) facilitating the analysis of interactions that are of interest.

Therefore chemical strategies of covalently attaching proteins through cysteine need to exploit the reactive properties of thiols. As mentioned in Section IIIa, thiols display chemical reactivity very similar to that of alcohols; thus one can use the vast knowledge of alcohol reactions to develop methods of attaching cysteine residues to the polymer. Three reactive thiol groups have been identified for this purpose: maleimides,[55, 56] vinyl sulfones,[55] and thiols (formation of disulfides).[55] The structures of these compounds are shown in Figure 7. It has been demonstrated that these functional groups are highly reactive toward the surface cysteine residues of proteins.[55, 56] When conducting these functionalization methods it is necessary to know the structure of the protein of interest, as the reaction is nonspecific in relation to where on the protein the attachment will occur—that is, which amino acid attaches to the polymer. For example, if the attachment happens to be in a binding pocket

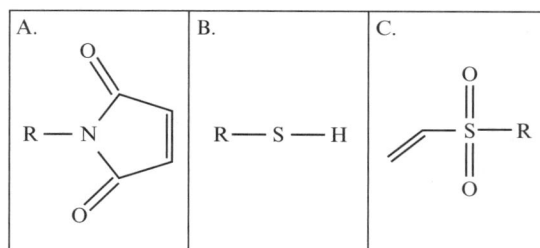

Figure 7. Structure of groups reactive to the thiol group of cysteine. **A.** Maleimides. **B.** Thiols. **C.** Vinyl sulfones.

the protein may be rendered inactive. An ideal protein candidate would be one in which all cysteine or lysine residues were located far from the active region of the protein. Attachment though the thiols of cysteine residues may offer an advantage over lysines in that cysteines are typically less abundant in proteins than amino groups (found in lysines). This results in more selective attachment of the target protein and also greater selective control over the position of attachment; therefore this reduces the likelihood of protein deactivation upon conjugation. The specific requirements (pH, ionic concentration, etc.) necessary for polymer-protein conjugation depends on which reactive groups (NHS, SPA, maleimides, vinyl sulfones, thiols, etc.) are used in the conjugation process. Most of this information is available in the Nektar advanced PEGylation catalogue, therefore the reader is encouraged to use this resource for more detail. An obvious additional requirement for this type of conjugation is functionality at both ends of the polymers, whether SAMs or other types of polymers (e.g., PEG). One end needs to contain the correct functionality to bind to the AFM surface (thiol for gold, NHS or SPA for amine-functionalized tips); the other end needs the correct functionality for covalent linkage to the protein.

IVb Single-Molecule Tip Functionalization

One of the most significant achievements of CFM was developing methods to conduct single-molecule force measurements for biologically relevant systems. A vast majority of molecular recognition events occurring at cellular membrane interfaces are initiated under a shear stress or applied force, including viral- and bacterial-host attachment, antibody-antigen recognition, cell-cell adhesion, and lectin agglutination. Molecular binding interactions for these biological processes have commonly been investigated through molecules free in solution[57] or directly affixed to a solid support.[58] In several of these processes the ligand and/or protein are anchored via a flexible or semi-flexible tether to a membrane, such as a cell, or a viral or bacterial membrane. Although equilibrium measurements have provided important insight in molecular recognition events, they largely ignore the fact that under physiological conditions these interactions are occurring at membrane interfaces, under non-equilibrium conditions, and within a limited interaction volume. It has been shown that equilibrium measurements, where at least one molecular entity is free in solution, does not scale well with the binding properties of membrane-bound molecules restricted to the two-dimensional plane of the membrane.[59, 60] *Ricinus communis* agglutinin (RCA), a known potent crosslinker of erythrocytes, and *Viscum album* agglutinin (VAA), a weak agglutinating lectin, were both shown to exhibit the same equilibrium off-rate, 1×10^{-3} s^{-1}; but under an applied force as measured through single-molecule force spectroscopy the rupture strengths were 1.5 times greater for RCA, 65.9 pN, than VAA, 43.5 pN.[5] This demonstrated that binding strengths cannot be predicted based on zero-force dissociation kinetics and equilibrium thermodynamics for biological recognition events occurring under an applied force or shear stress. Therefore, biological processes that occur under an applied force, such as viral-host cell adhesion and agglutination of cells, are better characterized using single-molecule force measurements under different loading rates to represent the physiological conditions more accurately.

The method of single-molecule force spectroscopy has only been possible through the developments in the techniques of tip functionalization presented in this chapter. Functionalization of tips with single molecules for applications in single-molecule force measurements requires a single polymer spacer. As discussed above (Section IVa), there is a wide range of nonspecific interactions that occur between the tip and substrate at close distance, making it difficult to determine the specific force of interest. This problem is amplified for single-molecule studies since the signal to noise ratio drastically drops, i.e., the rupture force of a single interaction is much lower than the force of rupture for several simultaneous interactions. Thus the nonspecific interactions dominate the rupture force. In fact, Ratto et al., and others, have recently used a double-tethered system to further delineate nonspecific interactions from specific interactions.[4, 7]

The main difficulty in conducting single force measurements is attaching a single molecule to the tip via a polymer spacer. For single-molecule tip functionalization polymer attachment is rarely done through thiol-gold adsorption. As mentioned in Section IIIb, this method allows very little control over the number of molecules that attach, since any thiol group can adsorb to gold. In addition, single-molecule studies on SAMs can prove difficult, as one cannot readily regulate the number of reactive polymers incorporated into SAMs. As a result, most single-molecule studies have used a functionalization scheme in which either silanization or esterification is used to attach an amine, followed by attachment of a polymer (generally PEG), followed by attachment of the biological molecule of interest (this reaction scheme was outlined in Section IVa without an emphasis on single-molecule attachment).

This then leads to the question, at what point is the chemistry adjusted to allow for single-molecule attachment? This typically occurs in the initial step of single-molecule tip functionalization during either silanization or esterification. *Therefore the key to single-molecule functionalization lies in the initial step of functionalization, attaching only one reactive amine to the tips during silanization or esterification.* Attempting to control single-molecule addition further down the functionalization process proves to be quite difficult, since the reaction involves highly reactive groups (NHS, SPA, maleimides, vinyl sulfones, thiols, etc). Controlling the addition of just one reactive group during silanization and esterification can be achieved by using a ratio of reactive amine molecules (e.g., APTES for silanization, ethanolamine for esterification) to nonreactive molecules, such as methyl-terminated molecules (e.g., methyltriethoxysilane (MTES) for silanization or methanol for esterification). If this ratio is adjusted properly, one may get attachment of a single amine-terminated group per tip; thus in the subsequent steps (attachment of polymer and biological molecule) the tip will contain a single biological molecule. Due to the small size of the AFM tip it is very difficult to characterize the tip after functionalization. Therefore in practice the molecular ratio of active amines to nonactive methyls in esterification and silanization is adjusted such that there is only one functional tip per 5–15 tips. This greatly increases the odds that the one tip displaying functionality contains only a single biological molecule. The most significant limitation of this method is that currently there is no independent method to determine that a tip bears a single functional biomolecule. Thus multiple control experiments, such as blocking the interaction with free ligands in solution, measuring interactions on samples known not to possess binding groups, etc., are required for high confidence experiments.

Figure 8. Chemical strategy for single molecule attachment of con-A. Chemical structures on the right are the
actual tip surface chemistry, images on the left are cartoon
representations of the chemistry occurring at the tip
surface. Initially tips are piranha etched for cleaning.
Vapor phase silanization is used to attach silanes with a
ratio of 1 APTES: 10 MTES in solution. This is followed
by attachment of SPA-PEG-SPA and then the Con A
protein. The chemistry involved in these last two steps is
covered in the text.

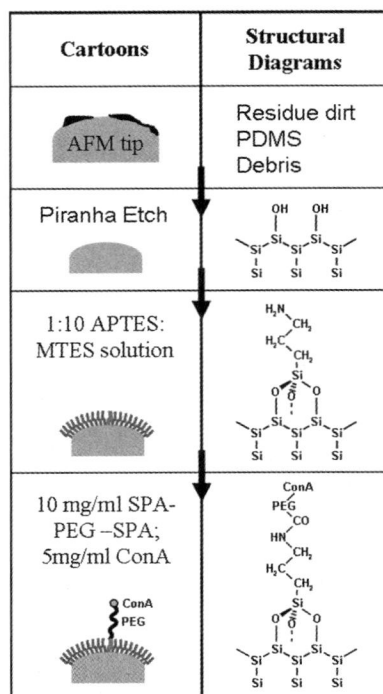

Cartoons	Structural Diagrams
AFM tip	Residue dirt PDMS Debris
Piranha Etch	OH OH ...
1:10 APTES: MTES solution	...
10 mg/ml SPA-PEG –SPA; 5mg/ml ConA	ConA ...

Figure 8 outlines a functionalization scheme that was shown to be successful for functionalizing AFM tips with a single concanavalin-A protein (con-A).[4, 7] The initial step in this
functionalization scheme was silanization from the vapor phase. To achieve a silane layer
containing a single amine group, the solution from which the vapor was produced contained
a molar ratio of 1 APTES:10 MTES. Due to the different vapor pressures exerted by each
silane, this results in a molar ratio of ~1 APTES:250 MTES in the vapor phase. This was followed by PEG attachment in chloroform at a PEG concentration of 10 mg/ml. Finally, con-A
was attached by incubating the tips in a phosphate buffer containing 2 mg/ml solution of con-
A at pH 8.0. It was determined from this functionalization scheme that ~ 1 out of 10 tips were
functionalized with con-A, and single-molecule attachment, as opposed to multiple-molecule
attachment, was determined through rupture distances and forces.

Single-molecule force spectroscopy is a relatively new field that contains a great deal of
potential. The largest hurdle to reproducibly conducting these studies lies in the chemistry of
tip functionalization. Through the methods presented in the previous chapters, one can readily
overcome this hurdle if careful attention is made to the method and chemical strategies used
during tip functionalization.

Conclusions

The main objective of this chapter was to give an overview of the different methods
used to functionalize AFM tips for applications in CFM. The importance of tip functionalization should not be underestimated, since the basis of all studies done with CFM start

with the chemical methods used to impart functionality to AFM tips. In this chapter we presented a wide range of methods to functionalize tips, starting with the initial stage of functionalization, followed by subsequent steps to build up layers of tip functionality. There is a wide range of different methods and chemical strategies to achieve the same type of tip functionality, and thus the method one chooses will depend primarily on the application (e.g., single-molecule properties vs. surface properties). In addition, there are alternative combinations of the chemical processes presented in this chapter that were not extensively covered, but the material should be sufficient for one to understand how different combinations may be effective. The intent of this chapter was to provide the essential information about tip functionalization for someone at the early stages of learning CFM, so that in the future they may contribute to this field through advances in both the method of experimentation and tip functionalization.

This work was performed under the auspices of the U.S. Department of Energy by the University of California, Lawrence Livermore National Laboratory, under Contract W-7405-Eng-48.

References

1. Dammer, U.; Hegner, M.; Anselmetti, D.; Wagner, P.; Dreier, M.; Huber, W.; Guntherodt, H. J., Specific antigen/antibody interactions measured by force microscopy. *Biophysical Journal* **1996,** 70, (5), 2437–2441.
2. Hinterdorfer, P.; Baumgartner, W.; Gruber, H. J.; Schilcher, K.; Schindler, H., Detection and localization of individual antibody-antigen recognition events by atomic force microscopy. *Proceedings of the National Academy of Sciences of the United States of America* **1996,** 93, (8), 3477–3481.
3. Allen, S.; Chen, X. Y.; Davies, J.; Davies, M. C.; Dawkes, A. C.; Edwards, J. C.; Roberts, C. J.; Sefton, J.; Tendler, S. J. B.; Williams, P. M., Detection of antigen-antibody binding events with the atomic force microscope. *Biochemistry* **1997,** 36, (24), 7457–7463.
4. Ratto, T. V.; Langry, K. C.; Rudd, R. E.; Balhorn, R. L.; Allen, M. J.; McElfresh, M. W., Force spectroscopy of the double-tethered concanavalin-A mannose bond. *Biophysical Journal* **2004,** 86, (4), 2430–2437.
5. Dettmann, W.; Grandbois, M.; Andre, S.; Benoit, M.; Wehle, A. K.; Kaltner, H.; Gabius, H. J.; Gaub, H. E., Differences in zero-force and force-driven kinetics of ligand dissociation from beta-galactoside-specific proteins (plant and animal lectins, immunoglobulin G) monitored by plasmon resonance and dynamic single molecule force microscopy. *Archives of Biochemistry and Biophysics* **2000,** 383, (2), 157–170.
6. Fritz, J.; Katopodis, A. G.; Kolbinger, F.; Anselmetti, D., Force-mediated kinetics of single P-selectin ligand complexes observed by atomic force microscopy. *Proceedings of the National Academy of Sciences of the United States of America* **1998,** 95, (21), 12283–12288.
7. Ratto, T. V.; Rudd, R. E.; Langry, K. C.; Balhorn, R. L.; McElfresh, M. W., Nonlinearly additive forces in multivalent ligand binding to a single protein revealed with force spectroscopy. *Langmuir* **2006,** 22, (4), 1749–1757.
8. Gourianova, S.; Willenbacher, N.; Kutschera, M., Chemical force microscopy study of adhesive properties of polypropylene films: Influence of surface polarity and medium. *Langmuir* **2005,** 21, (12), 5429–5438.
9. Tormoen, G. W.; Drelich, J.; Beach, E. R., Analysis of atomic force microscope pull-off forces for gold surfaces portraying nanoscale roughness and specific chemical functionality. *Journal of Adhesion Science and Technology* **2004,** 18, (1), 1–17.
10. Sheng, X. X.; Ward, M. D.; Wesson, J. A., Adhesion between molecules and calcium oxalate crystals: Critical interactions in kidney stone formation. *Journal of the American Chemical Society* **2003,** 125, (10), 2854–2855.
11. Noy, A.; Vezenov, D. V.; Kayyem, J. F.; Meade, T. J.; Lieber, C. M., Stretching and breaking duplex DNA by chemical force microscopy. *Chemistry & Biology* **1997,** 4, (7), 519–527.
12. Knapp, H. F.; Stemmer, A., Preparation, comparison and performance of hydrophobic AFM tips. *Surface and Interface Analysis* **1999,** 27, (5–6), 324–331.
13. Piramowicz, M. D.; Czuba, P.; Targosz, M.; Burda, K.; Szymonski, M., Dynamic force measurements of avidin-biotin and streptavdin-biotin interactions using AFM. *Acta Biochimica Polonica* **2006,** 53, (1), 93–100.
14. Desmeules, P.; Grandbois, M.; Bondarenko, V. A.; Yamazaki, A.; Salesse, C., Measurement of membrane binding between recoverin, a calcium-myristoyl switch protein, and lipid bilayers by AFM-based force spectroscopy. *Biophysical Journal* **2002,** 82, (6), 3343–3350.
15. Langry, K. C.; Ratto, T. V.; Rudd, R. E.; McElfresh, M. W., The AFM measured force required to rupture the dithiolate linkage of thioctic acid to gold is less than the rupture force of a simple gold-alkyl thiolate bond. *Langmuir* **2005,** 21, (26), 12064–12067.

16. Klein, D. C. G.; Stroh, C. M.; Jensenius, H.; van Es, M.; Kamruzzahan, A. S. M.; Stamouli, A.; Gruber, H. J.; Oosterkamp, T. H.; Hinterdorfer, P., Covalent immobilization of single proteins on mica for molecular recognition force microscopy. *Chemphyschem* **2003**, 4, (12), 1367–1371.

17. Kienberger, F.; Kada, G., Gruber, H.J., Pastushenko, V.,Riener, C., Trieb, M., Knaus,H-G., Schindler, H., Hinterdorfer, P., Recognition Force Spectroscopy Studies of the NTA-His6 Bond. *Single Molecule* **2000**, 1, (1), 59–65.

18. Bonanni, B.; Kamruzzahan, A. S. M.; Bizzarri, A. R.; Rankl, C.; Gruber, H. J.; Hinterdorfer, P.; Cannistraro, S., Single molecule recognition between Cytochrome C 551 gold-immobilized Azurin by force spectroscopy. *Biophysical Journal* **2005**, 89, (4), 2783–2791.

19. Kienberger, F.; Kada, G.; Mueller, H.; Hinterdorfer, P., Single molecule studies of antibody-antigen interaction strength versus intra-molecular antigen stability. *Journal of Molecular Biology* **2005**, 347, (3), 597–606.

20. Hinterdorfer, P.; Gruber, H. J.; Kienberger, F.; Kada, G.; Riener, C.; Borken, C.; Schindler, H., Surface attachment of ligands and receptors for molecular recognition force microscopy. *Colloids and Surfaces B-Biointerfaces* **2002**, 23, (2–3), 115–123.

21. Harada, Y.; Kuroda, M.; Ishida, A., Specific and quantized antigen-antibody interaction measured by atomic force microscopy. *Langmuir* **2000**, 16, (2), 708–715.

22. Touhami, A.; Hoffmann, B.; Vasella, A.; Denis, F. A.; Dufrene, Y. F., Probing specific lectin-carbohydrate interactions using atomic force microscopy imaging and force measurements. *Langmuir* **2003**, 19, (5), 1745–1751.

23. Bustanji, Y.; Arciola, C. R.; Conti, M.; Mandello, E.; Montanaro, L.; Samori, B., Dynamics of the interaction between a fibronectin molecule and a living bacterium under mechanical force. *Proceedings of the National Academy of Sciences of the United States of America* **2003**, 100, (23), 13292–13297.

24. Lee, G. U.; Chrisey, L. A.; Colton, R. J., Direct Measurement of the Forces between Complementary Strands of DNA. *Science* **1994**, 266, (5186), 771–773.

25. Hugel, T.; Holland, N. B.; Cattani, A.; Moroder, L.; Seitz, M.; Gaub, H. E., Single-molecule optomechanical cycle. *Science* **2002**, 296, (5570), 1103–1106.

26. Hegner, M.; Wagner, P.; Semenza, G., Immobilizing DNA on Gold Via Thiol Modification for Atomic-Force Microscopy Imaging in Buffer Solutions. *Febs Letters* **1993**, 336, (3), 452–456.

27. Conti, M.; Bustanji, Y.; Falini, G.; Ferruti, P.; Stefoni, S.; Samori, B., The desorption process of macromolecules adsorbed on interfaces: The force spectroscopy approach. *Chemphyschem* **2001**, 2, (10), 610–613.

28. Wal, M. V.; Kamper, S.; Headley, J.; Sinniah, K., Effects of contact force and salt concentration on the unbinding of a DNA duplex by force spectroscopy. *Langmuir* **2006**, 22, (3), 882–886.

29. Vezenov, D. V.; Noy, A.; Rozsnyai, L. F.; Lieber, C. M., Force titrations and ionization state sensitive imaging of functional groups in aqueous solutions by chemical force microscopy. *Journal of the American Chemical Society* **1997**, 119, (8), 2006–2015.

30. Thomas, R. C.; Tangyunyong, P.; Houston, J. E.; Michalske, T. A.; Crooks, R. M., Chemically-Sensitive Interfacial Force Microscopy - Contact Potential Measurements of Self-Assembling Monolayer Films. *Journal of Physical Chemistry* **1994**, 98, (17), 4493–4494.

31. Barrat, A.; Silberzan, P.; Bourdieu, L.; Chatenay, D., How Are the Wetting Properties of Silanated Surfaces Affected by Their Structure - an Atomic-Force Microscopy Study. *Europhysics Letters* **1992**, 20, (7), 633–638.

32. Friedsam, C.; Becares, A. D.; Jonas, U.; Gaub, H. F.; Seitz, M., Polymer functionalized AFM tips for long-term measurements in single-molecule force spectroscopy. *Chemphyschem* **2004**, 5, (3), 388–393.

33. Hinterdorfer, P.; Dufrene, Y. F., Detection and localization of single molecular recognition events using atomic force microscopy. *Nature Methods* **2006**, 3, (5), 347–355.

34. Lo, Y. S.; Huefner, N. D.; Chan, W. S.; Dryden, P.; Hagenhoff, B.; Beebe, T. P., Organic and inorganic contamination on commercial AFM cantilevers. *Langmuir* **1999**, 15, (19), 6522–6526.

35. Ossenkamp, G. C.; Kemmitt, T.; Johnston, J. H., Toward functionalized surfaces through surface esterification of silica. *Langmuir* **2002**, 18, (15), 5749–5754.

36. Riener, C. K.; Stroh, C. M.; Ebner, A.; Klampfl, C.; Gall, A. A.; Romanin, C.; Lyubchenko, Y. L.; Hinterdorfer, P.; Gruber, H. J., Simple test system for single molecule recognition force microscopy. *Analytica Chimica Acta* **2003**, 479, (1), 59–75.

37. Green, J. B. D.; Lee, G. U., Atomic force microscopy with patterned cantilevers and tip arrays: Force measurements with chemical arrays. *Langmuir* **2000**, 16, (8), 4009–4015.

38. Awada, H.; Castelein, G.; Brogly, M., Use of chemically modified AFM tips as a powerful tool for the determination of surface energy of functionalised surfaces. *Journal De Physique Iv* **2005**, 124, 129–134.

39. Hu, D. H.; Micic, M.; Klymyshyn, N.; Suh, Y. D.; Lu, H. P., Correlated topographic and spectroscopic imaging beyond diffraction limit by atomic force microscopy metallic tip-enhanced near-field fluorescence lifetime microscopy. *Review of Scientific Instruments* **2003**, 74, (7), 3347–3355.

40. Micic, M.; Chen, A.; Leblanc, R. M.; Moy, V. T., Scanning electron microscopy studies of protein-functionalized atomic force microscopy cantilever tips. *Scanning* **1999**, 21, (6), 394–397.

41. Vezenov, D. V.; Zhuk, A. V.; Whitesides, G. M.; Lieber, C. M., Chemical force spectroscopy in heterogeneous systems: Intermolecular interactions involving epoxy polymer, mixed monolayers, and polar solvents. *Journal of the American Chemical Society* **2002,** 124, (35), 10578–10588.

42. Mahapatro, M.; Gibson, C.; Abell, C.; Rayment, T., Chiral discrimination of basic and hydrophobic molecules by chemical force spectroscopy. *Ultramicroscopy* **2003,** 97, (1–4), 297–301.

43. Brant, J. A.; Johnson, K. M.; Childress, A. E., Characterizing NF and RO membrane surface heterogeneity using chemical force microscopy. *Colloids and Surfaces a-Physicochemical and Engineering Aspects* **2006,** 280, (1–3), 45–57.

44. Strunz, T.; Oroszlan, K.; Schafer, R.; Guntherodt, H. J., Dynamic force spectroscopy of single DNA molecules. *Proceedings of the National Academy of Sciences of the United States of America* **1999,** 96, (20), 11277–11282.

45. Sulchek, T.; Friddle, R. W.; Noy, A., Strength of multiple parallel biological bonds. *Biophysical Journal* **2006,** 90, (12), 4686–4691.

46. Raible, M.; Evstigneev, M.; Bartels, F. W.; Eckel, R.; Nguyen-Duong, M.; Merkel, R.; Ros, R.; Anselmetti, D.; Reimann, P., Theoretical analysis of single-molecule force spectroscopy experiments: Heterogeneity of chemical bonds. *Biophysical Journal* **2006,** 90, (11), 3851–3864.

47. Ellis, J. S.; Allen, S.; Chim, Y. T. A.; Roberts, C. J.; Tendler, S. J. B.; Davies, M. C., Molecular-scale studies on biopolymers using atomic force microscopy. *Polymer Therapeutics Ii: Polymers as Drugs, Conjugates and Gene Delivery Systems* **2006,** 193, 123–172.

48. Zhang, W.; Barbagallo, R.; Madden, C.; Roberts, C. J.; Woolford, A.; Allen, S., Progressing single biomolecule force spectroscopy measurements for the screening of DNA binding agents. *Nanotechnology* **2005,** 16, (10), 2325–2333.

49. Rief, M.; Clausen-Schaumann, H.; Gaub, H. E., Sequence-dependent mechanics of single DNA molecules. *Nature Structural Biology* **1999,** 6, (4), 346–349.

50. Krautbauer, R.; Pope, L. H.; Schrader, T. E.; Allen, S.; Gaub, H. E., Discriminating small molecule DNA binding modes by single molecule force spectroscopy. *Febs Letters* **2002,** 510, (3), 154–158.

51. Fujihira, M.; Tani, Y.; Furugori, M.; Akiba, U.; Okabe, Y., Chemical force microscopy of self-assembled monolayers on sputtered gold films patterned by phase separation. *Ultramicroscopy* **2001,** 86, (1–2), 63–73.

52. Ling, L. S.; Butt, H. J.; Berger, R., Rupture force between the third strand and the double strand within a triplex DNA. *Journal of the American Chemical Society* **2004,** 126, (43), 13992–13997.

53. Hards, A.; Zhou, C. Q.; Seitz, M.; Brauchle, C.; Zumbusch, A., Simultaneous AFM manipulation and fluorescence imaging of single DNA strands. *Chemphyschem* **2005,** 6, (3), 534–540.

54. Morii, T.; Mizuno, R.; Haruta, H.; Okada, T., An AFM study of the elasticity of DNA molecules. *Thin Solid Films* **2004,** 464–65, 456–458.

55. Herman, S.; Loccufier, J.; Schacht, E., End-Group Modification of Alpha-Hydro-Omega-Methoxy-Poly(Oxyethylene).3. Facile Methods for the Introduction of a Thiol-Selective Reactive End-Group. *Macromolecular Chemistry and Physics* **1994,** 195, (1), 203–209.

56. Goodson, R. J.; Katre, N. V., Site-Directed Pegylation of Recombinant Interleukin-2 at Its Glycosylation Site. *Bio-Technology* **1990,** 8, (4), 343–346.

57. Lavigne, J. J.; Anslyn, E. V., Sensing a paradigm shift in the field of molecular recognition: From selective to differential receptors. *Angewandte Chemie-International Edition* **2001,** 40, (17), 3119–3130.

58. McDonnell, J. M., Surface plasmon resonance: towards an understanding of the mechanisms of biological molecular recognition. *Current Opinion in Chemical Biology* **2001,** 5, (5), 572–577.

59. Chang, K. C.; Hammer, D. A., The forward rate of binding of surface-tethered reactants: Effect of relative motion between two surfaces. *Biophysical Journal* **1999,** 76, (3), 1280–1292.

60. Riper, J. W.; Swerlick, R. A.; Zhu, C., Determining force dependence of two-dimensional receptor-ligand binding affinity by centrifugation. *Biophysical Journal* **1998,** 74, (1), 492–513.

8

The Dynamical Response of Proteins Under Force

Kirstine L. Anderson, Sheena E. Radford, D. Alastair Smith, and David J. Brockwell

1 Introduction

At the macroscopic level, it is well understood that ordered assemblies of proteins are used extensively throughout biology to provide structures that are mechanically strong and yet not brittle. Examples include the triple coiled coil of collagen, the β-sheets of silk, and the coiled coil rods of keratin, a protein found in nails, claws, and skin tissue. More recently, it has been realised that at a cellular and subcellular level, there are also many proteins whose function requires them to resist mechanical deformation. Indeed at the subcellular (nanoscale) level, force is ubiquitous and is not only important in systems with a clear mechanical function such as processive motors that run on tracks (e.g., myosin-actin and kinesin-microtubules) but is also thought to play a role in mechano-signalling[1], fibrillogenesis[1], and protein degradation[2]. At this length scale, many protein systems that react to force are either monomeric or are expressed as tandem arrays of domains with closely related topologies.

Until recently the mechanical properties of single protein molecules were unknown, as techniques that allowed the manipulation and accurate quantification of their properties were unavailable. The development of robust atomic force microscope (AFM) and tweezer methodologies (controlled by either a magnetic field or a light source) has meant that single molecule manipulation of both nucleic acids and proteins can now be performed routinely. The availability of 'off the shelf' hardware, together with the development of efficient protein engineering techniques has resulted in a revolution in our understanding of the mechanical properties of proteins at the nanoscopic level over the last eight years. Force-mode AFM, the subject of this chapter, is an important technique that has been used extensively to characterise the mechanical and dynamical response of many different proteins at the single molecule level. The first aim of this chapter is to describe to the nonspecialist how these experiments are executed and how the resulting data are analysed. To this end, the molecular biology protocols used to construct biomolecules amenable to study, the forced unfolding methods used, and the techniques to interpret the data will be described. The second aim of this chapter, with reference to the literature, is to discuss the determinants of a protein's mechanical resistance and how nature has exploited these principles to generate a wide variety of mechanical phenotypes. An exciting aspect of this field of research is the continual development of instrumentation that allows previously impossible or impracticable experiments to be undertaken. As a consequence, the final aim of this chapter is to briefly discuss newly emerging techniques that are related to force-mode AFM and discuss how these can be applied to investigate the biophysical properties of proteins.

1.1 Force Perturbs the Energy Landscape

In mechanical unfolding experiments using the AFM, the 'strength' of a protein is measured either by increasing the distance between fixed points on the protein at a constant rate and measuring the force applied by the polypeptide chain onto a force sensor (a technique known as constant velocity mechanical unfolding), or by applying a constant or increasing force at fixed points on the protein and measuring the time over which each molecule remains in its native state (constant force or force-ramp mechanical unfolding). As many of the proteins that are studied by this technique readily unfold on a millisecond–second timescale under force but remain in a folded state over days in the absence of force[3], it is clear that the application of force onto a protein dramatically increases its probability of unfolding. Before the effects of force on a protein are discussed more fully, terms such as 'energy landscape' and 'intrinsic unfolding rate constant' that are used throughout this chapter will be introduced by reference, not to unfolding, but to protein folding.

Protein folding from a denatured state to the native state is a complex process governed by both thermodynamic and kinetic parameters. The process is often visualised in terms of 'free energy diagrams' (see Figure 1). In this two-state free energy diagram only the fully folded native state and the unfolded state are populated. Under equilibrium conditions, the largest proportion of the protein will be found in the lowest free energy region of the landscape which, for the protein shown in Figure 1, is the native state. The difference in free energy between the native and unfolded states is termed the thermodynamic stability (ΔG_{UN}).[1] However, the rate at which the protein folds from its unfolded to its native state will depend upon the height of the free-energy barrier to folding, i.e. the difference in free

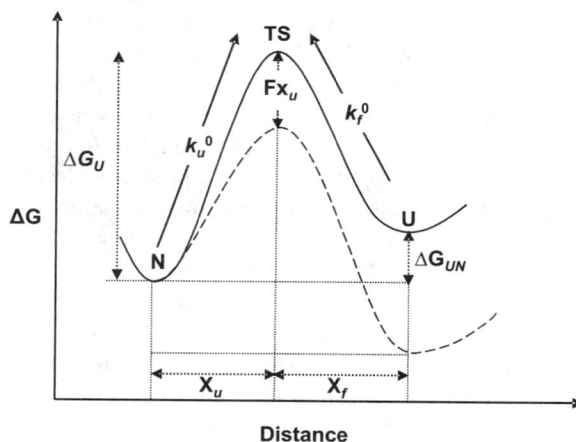

Distance

Figure 1. Free energy diagram of the unfolding landscape of a protein in the absence (solid line) and the presence (dashed line) of force, F. Upon application of a force, the energy landscape is tilted by $F(cos\theta)x$, where x is the distance reaction coordinate and θ is the angle between this coordinate and the geometry of the applied force. This stabilises the transition state (TS) and unfolded state (U), but has no effect on the native state (N). Reduction of the energy barrier to unfolding (ΔG_u) by Fx_u exponentially increases the unfolding rate constant (k_u^0), but exponentially decreases the folding rate constant (k_f^0). x_u and x_f are the distance to the transition state along the reaction coordinate from the unfolded and folded states, respectively.

---

[1] All of the free energy terms discussed in this Chapter relate to differences in standard free energies which are normally referred to as $\Delta G°$. However, for simplicity, differences in standard free energy will be referred to in this Chapter as ΔG.

energy of the transition and denatured states. This rate, which is usually measured by stopped flow techniques, is termed the intrinsic folding rate constant or k_f^0. In the case of unfolding proteins, the height of the transition state barrier to be traversed is defined by the free energy difference between the native and transition states, which is termed the intrinsic unfolding rate constant or k_u^0.

The effect of force on the unfolding of a protein can be thought of in terms of the free energy diagram discussed above (see Figure 1). With no applied force, the lowest energy state of the protein is the native state. To reach an unfolded state, the protein must first pass the high-energy transition state barrier. The height of this barrier (ΔG_u) defines the rate at which the protein will unfold due to thermal fluctuations. Upon application of a force F, the energy landscape is tilted by a function $-Fx(\cos\theta)$, where x represents the mechanical reaction coordinate of distance from the native well and θ is the angle between the mechanical reaction coordinate and the applied force. The transition state barrier will thus be reduced to ΔG_u-Fx_u, where x_u is the distance between the native well and the transition state. This increases the probability that the protein will pass over the transition state and become unfolded. The exponential increase in the unfolding rate constant under the application of force ($k_u^{(F)}$) can be described as:

$$k_u^{(F)} = \omega e^{-\frac{\Delta G_u - Fx_u}{k_B T}} \tag{1}$$

where ω is the pre-exponential factor which describes the diffusivity of states in the over-damped limit, k_B is Boltzmann's constant, and T is the temperature. This can be rewritten[4] as:

$$k_u^{(F)} = k_u^{(0F)} e^{\frac{Fx_u}{k_B T}} \tag{2}$$

In this equation, the intrinsic unfolding rate constant is that for the process under *zero applied force* ($k_u^{(0F)}$). If mechanical and chemical unfolding experiments measure the same barrier to unfolding, then the intrinsic unfolding rate constant measured by each technique should be equal. However, as we shall see later, this is rarely observed and it is now thought that mechanical and chemical denaturation probe distinct areas of the free energy landscape.

1.2 Methods of Mechanical Unfolding

The mechanical properties of single biomolecules and their complexes can be charac-terised by a variety of techniques that include AFM[5], laser[6] or magnetic tweezers[7], and the biomembrane force probe (BFP)[8]. However, since the underlying method by which each tech-nique exerts forces onto single molecules is different, each method is optimal over different force and distance scales. Consequently, each technique has mostly been applied to different biomolecular systems whose dimensions or force properties are most suited to that technique. For example, laser tweezers which can only apply small forces have been used extensively to measure the force generation and kinetics of load bearing, processive motors such as myosin and kinesin and the unfolding of RNA. Magnetic tweezers are capable of applying slightly larger forces and are particularly suited to the study of proteins that interact with DNA, as the effect of supercoiling can easily be investigated. In addition, this technique has also been used to investigate the viscoelastic properties of intracellular materials[9]. However, whilst these systems are ideal for measuring relatively small forces between points separated by microme-tres, they cannot measure relatively high forces exerted over smaller (nm-μm) distances. For these systems, AFM and the BFP are ideal. Of these techniques, the BFP has the largest dynamic range and has been used to characterise the dynamic force response of protein:lig-and systems. These include the high affinity interaction between biotin and (strept)avidin that is used extensively in biotechnological applications[8] and the interaction between P-selectin and its ligand (PSGL-1), which anchors leukocytes to the endothelial cell wall, allowing the extravasation of leukocytes to sites of inflammation[10, 11]. Despite the advantages of measuring force response by the BFP, the technique is currently rarely used because the instrument is not yet commercially available and needs to be constructed and operated by a specialist. Laser tweezers, magnetic tweezers, and the BFP are each addressed elsewhere in this book and will not be discussed further here. AFM has proven to be highly suitable as a tool for studying the mechanical properties of proteins, as this technique can measure forces from ~10 pN to many nN with pN sensitivity and also monitor changes in distance with sub-nm resolution. The robustness of the available commercial (or home-built) systems and the simplicity of the immobilisation methods have allowed many biomolecules and their complexes to be analysed by AFM. Notable examples are studies of the mechanical properties of carbohydrates[12–14] and nucleic acids[15]; the measurement of the dynamic force spectra of non-covalent interactions such as those between biotin:streptavidin[16], antigen:antibodies[17, 18], and nucleic acids[19, 20]; and the determination of the mechanical properties of proteins under force, to which we now turn our attention fully.

2   Mechanically Unfolding Proteins Using the AFM

2.1   Hardware: AFM in Force-Mode

Since its conception in 1986[21], AFM has been applied to a wide variety of imaging applications in material science, chemistry, and biology. The imaging capability of AFM has been complemented by the development of force-mode AFM[22], in which the force being exerted onto the cantilever is measured as a function of distance from the sample surface. This is possible because the cantilever acts as a Hookian spring and, as the spring constant is known (measured by the method of Hutter[23] for example), it is possible to calculate the force applied to the cantilever by monitoring its displacement. Commercial instruments specifically designed for force-mode AFM are available from Asylum Research (the MFP-1D) and Veeco (Picoforce). These specially designed instruments provide important advantages over imaging mode AFMs when measuring force-extension data such as the use of a low coherence light source to prevent optical interference effects at the detector (arising from reflections from the back of the cantilever and sample surface). These instruments are now complemented by next-generation instruments that allow both imaging and force experiments to be performed on the same sample—the MFP-3D (Asylum Research) and the Bioscope II (Veeco).

Single molecule force spectroscopy has been applied to measure the mechanical properties of both soluble proteins and proteins integrated into membranes. These latter experiments, performed on membranes containing the protein of interest immobilised onto a substrate, have allowed not only the unfolding pathway for these proteins to be delineated[24–27], but also the order in which the extracted secondary structural elements (usually α-helices in these studies) re-insert into the membrane upon a reduction in the applied force[28, 29]. This type of protein mechanical unfolding, which can give detailed insight into the forces that stabilise secondary structural elements within membranes, will not be discussed further in this chapter and the reader is directed to the endnotes[30, 31] for references.

A typical experimental setup for the analysis of the mechanical properties of soluble proteins is shown in Figure 2. The protein of interest, which for reasons described below is usually a polymeric chain of identical domains, is immobilised onto a substrate mounted beneath the AFM. This substrate is generally gold–coated, and therefore the protein can be specifically attached to the substrate via the strong, rapidly formed interaction between the

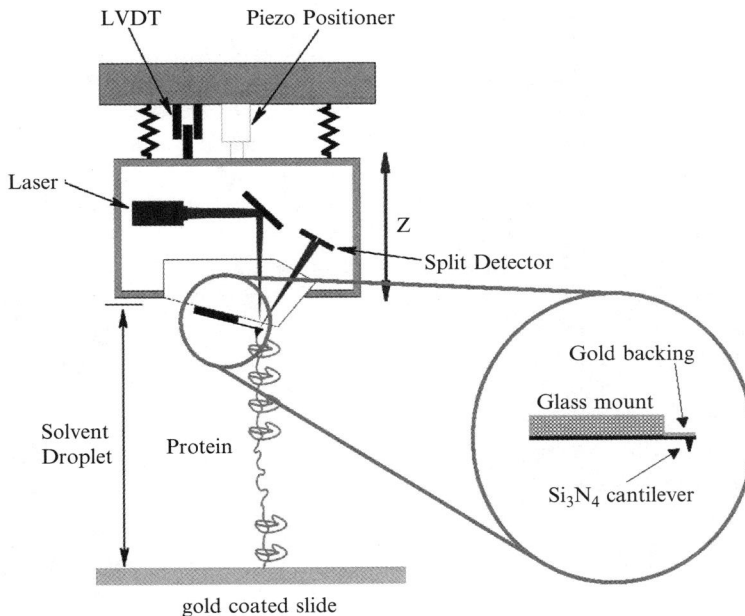

Figure 2. The experimental setup of a typical protein mechanical unfolding experiment. A polymeric protein in aqueous solvent (usually phosphate buffered saline) is immobilised at the C-terminus onto a gold coated substrate via the strong interaction that occurs between the sulphydryl groups of cysteine residues specifically placed there by molecular biological methods and the gold substrate. The cantilever (inset, shown magnified) is brought close to the surface by a piezo-positioner (the movement of which is measured, in this case, by a linear variable differential transformer [LVDT]), and sometimes (~ 2–5 %) a non-specific protein-tip interaction occurs. Upon retraction of the cantilever, the force being exerted onto the protein is measured (using the split detector) by monitoring the change in deflection of a laser which is incident to the gold-covered top of the cantilever. Reproduced by permission of The Royal Society of Chemistry[119].

gold substrate and the sulphur atoms of two cysteine residues that are engineered at the C-terminus of the protein construct (see below). Protein that is immobilised on this surface is attached to the cantilever (which is 20–300 μm in length and has a sharpened tip made from Si_3N_4) by repeatedly approaching and contacting the surface at a fixed rate and for a defined time. This is called the approach phase. Upon retraction of the cantilever at a fixed rate away from the surface (known as the pulling speed or extension rate), successful binding events of single proteins are identified by their distinctive force-extension profiles. The probability of a single approach-retract cycle bearing fruitful data is necessarily low and consequently hundreds or even thousands of approach-retract cycles are usually required in order to obtain a statistically valid dataset. As the protein attaches to the cantilever by a poorly understood, nonspecific mechanism termed physisorption, the cantilever may pick up the protein at any point along its length. This means that for polymeric proteins, the unfolding of all the domains is not necessarily observed in every pulling trace.

The dynamic range of rates over which a protein can be mechanically unfolded using the AFM is relatively narrow (~50–2000 nms⁻¹). At slow extension rates, thermal drift of the cantilever becomes highly problematic in that linear baselines are difficult to achieve (note: this drift occurs at all pulling speeds but has a larger effect at slow extension rates, as the time taken to carry out one approach-retract cycle is longer). At higher speeds, viscous drag of the cantilever in the solvent becomes significant, which causes the cantilever to bend, hence increasing the apparent force. It may be possible to widen this dynamic range by improvements in hardware. For example, the use of closed cell, temperature controlled systems with closed loop piezoelectric scanners in the xy dimension may allow slower speeds to be undertaken in the near future. Also, as viscous drag is a function of a cantilever's dimensions, it is possible to reduce this problem by using smaller cantilevers. Many cantilevers made by different companies are available and these cantilevers differ markedly in their spring constants (which can vary from tens of pNnm^{-1} to tens of nNnm^{-1}) and in their coating (e.g., whether the tip is bare silicon nitride or coated with a thin layer or gold). Cantilever choice is informed by the need to derivatise the tip to allow some kind of specific immobilisation, and also by the strength of the interaction that is to be measured, which for mechanically unfolding proteins is usually in the 15–300 pN range. Typically, cantilever spring constants used for these sorts of experiments vary between 30 and 100 pNnm^{-1}; however, cantilevers with spring constants as low as 6 pNnm^{-1} are available (Olympus Biolever, Japan). Intuitively, one would select a soft cantilever (a small spring constant), since this would allow a larger deflection to be measured at lower forces, thus improving force sensitivity. However, use of a soft cantilever suffers from greater positional noise due to thermally induced cantilever oscillations as described by the equipartition theorem:

$$\frac{1}{2}k_BT = \frac{1}{2}K\langle x^2 \rangle$$

(3)

where $\langle x^2 \rangle$ is the thermally driven mean square deflection amplitude of a cantilever with spring constant K Nm^{-1}. Generally, cantilevers with spring constants of ~40 pNnm^{-1} are used in these studies and the thermal noise limit is thus ~13 pN. As most proteins studied by this technique unfold at forces greater than 50 pN, these cantilevers give a good signal to noise ratio. Greater force sensitivity (sub-picoNewton AFM) may be achieved by damping the thermally induced motions of a soft cantilever using either radiation pressure[32] or an applied magnetic field[33].

2.2 Construction of Polymeric Proteins

Initially, mechanical unfolding experiments were performed either on monomeric proteins[34] or heterogeneous polymeric proteins that are naturally expressed as tandem arrays, such as titin[35] or tenascin[36]. However, neither of these situations is ideal; analysis of monomers may result in difficulty in ascertaining whether a genuine protein unfolding event has given rise to the observed data; for heteropolymeric proteins it is impossible to assign a certain mechanical property to an individual domain (titin, for example, consists of approximately 250 different fibronectin type III (FNIII) and immunoglobulin [Ig] domains. Construction of artificial homopolymers comprising 4–12 copies of a single domain linked by their C- and N-termini via a sequence of linker amino acids obviates these problems. If the domain under study is reasonably mechanically stable, each copy within the polymer unfolds discretely. The result is a highly distinctive force-extension profile consisting of evenly spaced, repetitive elements (or 'sawteeth'), allowing these events to be unequivocally assigned to real unfolding events (see Figure 3). Furthermore, as a polyprotein gives rise to several unfolding events per approach-retract cycle, the amount of time taken to accumulate sufficient data is decreased. As well as homopolyproteins, experiments are sometimes carried out on heteropolyproteins[38-41], which is viable as long as the domains are of significantly different size or of different mechanical character to allow each unfolding event to be assigned unequivocally to each domain type within the polyprotein (see below).

Covalently-linked polyproteins have been constructed by different laboratories using several different molecular biological techniques (see below). At the protein level, however,

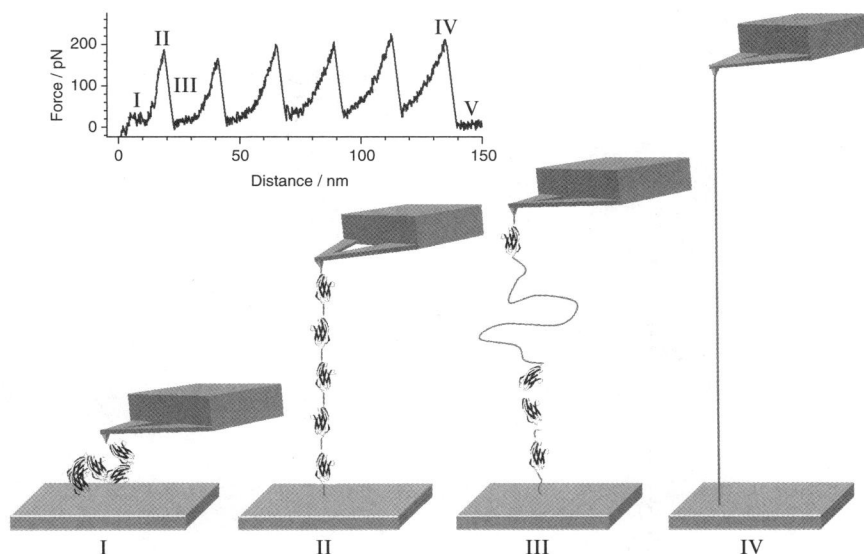

Figure 3. A typical force-extension profile for an approach-retract cycle during which a polymeric protein has been unfolded. (I) The cantilever approaches the surface and non-specifically binds, in this case to the N-terminus of a pentameric homopolymer. The cantilever is retracted at a set rate, extending the linker regions between the protein domains which then resist further extension; hence the cantilever is deflected, indicating an increase in force (II). At some point, one of the protein domains spontaneously unfolds due to thermal fluctuations (III). The force then suddenly drops because of the drastic increase in the end-to-end length of the polyprotein. This process is repeated until all the domains have been unfolded, leaving an unstructured polypeptide chain which is extended (IV) until the protein-tip interaction is broken (~1–4 nN) and the force returns to the baseline thermal noise value (V).

many of these constructs have similar features. As an example, the first concatenated protein to be made in our laboratory, $(I27)_5*$, is shown in Figure 4. $(I27)_5*$ consists of four elements: an N-terminal polyhistidine tag to allow facile purification by Ni^{2+}-nitrilotriacetic acid (NTA) affinity chromatography, two C-terminal cysteines to allow specific attachment to the gold surface via their sulphydryl side-chains, the repeated domain of interest, and a series of 'linker' regions. These linkers, which are typically 5–8 residues in length, are introduced to separate the domains, hence preventing aberrant interdomain interactions and allowing sterically unhindered movement when the polyprotein is extended. There is, however, some debate as to whether the nature of these linkers affects the results of mechanical unfolding experiments and whether or not they prevent domain interactions[42]. Whilst these features are generic for many polyproteins, it should be noted that both the number of domains and the length of the linkers between each domain varies considerably, even between constructs made in the same laboratory. For reasons discussed later, differences in these parameters can make direct comparison of a protein's mechanical strength difficult to assess.

Three molecular biology schemes have been devised for the creation of tandem arrays of polyproteins. Each of these methods has advantages and disadvantages, hence which one is the most appropriate to use depends on the specific polyprotein that is being produced.

1. *The modular cassette method.* In this approach, individual domains are amplified by the polymerase chain reaction (PCR) using forward and reverse primers with 5′ overhangs that, at the DNA level, place a specific, unique restriction site onto each end of the gene and, at the amino-acid level, encode the linkers. The PCR products are directly ligated into predigested 3′ thymidine-tailed vectors such as pGEM-T via the adenine-overhang left by *Taq*

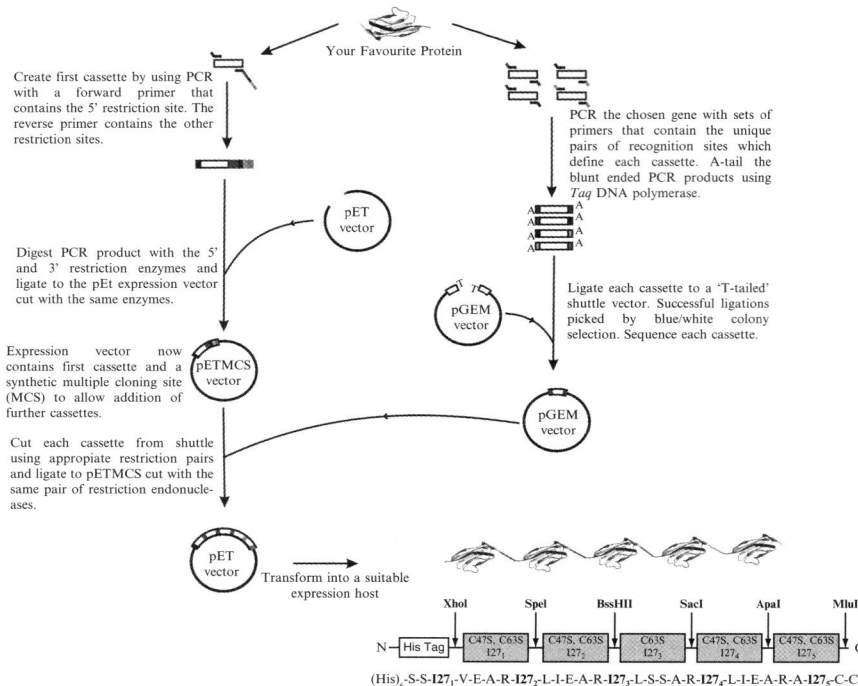

Figure 4. Flow diagram showing the construction of a pentameric polyprotein. PCR is used to introduce a unique restriction site (black rectangle) and a synthetic multiple cloning site (containing 5 unique restriction sites, represented by shaded blocks) 5′ and 3′ to the DNA encoding 'Your Favourite Protein' (white rectangle), respectively. This cassette is then ligated directly into the expression vector (such as pET) and sequenced. The other cassettes are generated by four separate PCR reactions using pairs of primers with 5′ and 3′ overhangs that code for the pairs of unique restriction sites that define each cassette and occur in the synthetic multiple cloning site. *Taq* polymerase is used to add a single A base onto each end of the PCR products and these are then ligated into a T-vector (such as pGEM, Promega) obviating the need to digest each product and host plasmid with separate pairs of restriction endonucleases. The use of such a 'shuttle vector' allows successful ligations to be rapidly identified and their sequence confirmed by DNA sequencing. These cassettes are then sequentially ligated into the expression vector (at least 3 cassettes can be added simultaneously). Note: to limit the probability of homologous recombination of these cassettes, the expression vector is always handled in a *rec*⁻ strain. Expression by a suitable *rec*⁻ expression host yields a polyprotein containing five copies of the domain, the details of which are shown below. The unique pairs of restriction endonucleases for each cassette of I27 are shown above the pentamer and the amino acid sequence of the linkers below. In this example, the polymer is constructed of different variants of I27 created by mutagenesis (C47S, C63S I27 and C63S I27). Reprinted with permission of Elsevier[43].

polymerase on PCR products (see Figure 4). After verification of their identity and fidelity by sequencing, the individual domains are then cut out of the pGEM vectors via the unique pair of restriction sites and sequentially ligated into a vector using a multiple cloning site introduced upon insertion of the first cassette (see Figure 4). Whilst this approach[43-45] seems laborious and repetitive, its strength is the ability to precisely control the identity of each cassette: any given domain in the concatamer can be swapped for another protein as long as the correct restriction sites are present at the termini of the respective genes. This allows the mechanical properties of putative model proteins to be rapidly assessed and has also proved useful in allowing the over-expression and purification of heteropolymeric protein constructs containing proteins that do not over-express as homopolyproteins[41, 46].

2. *The Ava*I *method.* This approach relies on the fact that the *Ava*I restriction sequence is not palindromic. PCR is used to produce a gene that is flanked on each side by this restriction site. After restriction with *Ava*I, these genes are ligated into a precut plasmid which, because of the non-palindromic nature of the site, is directional. Multiple ligation events can be carried out in this way to produce a polymeric protein. This method[47] is much quicker than the cassette approach and is useful for constructing homopolymers, although the correct number of gene insertions must be selected for. However, it is not ideal for making heteropolymers because the position in which the gene inserts is nonselective.

3. *The Bam*HI/*Bgl*II *method*: This procedure is essentially a hybrid of the above methods. PCR is used to introduce a *Bam*HI and *Bgl*II restriction site 5′ and 3′ to the gene to be concatenated, respectively; and the gene is then ligated into a shuttle vector. As these restriction enzymes produce complementary sticky ends (note that the restriction sequence is different for each enzyme), pairs of these cassettes can be ligated into an expression vector and both restriction sequences are interrupted upon ligation. A polyprotein of $(2)n$ domains is made by iterative rounds of ligating dimers to dimers, tetramers to tetramers, etc. This method[47] is advantageous, as it allows for some heterogeneity (e.g., alternating domain types) but is faster than the modular cassette method.

2.3 Experimental Conditions and Their Optimisation

The hit rate in force-mode AFM experiments (the probability that an approach-retract cycle will yield clean protein unfolding events) is necessarily low in order to ensure that truly single molecule events are being measured. The hit rate can vary but is usually ~ 4–10% and is strongly affected by the purity of the protein being studied and the quantity of protein introduced onto the gold surface. In our laboratory, we have found that polyproteins subjected to a two-step purification strategy (Ni^{2+}-NTA affinity chromatography followed by size exclusion chromatography) give excellent results. The protein concentration in a sample can affect both the quality of the force-extension data and hit rate of the experiment. For example, if too much protein is present on the gold surface, the chance that the cantilever will pick up multiple molecules is increased and useful data will not be obtained. On the other hand, if the protein concentration is too low, then the number of force-extension curves in which a polyprotein is being unfolded will also be low, meaning that the experimental time will be large. This is problematic because not only does it decrease the efficiency of data acquisition, but it also reduces the probability of obtaining a sufficiently large dataset as evaporation of the solvent can be significant, setting an upper limit on accumulation time. Previous studies have used protein concentrations ranging from 0.5 to 150 μM, but a range of concentrations for each protein is usually tested studied to find the best conditions before acquiring datasets.

Similarly to studies of protein (un)folding using chemical denaturants, mechanical unfolding studies on different proteins are often carried out under different buffer conditions and different temperatures. However, many groups use phosphate buffered saline (PBS) as this relatively simple buffer has a pH and ionic strength that is considered 'physiological'. The moderate concentration of salts may also help to screen the protein from electrostatic interactions at the substrate surface, thereby reducing non-specific attachment of the protein. Although closed cell AFMs are available in which the temperature can be independently regulated, many of the current dedicated force-mode AFM machines do not have temperature control and hence most mechanical unfolding studies are performed at room temperature.

2.4 Constant Velocity Mechanical Unfolding Experiments

In constant velocity experiments, the cantilever is retracted from the substrate at a constant speed and the deflection of the cantilever recorded. These experiments are typically carried out under a far from equilibrium regime and, under these conditions, the unfolding of a mechanically strong polyprotein results in a 'sawtoothed' force-extension profile (see Figure 3) in which the height of each tooth corresponds to the force at which each protein domain unfolds. Each unfolding event is highly asymmetric: the rising edge is nonlinear, showing an exponential-like relationship between extension and force which is followed by a rapid linear decay in force after unfolding. These effects occur because the polypeptide chain is behaving as an entropic spring that resists extension and can be explained by reference to Figure 3. In Figure 3II, each domain in the polyprotein is still natively folded, and the force rapidly increases since only the linker regions can be easily extended. At some point, one of the domains unfolds, releasing a large length of previously 'hidden' polypeptide chain, increasing the length of the unfolded polypeptide chain. As this release occurs at a rate faster than the extension rate, the force rapidly drops to zero. In order to unfold the next domain, this excess of polypeptide chain needs to be extended to a point where the force exerted onto the remaining folded domains reaches a level sufficient to allow thermally activated transitions.

The force necessary to extend the end-to-end length of an unfolded polypeptide can be modelled by the worm-like chain model (WLC), Equation 4[48, 49]. This model describes the elasticity of a polymer that is assumed to be made of units that have fixed inter-unit angles but are free to rotate:

$$F(x) = \frac{k_B T}{p}\left(0.25\left(1 - \frac{x}{L_c}\right)^{-2} - 0.25 + \frac{x}{L_c}\right) \tag{4}$$

where L_c is the contour length (discussed below), F is the force, x is the extension of the molecule, and p is the persistence length, which is a measure of the stiffness of the polymer and can be thought of as the length of the individually rotating units in the polymer (generally around 0.4 nm in proteins[44, 50–52]). A series of WLC fits to a pentameric construct of protein L (the B1 domain of Protein L from *Peptostreptococcus magnus*), a model protein studied in our laboratory, is shown in Figure 5.

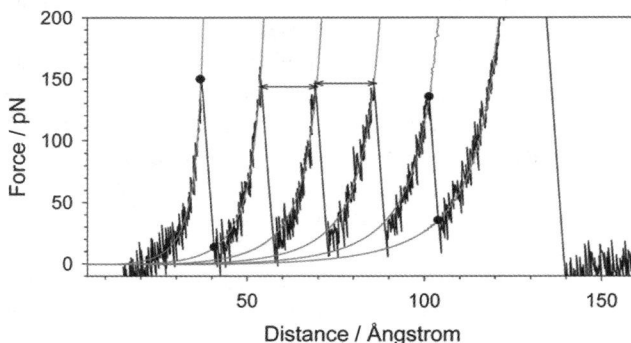

Figure 5. Force-extension profile for a pentameric construct of protein L showing all five unfolding events and the protein-tip detachment at high force. Black arrows show the inter-peak distance. Grey lines are worm-like chain fits to the rising edge of each unfolding event, showing that the length of polypeptide chain released upon unfolding behaves like an entropic spring with a persistence length of 0.4 nm and a contour length that increases by ~19 nm (determined from the WLC fit) at each unfolding event. An important effect of the WLC-like behaviour of the unfolded polypeptide chain is that as each domain unfolds, the relative increase in contour length decreases and, consequently, the entropic restoring force applied onto the polypeptide chain increases. This means that the difference in force between the point of domain unfolding and the point immediately after decreases at each successive event (shown by pairs of black filled circles on the first and fifth unfolding event). The unfolding force must therefore be measured relative to a baseline after the protein-tip interaction is broken.

By fitting the rising edge of each unfolding event to a WLC, it is possible to estimate the end-to-end length of the molecule extrapolated to full extension at each unfolding step—a parameter known as the contour length (L_c). Quantification of this parameter is important because the change in contour length, ΔL_c, is equal to the difference in length between a folded and unfolded domain. Consequently, if a high resolution three-dimensional structure of the protein under study exists, the length between successive unfolding events of polypeptide released in an unfolding event can be compared with that expected if the entire domain unfolded. Discrepancies may reveal regions of the protein that unfold at a low force or, in more complex systems, can suggest three-state unfolding (i.e., unfolding via intermediates). In the latter case, measuring the ΔL_c for each unfolding step allows the number of amino acids released from the folded core at each stage to be estimated. This information, in conjunction with knowledge of the protein's structure, may allow the unfolding pathway of these three-state proteins to be visualised and will be discussed in more detail below. However, if the polyprotein under study has been previously characterised by force-mode AFM, simple measurement of the distance between each peak can be used to rapidly filter the data. It is important to note that ΔL_c is not equal to the observed distance between unfolding events, because the force at which a protein unfolds at a given retraction speed is distributed widely due to the random nature of thermal fluctuations. The WLC chain fit to the rising edge of each sawtooth predicts how force will be loaded onto the polypeptide chain as a function of distance; and the next domain to unfold can, in principle, unfold at any point along this line. If, by chance, this domain were to unfold at a low force (and therefore low polypeptide chain extension), the distance to the next unfolding event (peak to peak distance) would be larger than average.

We have seen that analysis of the force-extension profiles of a homopolymer can directly yield information on the size of the mechanically resistant structure in a protein (in terms of the number of amino acids) and also the mechanical strength of this structure. In addition to these parameters, it is possible to characterise basic features of the underlying energy landscape, such as the height of the barrier to unfolding ($k_u^{(0F)}$) and the distance from the native well to this barrier (x_u), by analysing how the observed unfolding force is affected by the rate at which force is loaded onto the protein. The loading rate (which has units Ns^{-1}) affects the mechanical strength of a protein because it governs the number of thermally activated barrier crossing attempts before the protein is unfolded. Under low loading rates, the protein will spend a longer amount of time experiencing a given force, meaning that the likelihood of thermal fluctuations driving it over the transition state at this force is increased. The relationship between the most likely unfolding force and the logarithm of the loading rate is linear:

$$f^* = \frac{k_B T}{x_u} \ln\left(\frac{x_u}{k_u^{0F}}\right) + \frac{k_B T}{x_u} \ln(r) \tag{5}$$

where f^* is the mode of unfolding forces and r is the loading rate[53]. In principle, a plot of f^* against $\ln(r)$ should allow the parameters $k_u^{(0F)}$ and x_u to be obtained. However, the loading rate that is applied onto the biomolecule is not simply the product of the cantilever spring constant and extension rate, and so does not remain constant throughout the experiment. Figure 6 shows how the force applied onto WLCs with different contour lengths varies as a function of time when extended at constant velocity. Examination of the gradients of each force-time profile shows that the instantaneous loading rate not only varies as a function of time for each polypeptide, but also that the applied loading rate is strongly affected by the contour length of the WLC. Consequently, the rate at which force is loaded onto each domain in a polyprotein

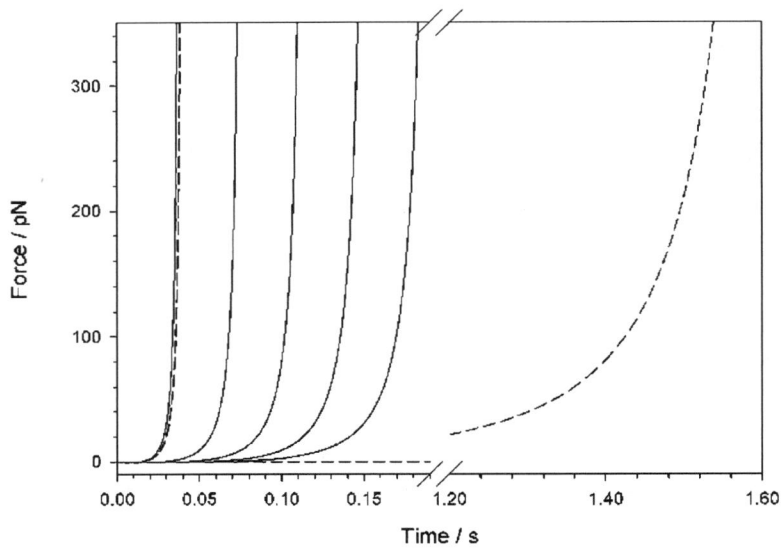

Figure 6. Effect of contour length (continuous lines) and extension rate (dashed lines) on the loading rate of an unfolded polypeptide chain that is assumed to behave as a WLC. By comparing the instantaneous gradient for polypeptides of different length (L_c = 28, 56, 84, 112 and 140 nm for continuous lines left to right) it can be seen that the increase in contour length that occurs upon domain unfolding causes a reduction in the loading rate that is applied onto the remaining folded domains, even though the extension rate remains constant (700 nms^{-1}). Dashed lines show the effect of extension rate (50 nms^{-1} and 2000 nms^{-1}, left and right, respectively) on the rate of force loading onto a WLC with a contour length of 84 nm.

not only varies during the extension of each unfolded domain but also varies as a function of event number. These complex changes in loading rate have two important consequences. Firstly, the loading rate at the point of rupture is highly variable; and secondly, the most probable unfolding force is expected to decrease as a function of event number due to the decrease in effective loading rate (known as a compliance effect; see Section 4.1).

In order to plot the dependence of the unfolding force on loading rate it is therefore necessary to calculate the instantaneous loading rate at the point of rupture for each unfolding event. This can be performed by fitting a WLC to the rising edge of each sawtooth in force-extension profiles that have not been corrected to account for tip deflection.[2] The force at which the domain unfolds can then be used to calculate the distance at which unfolding occurred (x in Equation 4), and this value along with L_c and p can be inserted into a differentiation of the WLC equation to obtain dF/dx which is then converted to a loading rate by multiplying this parameter by the retraction speed at which the data was taken. However, this method is rarely used in the field of protein mechanical unfolding[54, 55] because the accuracy of the fitting procedure (which can be automated) is dependent on the ability to fit the WLC to the rising edge of each sawtooth, which in practice is sometimes difficult to achieve. Furthermore, in contrast to dynamic force studies performed on protein:protein and protein:ligand interactions, where this

---

[2] In order to calculate the increase in end-to-end length of a biomolecule immobilised between the cantilever tip and substrate, the distance change due to the deflection of the cantilever is normally subtracted from each force-extension profile during data processing. However, as the cantilever is also compliant, its effects must be included in the measurement of instantaneous loading rate.

fitting procedure is more commonly used, more than one unfolding event per approach-retract cycle is being measured. This means that not only does the compliance change throughout the time course of the unfolding experiment, but also that the number of folded domains changes. As the forced denaturation of proteins is a stochastic, thermally activated process, the number of domains that are available to unfold affects the most probable unfolding force, since a decrease in the domain number results in a lower total unfolding probability (less total barrier crossing attempts per unit time) and hence increased unfolding forces. In contrast to the compliance effects discussed above, the domain number effect would thus be expected to increase the most probable unfolding force as a function of event number. The relative importance of these counteracting effects varies between constructs and this means that the apparent mechanical properties of a protein can change markedly between different polymeric constructs. This is discussed more fully in Section 4.1. As a result of these difficulties, many protein mechanical unfolding studies measure how the most probable unfolding force varies as a function of extension rate (see Figure 7a) and a Monte Carlo model or analytical solution that models the effects that compliance and domain number have on the underlying process is then iteratively fitted to this data to extract the parameters $k_u^{(0F)}$ and x_u. This procedure is described briefly below.

In order to extract the parameters that describe the basic features of the energy landscape, the extension speed dependence of the most probable unfolding force for the protein under study must first be determined. This is achieved by performing mechanical unfolding experiments over the widest dynamic range possible with the instrument ($\sim$50–2000 nms^{-1}). As the unfolding of each domain within the polyprotein is a stochastic event, in order to estimate the most likely unfolding force of a polyprotein at each speed it is necessary to collect at least fifty (ideally 100–200) unfolding events per data set. This then allows a force-frequency histogram to be constructed and the mode calculated. This process is performed in triplicate and the mean of the modes is calculated to minimise the effects of experimental error. The experimentally derived pulling speed dependence of the mean unfolding force calculated as described above is shown for three proteins in Figure 7a. Values for $k_u^{(0F)}$ and x_u are then estimated by fitting these data to simulated data obtained from either a Monte Carlo process or an analytical solution. As both of these procedures are based on the same assumptions and give identical results, only one method, Monte Carlo simulations, will be discussed in detail. Further details on the implementation of an analytical approach can be found elsewhere[53, 56].

The Monte Carlo simulation is a computational process that can be used to model a set of data and is often used to obtain parameters for $k_u^{(0F)}$ and x_u. In a Monte Carlo simulation, a number (N_f) of folded domains of length L_f, which are separated by unstructured polypeptide linkers of length L_u, are placed in series with a cantilever of known spring constant. These parameters will vary according to the polymeric protein construct that is under study. The cantilever is then retracted at a constant rate. The force applied onto the polyprotein after a small time increment, Δt, is calculated from the WLC model (Equation 4) and this force is then used to calculate the unfolding rate at this extension ($k_u^{(F)}$) by Equation 2 using arbitrary values of $k_u^{(0F)}$ and x_u. The probability of a domain unfolding at each time step is then calculated:

$$P_u = N_f k_u^{(F)} \Delta t \tag{6}$$

(Note: because force exponentially increases the height of the folding barrier, the probability of refolding is assumed to be zero.) To replicate the stochastic nature of the process, unfolding occurs if the calculated probability is greater than a randomly generated number between 0 and 1. If unfolding occurs, the contour length of the polymer is increased by L_u-L_f and the number of folded domains decreased by 1. If unfolding does not occur, the time is incremented and the cycle repeated. This process is continued until all the domains in the polyprotein are unfolded. The whole process is then repeated thousands of times at one

(a)

(b)

(c)

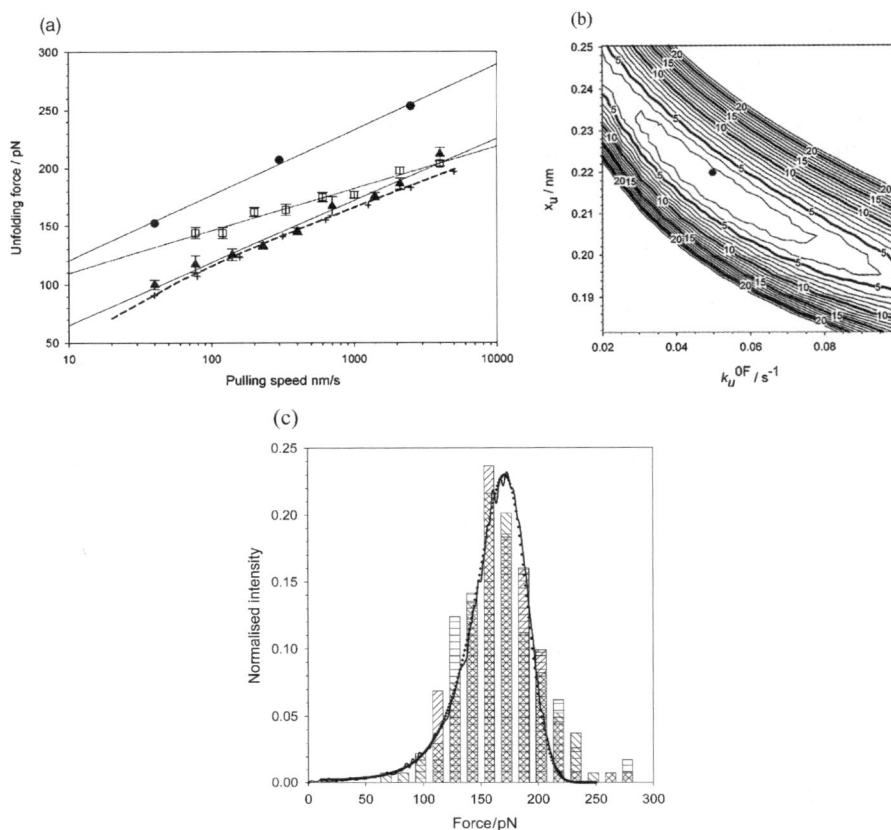

Figure 7. (A) Comparison of the pulling speed dependence of the unfolding forces of (protein L)$_5$ (triangles), (I27)$_5^*$ (squares), and (ubiquitin)$_9$ (circles). Solid lines through each data set are a best fit to guide the eye. Data taken from Brockwell et al.[54], Brockwell et al.[44] and Carrion-Vazquez et al.[61], respectively. Fitting the data for protein L to an analytical solution (dashed line) estimates that the height and the position of the unfolding barrier relative to the native state is smaller and shorter ($k_u^{(0F)}$ = 0.05 s^{-1}; x_u = 0.22 nm) than that obtained for (I27)$_5^*$ ($k_u^{(0F)}$ = 0.002 s^{-1}; x_u = 0.29 nm). Monte Carlo simulations, using the best fit parameters for protein L obtained above, give identical modal values (crosshairs) to those predicted by the analytical model. (B) Error analysis of parameter pairs reveals degeneracy in the fit of $k_u^{(0F)}$ and x_u to the observed experimental data for (protein L)$_5$. Contour lines link parameter pairs calculated to have equal χ^2 error. (C) Experimental force frequency distributions are consistent with those predicted by the analytical model (dotted lines) and Monte Carlo simulation (solid black line) using the parameter pair marked by a solid circle in (B) [54]. Used by permission.

speed and a force frequency histogram is constructed to calculate the most probable unfolding force at that retraction speed. This process is carried out over the same range of speeds as used experimentally, resulting in a simulated pulling speed dependence of the unfolding force for a protein with a certain $k_u^{(0F)}$ and x_u value. This simulated pulling speed dependence of the unfolding force is then compared with that obtained experimentally and the quality of the fit estimated by calculating the χ^2 error[54, 57]. The whole process is repeated for a matrix of $k_u^{(0F)}$ and x_u values, so that the data can be used to estimate the values of these parameters that best fit the experimental data by constructing contour plots of χ^2 error over a matrix of $k_u^{(0F)}$ and x_u values. Fitting the experimental speed dependence of the unfolding force using

Monte Carlo methods is computationally expensive. Whilst more difficult to implement for the nonspecialist, the analytical approach is far faster, since the probability distribution is calculated directly at each speed for a pair of $k_u^{(0F)}$ and x_u values[53, 54, 56]. From this, the mode unfolding force at each speed can be calculated, which then yields the speed dependence of the unfolding force. Parameters that best fit the data are then chosen by construction of a χ^2 error matrix, as described for the Monte Carlo method above.

Since both methods are based on the same principles, identical results are obtained if implemented correctly. Figure 7a shows the result of Monte Carlo (cross hairs) and analytical (dashed line) fits for a protein studied in our laboratory (protein L). From the χ^2 error analysis (Figure 7b), it can be seen that many pairs of $k_u^{(0F)}$ and x_u fit the data equally well; and whilst this leads to relatively small errors on x_u, the errors on $k_u^{(0F)}$ are very significant. As the shape and modal force of the unfolding force frequency histograms at a single speed are affected by the values of $k_u^{(0F)}$ and x_u[58], it is theoretically possible to fit the experimental distributions directly. However, as the experimental variability in the modal unfolding force of a protein at a fixed extension rate is relatively high due to errors in the calculation of the spring constant of the cantilever, this method is inferior to fitting the speed dependence of the unfolding force where multiple experiments have been averaged. Comparing the experimentally derived force-frequency distributions with those obtained by Monte Carlo or analytical methods using the optimal parameters derived from the fitting of the speed dependence of unfolding force can, however, be used to assess the quality of the experimental data (see Figure 7c).

2.5 Constant Force Mechanical Unfolding Experiments

As discussed above, the interpretation of constant velocity experiments is complicated by the fact that, as the protein is stretched, the force being applied to it is constantly changing in a non-trivial manner. A more direct way of measuring the force-dependence of protein unfolding is to hold the protein at a constant force (force-clamp) or to apply a force that increases at a fixed rate (force-ramp) and measure the rate at which domains unfold[59, 60]. The advantage of this technique is that the loading rate being applied onto the polyprotein is always controlled, obviating the effects of compliance and variable loading rates described above.

The setup for constant force experiments is similar to that used in constant velocity experiments, with the addition of a feedback system that compares the force applied onto the polyprotein (calculated from the deflection of the cantilever) to the desired preset value. The difference between these values is then fed to an amplifier that is output to the piezoelectric sample positioner, which raises or lowers the sample to control the deflection of the cantilever. The typical time response of the repositioning is approximately 4–10 ms, meaning that any changes in protein length occurring on a timescale shorter than this cannot easily be resolved. Typically, data acquired from a force-clamp experiment show that the protein length increases in a stepwise manner as a function of time (see Figure 8a). This is because upon the unfolding of a domain, the applied force dramatically drops. The feedback system returns the system to the set force by increasing the distance between the piezopositioner and the substrate. In a two-state mechanical unfolding reaction, these steps will thus correspond to the unfolding of one domain in the polyprotein. Similar to the distance between peaks in constant velocity experiments, the size of each step corresponds to the difference in length between a folded and an unfolded domain, and this length depends on the force with which the protein is being stretched.

The application of force-clamp and force-ramp techniques to protein mechanical unfolding studies have been developed by the Fernandez group and only limited data are available[59, 60]. Despite differences in how force is loaded onto the system, the same underlying theory as that described for constant velocity experiments is used to interpret the data and extract the parameters $k_u^{(0F)}$ and x_u. Equation 2 shows that the unfolding rate of the protein is exponentially

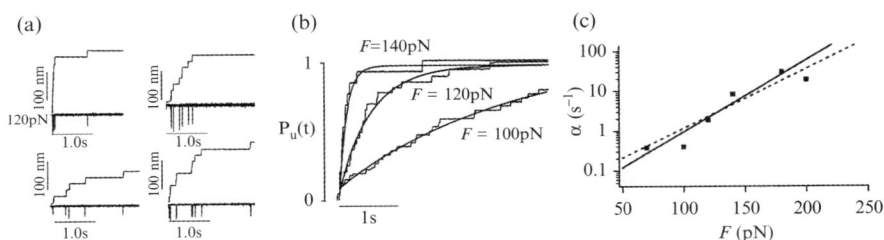

Figure 8. Constant force unfolding data for ubiquitin. (A) Four different examples under an applied force of ~ 120 pN. Each dataset shows the end-to-end length (top trace) and force (bottom trace) as a function of time for (ubiquitin)$_9$ homopolymers. Note the 20 nm staircases in the top traces that signify sequential unfolding of single domains. The spikes in the force-time plots reflect the time response of the force clamp feedback system and signify an unfolding event. (B) The unfolding rate constant (denoted α in this figure) under each applied force can be estimated by fitting a single exponential function (curvilinear line) to averaged and normalised time courses accumulated at the same applied force. (C) Fitting the relationship between the applied force and unfolding rate constant (squares) to Equation 2 (solid line) allows the intrinsic unfolding rate constant at zero applied force to be measured. The dashed line corresponds to Equation 2 evaluated with values for $k_u^{(0F)}$ and x_u obtained from force ramp experiments (see text) [60]. Reprinted with permission of the National Academy of Sciences, U.S.A.

related to the force being applied to it. Similarly to the measured unfolding forces in constant velocity experiments, the lifetimes ($1/k_u^{(F)}$) of each domain within a homopolymer are highly variable (see Figure 8a), demonstrating the stochastic nature of this thermally activated process. As these experiments use homopolymeric constructs of ten domains, the probability of a domain unfolding per unit of time decreases after each unfolding event. Summing and normalising several data sets at a constant force results in an exponential time-course from which the lifetime and therefore the unfolding rate at force F can be directly measured. By performing this process at several forces, the data can be fitted directly to Equation 2 to obtain the parameters $k_u^{(0F)}$ and x_u (see Figure 8b and 8c).

In a force-ramp experiment, the force loaded onto the protein is increased linearly. The raw data obtained from a force-ramp shows how the polyprotein length varies as a function of time, but can be converted easily into length as a function of force. By pooling many of these 'unfolding staircases' it is possible to construct a frequency histogram of the force at which unfolding events occur. The unfolding probability distribution (a sigmoidal function) is then obtained by integration and normalisation of the force frequency histogram (so that the cumulative probability is unity). This is then fit to:

$$P_u(F) = 1 - e^{-\frac{k_u^{(0F)}k_BT}{rx_u}(e^{\frac{Fx_u}{k_BT}}-1)} \tag{7}$$

(where r is the loading rate in pN/s), to obtain values for $k_u^{(0F)}$ and x_u. To check these values Fernandez and co-workers used the values of $k_u^{(0F)}$ and x_u obtained by this procedure to compare a simulated force-frequency histogram to the experimentally derived one. This is similar to the method described for constant speed experiments, where the validity of the estimated values of $k_u^{(0F)}$ and x_u can be assessed by comparison of the simulated and experimental force-frequency histograms.

The relative merits of constant force experiments over those of the more established constant velocity experiments is difficult to assess, as only a few proteins have been studied by constant force methods (I27[59], I28[59], and ubiquitin[60]). The intrinsic unfolding rate constants under zero force estimated by each technique are very similar for I27 but significantly different for I28[59]. For ubiquitin, comparison of the parameters obtained by each technique has not been performed, since in the constant velocity experiments $k_u^{(0F)}$ was not obtained by a fit to the speed dependence of the unfolding force, but was instead assumed to be identical to that obtained from chemical unfolding studies[61]. Such an assumption is now thought to be invalid. However, it is possible to compare the values of $k_u^{(0F)}$ and x_u obtained for ubiquitin when using force-clamp and force-ramp methods which were found to be 0.015s^{-1} and $0.17\,\text{nm}$, and 0.0375s^{-1} and $0.14\,\text{nm}$, respectively. The close agreement between the x_u values and the relative large variation in $k_u^{(0F)}$ is similar to the situation for constant velocity experiments (see Figure 7b), and may suggest that the degeneracy of the fit parameters observed in constant velocity experiments may also be problematic in the analysis of constant force data.

3 Determining the Mechanism of Mechanical Unfolding

In previous sections we have seen how, by unfolding polymeric protein constructs over a range of constant velocities or forces, it is possible to estimate the size of the unfolding unit, to gain some idea of each protein's mechanical strength, and to obtain parameters that describe basic features of the unfolding energy landscape. However, these data alone are unable to give any detailed information about the structural mechanism of mechanical unfolding at the molecular level. Two techniques used extensively in protein folding studies, computational simulations and Φ-value analysis, have been shown to be extremely powerful tools to investigate the mechanical unfolding process, allowing the experimentalist to identify the specific protein interactions that limit the rate of mechanical unfolding (the transition state), and to delineate the features of the protein that allow it to resist force.

3.1 Molecular Dynamics Simulations

Steered molecular dynamics (SMD, also called constant velocity molecular dynamics or CVMD) and constant force molecular dynamics (CFMD) simulations are designed specifically to simulate the response of proteins to an applied extension or force. SMD simulations are analogous to constant velocity experiments. These are carried out by fixing the position of one point of an all-atom model of a monomeric protein and attaching another point to a harmonic spring which is pulled away from the fixed point at a constant velocity[62]. CFMD simulations, on the other hand, are analogous to constant force experiments. In these simulations, an all-atom model of the protein is held at a constant force that is applied between two points and the distance between these points monitored as a function of time. This latter technique has the advantage that any metastable states populated in the unfolding trajectory are easily identifiable as plateaus where the distance between extension points remains constant for a period of time. Application of a suitable constant force, therefore, allows any thermodynamically stable state, such as an intermediate, to be stabilised and held in this state long enough for this species to be characterised structurally.

CVMD and CFMD simulations obtained for protein L, a model protein studied in our laboratory, are shown in Figure 9. The constant velocity simulation (Figure 9a) demonstrates that this protein is mechanically brittle: the force applied onto the protein increases rapidly with little gain in extension. At some force (~ 1700 pN in these simulations) key interactions are broken, the protein begins to unravel, and consequently the force drops. By repeating the simulation many times (each starting from a different configuration obtained by minimising the solution or crystallographic structure under native conditions), it is

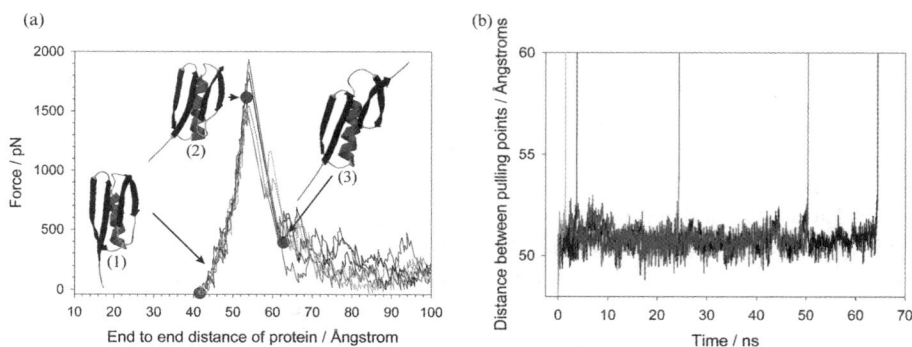

Figure 9. MD simulations of the mechanical unfolding of protein L. Both (A) constant velocity and (B) constant force MD techniques show that protein L is highly resistant to force and unfolds via a well-defined pathway (multiple simulations produce almost identical force-extension plots in CVMD or show the same end-to-end length at the transition state in CFMD). Structures at selected time-points are shown inset in (A) and demonstrate that the unfolding transition state for protein L is highly native (compare structures [2] and [3]; [1] depicts the native state) [54]. Used by permission.

possible to investigate the heterogeneity of the unfolding pathway. For protein L, independent simulations result in almost identical force-extension profiles that differ only in their unfolding force (similar to experimental results), indicating that this protein unfolds via a well defined transition state described by a narrow structural ensemble[54]. By contrast, simulations performed on a tandem array of spectrin repeats (α-actinin) suggested that these proteins could unfold by two distinct pathways[63]. The route taken was found to depend on the behaviour of the helical linker between each repeat. In one pathway, the linker was found to unfold before each domain, resulting in mechanically independent units. In the second pathway, the linker was found to remain intact, leading to simultaneous unfolding of the tandem repeat. This observation accords with previous experimental results that had demonstrated, by analysis of the distance between unfolding events, that serial spectrin repeats could unfold by single or tandem repeat unfolding[64].

The constant force simulations for protein L (Figure 9b) yield similar conclusions to those performed under constant velocity. Under application of a constant force of 400 pN, the distance between the N- and C-terminal residues increases almost instantaneously to a value of ~51 Å that remains constant for a time that varies between replicate simulations. This metastable state is identical to the highly force-resistant state just before unfolding in the constant velocity simulations. The variability in the lifetime of this metastable state (i.e., the width of the plateau in Figure 9b) once again illustrates the stochastic nature of the mechanical unfolding process. However, if enough replicates are performed, the distribution of lifetimes for each metastable state can be used to estimate its kinetic stability[46]. In simulations of protein L unfolding at constant force, the species corresponding to the extended native state remains folded over nanosecond timescales before unfolding rapidly without the population of intermediates. The transition state in this simple two-state unfolding mechanism is assumed to be an ensemble average of the protein's configuration at the force maximum for CVMD simulations and the last few structures before the rapid increase in end-to-end distance that occurs in the CFMD simulations.

Whilst describing the heterogeneity of the unfolding pathway, simply monitoring the force as a function of the distance between pulling points or the end-to-end distance of a protein as a function of time gives no more insight into the mechanism of unfolding than that obtained by experiment. However, as these simulations are atomistic (they use the three-dimensional atomic coordinates contained in a Protein DataBank file), each trajectory describes the behaviour of every atom in the protein as a function of time. Identification of the specific perturbations that lead to mechanical unfolding in a background in which every atom fluctuates around a mean position even at equilibrium can be difficult, especially for simulations of proteins with force-extension profiles more complex than protein L. The power of all atom simulations of mechanical unfolding is that any number of order parameters can be quantified. This analysis can be performed globally to locate transition states for example Once regions important to the unfolding process have been identified, the flucutuations of a single residue can then be monitored as a function of time single residue as a function of time. For example, in constant force simulations of the mechanical unfolding of the third FNIII domain of human tenascin (TNfn3, see Figure 10a), Clarke and co-workers[65] first identified a putative transition state by calculating the difference in C_α RMSD values for each timestep and plotting this against the distance between the N- and C-termini (Figure 10b). By analysing the trajectory in this way, metastable states that are populated for a significant time form easily identifiable clusters. These clusters describe the ensemble of configurations that each state visits as a consequence of thermal energy and are separated by a region of parameter space rarely visited by the molecule—the transition state (circled in Figure 10). Once the transition state was identified, the role that each residue played in the mechanical unfolding process was assessed by plotting either the fraction of native contacts or the effective energy as a function of the distance between the N- and C-termini for each residue. Other groups use different order parameters, such as solvent accessible surface area as a function of extension[63, 66] or the energy of individual hydrogen bonds as a function of time or extension[67, 68].

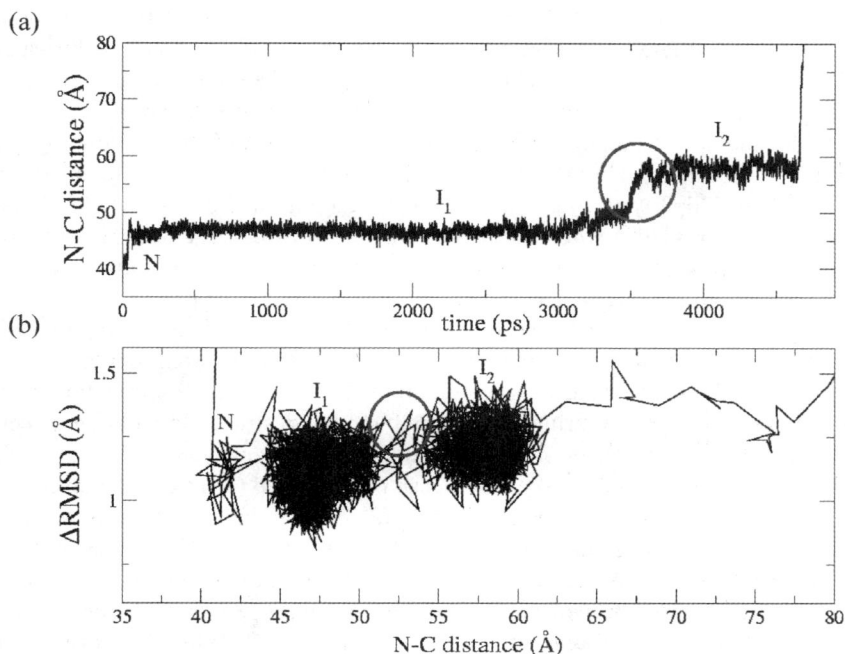

Figure 10. (A) A constant force trajectory for TNfn3. N denotes the native state, I_1 intermediate 1, and I_2 intermediate 2. (B) Cluster analysis can be used to identify the rate-limiting transition state. By calculating the difference in root mean square standard deviation (ΔRMSD) of the structures at consecutive time-steps, areas of configurational space rarely populated (the transition state, circled) can be identified[65]. Reprinted with permission of Elsevier.

(a)

(b)

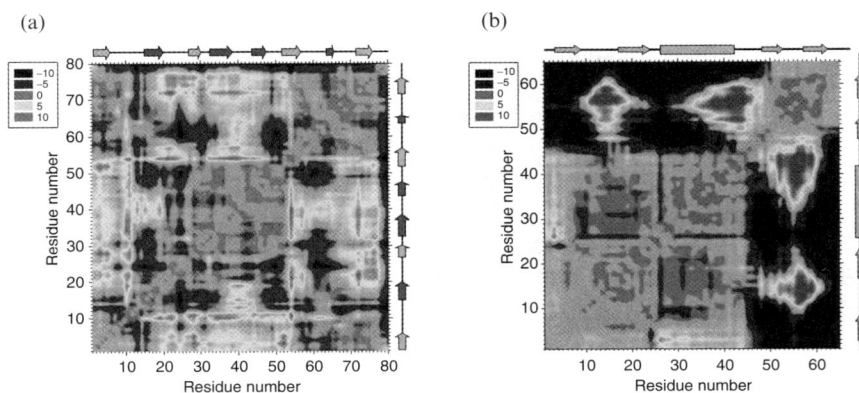

Figure 11. Difference distance maps for (A) E2lip3 when extended between the N-terminus and lys 41, and (B) protein L when extended by the N and C-termini. The maps were calculated by subtraction of the distance between each residue in a representative structure (from CVMD simulations) just after the transition state from the distance between each residue in a representative structure just before the transition state. An increase in the distance between pairs of amino acids is shown in blue, a reduction in red, and areas that remain constant in green. The position of secondary structural elements within each protein in the native state is shown top and right of each map. Inspection of each map suggests that upon extension, force is transmitted globally through E2lip3, extensively deforming the protein from its native state. protein L, however, shows a highly local response that results in the shearing of two mechanical sub-units past each other. (*See Color Plates*)

A particularly simple method to identify the areas of a protein most perturbed upon unfolding is to calculate difference distance or energy maps. These are obtained by calculating the difference in distance (or energy) between every pair of amino-acids in a representative (or averaged) structure just after the transition state, and in a representative structure just before the transition state[54, 65]. Comparison of such maps for proteins of distinct topology (protein L and E2lip3, see Figure 11) clearly shows that these proteins have different mechanical responses. When E2lip3 is extended between its N-terminus and the side chain of lysine 41, force is applied across the whole structure, causing diffuse structural perturbations. By contrast, most of protein L remains native-like in the mechanical unfolding transition state ensemble, apart from the interface between two subdomains of the protein.

As well as being an invaluable tool for visualising the unfolding process of experimentally measured systems, MD simulations are an important research tool that can provide insights into the effects of mechanical perturbation on the structure and function of proteins that are difficult to express experimentally as polymeric constructs. In addition, experiments that are very difficult in practice can be performed relatively easily *in silico*. As we shall see later, the rapidity with which these simulations can be performed (relative to the time-consuming experiments) means that a wide variety of proteins can be studied, allowing the effects of topology and sequence to be delineated[67]. The higher resolution and choice of reaction coordinate also reveals features of the unfolding process that cannot be measured experimentally. For example, the group led by Ikai studied the mechanical properties of monomers of the 259 amino acid residue bovine carbonic anhydrase using the AFM[69]. When extended, this protein typically shows an unfolding peak around 40–60 nm. The presence of only one measurable unfolding barrier and the complex topology of this protein, however, made the elucidation of the unfolding pathway extremely difficult. SMD simulations carried out in the presence of explicit or implicit water suggested that three major transitions occurred[70]. The first unfolding event was thought to be obscured in the experiment by either the binding of the tip to the protein or nonspecific tip-protein interactions (since a monomeric construct was used). The second and third peaks in the simulations corresponded to the disruption of β-sheet structures within the core of the molecule (note that the second peak was not observed in the original AFM study but identified in a later study[71]). Interestingly, a zinc ion coordinated between three histidine residues on strands S5 and S6 remained bound until these strands separated at a high force. It is possible that the presence of a coordination bond across regions that are sheared past each other may endow a mechanical strength onto this region that is equivalent to that of a covalent bond. A second example of the power of mechanical unfolding simulations is that of cryptic RGD domains in fibronectin type III domains. Fibronectin (FN) is a multimodular protein composed of mainly FN type I, II, and III domains that forms part of the extracellular cell matrix. It was thought that mechanical forces may play a critical role in controlling FN functions such as the binding of cellular integrins. Simulations on the tenth FNIII domain in fibronectin[72] revealed that upon application of force to the termini of this protein, the region to which integrins bind (a region containing the residues arginine, glycine, and aspartic acid known as the RGD loop) is deformed, but the rest of the protein remains largely native-like. As the conformation of this loop strongly affects its affinity for integrins, the authors concluded that this interaction could act as a mechanosensitive regulator.

Despite the potential power and widespread use of MD to simulate mechanical unfolding, the validity of using such techniques to interpret experimental data remains controversial. The main problem in comparing MD and experimental AFM data arises from the difference in extension rates that are used in each technique (10^6–10^{10} and 10^1–10^3 nms^{-1} for simulation and experiment, respectively). This huge difference is largely due to the computational expense of running fine-grained atomistic simulations. Consequently, the extension rate for *in silico* mechanical protein unfolding is usually one that will induce a transition in the longest time that can be simulated over a reasonable timescale (a few nanoseconds). The unfolding forces observed in simulations are thus considerably higher than those observed in the analogous experiment and are affected by dissipative forces that are experimentally negligible. The effect that the pulling speed has on the observed unfolding force is also different for the experimental and simulated process [73, 74], making it difficult to extrapolate the results of the simulation to the experimental timescale because of 'hidden' outer barriers that may become rate-limiting only at lower forces. This means that experimental data and simulations may not probe the same transition state between folding and unfolding. Methods to reduce the difference in timescale between MD simulations and experiment include subtraction of the dissipative forces to obtain the equilibrium potential of mean force[75] and derivation of the potential of mean force from a series of equilibrium MD simulations by using either umbrella sampling[73] or computationally efficient 'course-grained' Gō models rather than all-atom models of the protein under study[76, 77].

From the discussion above it can be seen that it is necessary to benchmark the simulated unfolding process by experiment. One way to test the identity between simulated and experimental mechanical unfolding is to compare the structural properties of the transition states crossed during each process. We have already seen how the transition state of the simulated process can be characterised. In order to characterise the transition state experimentally, it is necessary to perform a process known as phi-value (Φ-value) analysis.

3.2 Mechanical Φ-Value Analysis

Φ-value analysis has been extensively employed in traditional protein folding experiments to infer the structure of intermediates and the transition state along the folding reaction coordinate[78, 79]. Before discussing the use of Φ-value analysis in mechanical unfolding experiments, it is important to explain how Φ-value analysis works and to note the information that can be gained from it. Essentially, the technique analyses the effect of a point mutation on the folding and unfolding kinetics of a protein to ascertain the relative effects of the mutation on the free energy of the transition state (TS) and ground states of the protein. In other words, a Φ-value (Φ_{TS}) is defined as:

$$\Phi_{TS} = \frac{\Delta\Delta G_{U-TS}}{\Delta\Delta G_{UN}} \quad \text{or} \quad \Phi_{TS} = 1 - \frac{\Delta\Delta G_{TS-N}}{\Delta\Delta G_{UN}} \tag{8}$$

where ΔG_{U-TS} is the free energy difference between U and TS, and ΔG_{TS-N} is the free energy difference between TS and N; hence $\Delta\Delta G_{U-TS}$ and $\Delta\Delta G_{TS-N}$ are the differences between the ΔG_{U-TS} and ΔG_{TS-N} values of the wild-type protein and a variant created by mutation of a single amino acid side chain. Similarly, $\Delta\Delta G_{UN}$ is the difference between the variant and wild-type ΔG_{UN} values. Consider an example whereby a conservative mutation (the truncation of a buried hydrophobic side chain that occurs when a valine residue is mutated to alanine in the hydrophobic core, for example) has destabilised the native and transition states equally, but U is unaffected (see Figure 12a). This will mean that the rate of *unfolding* will be unaffected because the free energy difference between N and TS is identical for the wild-type and variant proteins, but the *folding* rate will decrease, because of the increased free energy difference between U and TS. This situation equates to a Φ_{TS} value of 1 and means that the contacts made by the side chains that are deleted by the mutation are 'native-like' in TS. In the converse scenario (see Figure 12b), the mutation destabilises N but has no effect on U or TS. In this case $\Phi_{TS} = 0$ as $\Delta\Delta G_{U-TS}$ is zero and the data are interpreted to mean that the side chain of interest is unstructured in both TS and U. Partial Φ-values are more complicated to interpret. In bulk experiments, they could mean that alternative pathways to folding exist, leading to an ensemble of different transition states, or they could signify a partial formation of native-like structure in a transition state ensemble that is well defined on a single reaction coordinate.

The principles of mechanical Φ-value analysis[57] are identical to those of the ensemble method, but the interpretation and relevance of the values obtained is less clear-cut, as shall be explained below. One advantage of applying Φ-value analysis to single molecule protein mechanical unfolding studies is in the interpretation of partial Φ-values ($\Phi \neq 0$ or 1), as these must be due to partial native-like structuring of the residue in the transition state unless very different unfolding paths are resolved by the single molecule experiment. By examination

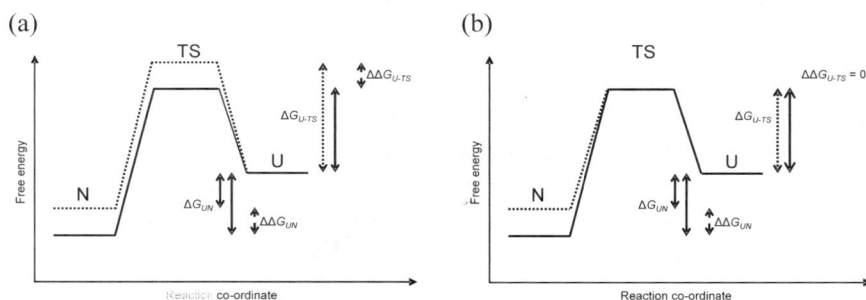

Figure 12. Free energy diagrams for the limits of Φ-value analysis. In (A), substitution of a single amino acid side chain destabilises the native (N) and transition states (TS) equally, relative to the unfolded state (U). In this case, the mutated residue is assumed to form native-like contacts in the transition state and $\Phi = 1$. In (B), a different amino acid is substituted, which destabilises the native state but does not affect the stability of the transition state. In this limit, the residue is inferred to be unstructured in the transition state and $\Phi = 0$.

of Equation 8 it is clear that not all of the data required for the calculation of mechanical Φ-values can be obtained by mechanical means. To provide Φ_{TS}^{mech}, $k_u^{(0F)}$ values are obtained from mechanical unfolding experiments, as described above, but $\Delta\Delta G_{UN}$ must be obtained from equilibrium denaturation experiments since, irrespective of whether constant force or constant velocity methods are used, mechanically unfolding proteins by the AFM is a far from equilibrium technique. Herein lies the controversy regarding the accuracy of mechanical Φ-values because in many cases the native state is not the ground state from which the protein crosses the transition state. To date, this technique has been used to characterise the mechanical unfolding transition state of two proteins (I27[80] and TNfn3[65]). In each case, the rate limiting transition was found not to be between the native state and the unfolded state, but between a force-stabilised intermediate and either the denatured state (I27) or another intermediate (TNfn3). The validity of normalising the effect of the mutation on the transition state by the effect that the mutation has on the native ground state, rather than the force activated 'ground state', is unclear. For I27, this problem was circumvented by measuring the effect of mutations (i.e., $\Delta\Delta G_{UN}$) in a variant that modelled the force stabilised intermediate (the entire A-strand was deleted). This was not possible for TNfn3. In this case, the authors suggested that referencing to the ground state may result in an overestimation of Φ-values. In the future it may be possible to directly measure the stability of these force-activated states using experimental constant force techniques under equilibrium conditions (i.e., the mechanically unfolded protein can also refold under the applied force[81]). Performing Φ-value analysis by single molecule mechanical unfolding methods is a time consuming undertaking as, for every mutation, a new homopolymer has to be engineered, purified, and then mechanically characterised over a wide range of pulling speeds. Consequently, whilst this technique is extremely powerful in visualising the unfolding transition state, it is rarely undertaken.

3.3 Visualising the Unfolding Process: the Synergy Between Experiment and Molecular Dynamics Simulation

A common approach in mechanical unfolding studies is to assume that the simulations replicate the experimental process and then, guided by these simulations, to alter the protein under study by protein engineering methods so that its unfolding properties change in a measurable way (i.e., either an alteration in the unfolding force or the distance between peaks). For example, as well as unfolding independently or as a tandem unit (see Section 3.1), other mechanical unfolding experiments on spectrin repeats suggested that each domain itself could also unfold by different pathways[82, 83]. Simulations suggested that kinking and partial melting of the central helix close to a proline residue led to the formation of unfolding intermediates[83]. Stabilisation of this helix by a double proline to alanine and glycine to alanine mutation successfully altered the unfolding pathway, ablating this intermediate.

Perhaps the best example of the synergy between experiment and simulation is the elucidation of the unfolding mechanism of I27, an Ig domain from titin (a multimodular protein found in muscle) whose three-dimensional structure and topology is shown in Figures 13a and 13b. I27 is mechanically strong (resisting forces of ~200 pN at a pulling speed of 700 nms^{-1})[47]. Interestingly, the rising edge of force-extension profiles of this protein sometimes shows a regular 'shoulder' or 'hump'-like deviation from the expected WLC behaviour (Figure 13c). The magnitude of this deviation, which decreases as each domain unfolds, is dependent on the total number of domains held between the tip and substrate. By plotting the difference in contour lengths between WLCs fitted to the top and bottom portion of the first unfolding event (see Figure 13c) as a function of the number of domains held between the tip and substrate, Fernandez and co-workers showed that at ~100 pN, this domain undergoes a structural transition that results in an extension of 6.6 Å per domain[84]. SMD simulations previously carried out

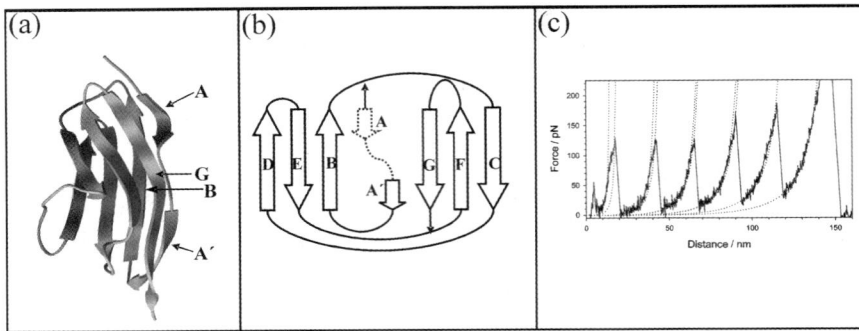

Figure 13. The paradigm for protein mechanical unfolding—I27. (A) Three dimensional structure of I27. The strands thought to be important for the mechanical strength of the native domain (A and B) and its unfolding intermediate (A' and G) are labelled. (B) Topology of I27 showing the points of extension when mechanically unfolded (black closed arrows); each strand is labelled A to G. The intermediate is formed by the detachment of strand A (dotted line) from the rest of the natively folded structure. (C) Force-extension profile of $(I27)_5$ [ref 44] shows significant deviations from a single WLC fit. Fitting separate WLC models to the rising edge of each sawtooth (dotted lines) before and after the 'hump' reveals that the magnitude of the deviation decreases upon each subsequent unfolding event.

on I27[66] suggested that the high mechanical stability of I27 results from interactions between the A and B and A' and G strands that formed a 'mechanical clamp', effectively isolating the rest of the structure from the applied extension. Constant force simulations were able to show, however, that the breakage of these interactions occurred in two distinct phases[68]. In the first transition, hydrogen bonds between the A and B strands were broken extending the end-to-end length of each domain by 4–7 Å, a distance remarkably similar to that derived experimentally (6.6 Å). The resulting intermediate, with the A strand completely detached from the core of the protein, was found to be very stable to force (having a lifetime of ~850 ps at an applied force of 750 pN compared with a lifetime of 200 ps for the native state). The rate-limiting barrier to unfolding was found to be the simultaneous breakage of six hydrogen bonds between the A' and G strands at an extension of only 3 Å longer than that achieved in the first phase. Thus, the synergy between experiment and simulation has resulted in a detailed picture of the unfolding pathway of I27.

The mechanical unfolding pathway of I27 has also been extensively tested experimentally using mutational analysis. For example, introduction of a highly disruptive proline residue into strand A (at residue 6) of I27 totally ablated the formation of the mechanical intermediate, probably because formation of the hydrogen bonds between stands A and B was prevented[85]. This interpretation was further tested by inserting five glycine residues either between strands A and A', between strands F and G, or C-terminal to the G strand. In agreement with the simulated unfolding pathway, a difference in the domain unfolding length was apparent only for the construct containing glycine insertions within the mechanically clamped core (i.e., between the F and G strands)[86]. Clarke and co-workers have also tested these simulations by combining AFM, protein engineering and MD simulations. Firstly, a mutant of I27 in which the entire A strand had been deleted was shown to be stable and predicted to have a structure very similar to the mechanical unfolding intermediate observed in simulations[87]. Secondly, mutating Val 4 which is located in the A strand was found not to affect the mechanical properties of full length I27, suggesting that this region does not contribute to the overall mechanical stability of the protein[87]. Thirdly, a mechanical Φ-value analysis of I27 showed that the transition state was highly native-like with only V13A (strand A') having a Φ-value of significantly less than one (0.6)[80].

These data thus show that, despite the huge difference in timescales, MD techniques can be used to visualise the mechanical unfolding pathway of at least I27. However, this conclusion may only be valid for mechanically strong proteins whose mechanical unfolding transition state is close to the native state (on a reaction coordinate of end-to-end distance) as these may be easier to simulate and less likely to suffer from the effects of the very fast extension rates used in simulations. In accord with this, the presence of an intermediate in the mechanical unfolding pathway of I27 has also been predicted using off-lattice, coarse grained Gō models[76] or even by calculation of the relative stabilisation (by tertiary contacts) of different blocks of secondary structure within each domain[88]. Further benchmarking is thus required on a wide variety of proteins that show disparate mechanical responses before a consensus on the general accuracy of simulations is reached.

4 What Governs a Protein's Mechanical Resistance?

Why are some proteins more able to resist mechanical deformation than others? Since the first report of forced unfolding some eight years ago[34], single molecule mechanical unfolding studies have been performed on many proteins. With such a database now available, it should be possible to discern commonalities between proteins of similar mechanical strength and consequently to develop rules which allow the mechanical properties of a protein to be predicted and eventually even tailored. However, identification of the determinants of the mechanical resistance of proteins is problematic due to several factors: many of the proteins studied have been analysed under different conditions by using polyproteins with different numbers of domains and different linkers, and the speed dependence of the unfolding force for many proteins has not been analysed in detail and can vary greatly between even closely related proteins. Finally, despite the number of proteins analysed, the diversity of their structures is rather narrow. In this section, we shall first discuss the care that must be taken when attempting to rank proteins by mechanical strength and then, by reference to the literature, we discuss the features that govern a protein's mechanical resistance.

4.1 Scaffold Effects and the Dynamic Force Spectrum

As discussed in Section 2.4 the most probable force at which a protein domain will unfold is dependent not only on the intrinsic properties of the domain under study, but also on the composition of the scaffold within which it is placed and the rate at which force is loaded onto the domains. For example, Figure 14 shows the experimentally derived relationship between extension rate and the unfolding force observed for I27 domains in two different polyprotein constructs, whose compositions are also shown. It can be seen that I27 in the heteropolymeric construct with E2lip3 unfolds at a significantly higher force relative to the homopolymer at all speeds tested, even though the I27 domains and the linkers that separate each domain are identical. The constructs shown in Figure 14 differ in two ways: the number of I27 domains (five and four, for the homo- and heteropolymer, respectively), and the total number of amino acids that are being extended between the tip and attachment point (476 and 428, respectively). The apparent increase in the mechanical resistance of I27 when in the heteropolymeric construct therefore is due both to the domain number effect (there are fewer I27 domains, hence a lower chance of an unfolding event per unit of time, which results in a higher measured unfolding force), and to a compliance effect (a shorter chain length leads to a greater loading rate at the same extension rate).

As both the compliance and number of folded domains change during the unfolding of a single protein polymer, the most probable unfolding force for each domain is not constant

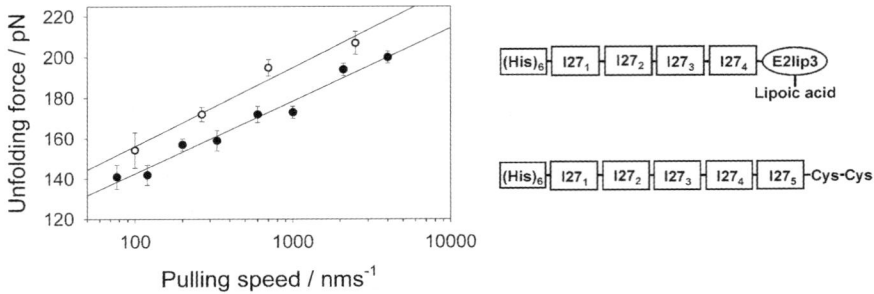

Figure 14. The mechanical stability of a protein is affected by the composition of the polyprotein. I27 domains unfold at a mean force of 173 pN at an extension rate of 700 nms^{-1} when in a homopolymer composed of five domains (closed circles, construct shown bottom right). However, the identical domain unfolds at a significantly higher force (189 pN) when extended at the same speed but in a heteropolymeric protein (I27)$_4$E2lip3 (open circles, construct shown top right). Error bars denote the standard error of the mean, and solid lines are linear fits to each dataset[105]. Used by permission.

but depends on the event number (i.e., the position at which the unfolding event occurs in the 'sawtoothed' profile). The competition between the effect of compliance and domain number results in a nonlinear relationship between unfolding force and event number that is small but measurable (see Figure 15a). In order to compare the mechanical strengths of different proteins in the context of a synthetic construct, it is necessary to quantify the effects of domain number, compliance, and loading rate. Unfortunately, it is impracticable to experimentally investigate the effect of domain number and compliance on the modal unfolding force gained from a mechanical unfolding dataset, since constructing and accumulating data for such a wide variety of constructs is infeasible. However, we have already seen that these effects can be readily investigated by Monte Carlo or analytical approaches. For example, Monte Carlo simulations using parameters for I27 based on its structure (i.e., unfolded length (L_u) = 28.0 nm, folded length (L_f) = 4.0 nm) and those previously obtained by fitting the speed dependence of unfolding force of this protein (i.e., $k_u^{(0F)}$ and x_u)[44] show good agreement with the experimentally derived event number dependence of the unfolding force of a pentameric I27 construct (see Figure 15a). This technique thus allows the roles that domain number, domain size, and scaffold stiffness play in modulating a protein's mechanical resistance to be investigated. Figure 15b shows that varying the number of domains in a homopolyprotein strongly affects the mean unfolding force observed for that protein. The differences in mean unfolding force are particularly large when increasing the number of domains from 5 to 15. Unfortunately, different research groups construct polyproteins that differ by both domain number (4 to 12 domains[82, 47]) and compliance (since the size of each domain, the length of linkers, and the choice of cantilever all affect this parameter). These differences suggest that it should be extremely difficult to compare the results of mechanical unfolding experiments performed on different constructs. However, as the effect of domain number can be accounted for quantitatively[44], and as relatively large changes in compliance are necessary to alter the mechanical properties of a protein significantly[89], it is possible to compare data obtained from different constructs.

However, care is still required when interpreting the functional relevance of results obtained from synthetic constructs, since many proteins studied by this technique occur naturally in complex modular proteins that consist of tandem arrays of sometimes hundreds of similar domains interspersed with other domains or extensive unfolded regions such as those found in titin[90] and myomesin[38, 39]. Some of these multimodular proteins are even expressed as different isoforms in different tissues; for example in titin, both the number of Ig domains

(a)   (b)

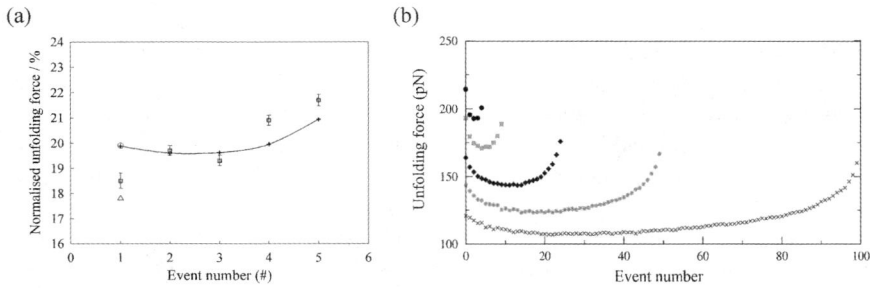

Figure 15. Monte Carlo (MC) simulations can be used to investigate compliance and domain number effects on the measured mechanical properties of a polyprotein. (A) MC simulation (cross hairs) and experimental unfolding forces (squares) for (C47S, C63S I27)$_5$ as a function of event number. The black line is a guide for the eye. The first event was found to be composed of real unfolding events (circle) and non-specific protein-protein or domain-surface interactions (triangle)[89]. (B) MC simulations showing the effect of varying the number of domains in the polyprotein on the simulated unfolding force (5, 10, 25, 50, and 100 domains for circles, squares, diamonds, asterisks, and crosses, respectively)[120]. Used by permission.

(37 and 90 in the I-band) and the length of the natively unfolded PEVK region (163 to 2174 residues) varies in cardiac and skeletal muscle, respectively[37]. The loading rates exerted on these domains and hence their apparent mechanical resistance, not only varies between natural and synthetic tandem arrays but also between different isoforms of the naturally expressed protein. The effects of domain number, length of linker or unfolded regions, and compliance of the surrounding structures play a key role in modulating the mechanical strength of proteins, allowing nature to tailor the mechanical strength of proteins and their interactions with ligands or receptors.

The second caveat in the comparison of the 'mechanical strength' of proteins is the different sensitivities that proteins display to the effect of the loading rate (or extension rate) on the most probable unfolding force. The speed dependence of the unfolding force of two proteins studied in this laboratory (I27 and protein L) is shown in Figure 7a. It can be seen that whilst both proteins display relatively high mechanical strength, their sensitivity to the extension rate differs markedly. The gradient of the speed dependence is inversely proportional to x_u, which is the distance between the native well and transition state along the mechanical coordinate (see Figure 1). The data suggest, therefore, that the mechanical transition state of protein L is more native-like than that of I27. The difference in extension rate dependency of a protein's unfolding force necessitates that the mechanical properties of a protein are characterised over the full dynamic range of the instrument, as a single measurement can be highly misleading. For example, the relative mechanical strengths of I27 and protein L at 50 and 4000 nms^{-1} are 135 and 102 pN, and 193 and 188 pN, respectively. Extrapolation of the relative mechanical strengths of proteins to extension rates outside the range of the instrument is also problematic if the underlying landscape is rugged, as different barriers along the reaction coordinate can become rate limiting at different loading rates (as seen for the dissociation of biotin from (strept)avidin[8, 91]). However, the pulling speed dependence of the unfolding force for all proteins to date shows that only a single barrier is being probed. This may be because unfolding proteins by extension of their termini dramatically reduces the possible choice of routes through which a protein can unfold. For mechanically resistant proteins, this may result in a large barrier to unfolding which remains rate limiting over all of the pulling speeds currently accessible to the AFM. At pulling speeds slower than the dynamic range of current AFM instruments, however, a previously hidden outer barrier may become rate limiting. In this case, extrapolation of the data obtained using the AFM to slower speeds would result in an underestimate of the protein's mechanical strength. Work by Williams et al.[92] suggests such behaviour for I27. Characterisation of the mechanical properties of a series of I27 mutants revealed that one, V86A, had a significantly different speed dependence to the wild-type protein (x_u = 5.5 and 3.3 Å for V86A and wild-type, respectively), suggesting that the position of the transition state along the mechanical unfolding reaction coordinate had altered upon mutation. A similar effect had been observed previously when studying the mechanical properties of a series of proline mutations introduced into this protein[85]. By modelling the data for wild-type and V86A to a three state model, it was shown that V86A unfolded not from the metastable intermediate, but directly from the native state. Consequently, the distance from the 'ground state' to the transition state (x_u) increased and the speed dependence decreased. Furthermore, this model suggested that at speeds slower than the dynamic range of the AFM, wild-type I27 would also unfold directly from the native state, changing the dynamic force spectrum at lower pulling speeds and significantly increasing the lifetime of I27 under physiological forces. Ideally, the mechanical behaviour of a protein would be characterised over the widest possible range of loading rates and, in principle, this can be achieved by using other devices such as the BFP or laser/magnetic tweezers in addition to the AFM. However, the experimental constraints on each method necessitate the engineering of different constructs (i.e., different linkers and attachment methods) which also introduce further complexities. A more ideal solution would be to increase the dynamic range of the AFM. This may be possible in the near future by the development of a sub-picoNewton AFM[32, 33] with a stable, precise piezo-electric positioner with closed loops in xyz. This would allow much lower loading rates to be applied as a result of the softer cantilevers and slower extension rates made accessible by this technique.

4.2 How Does the Equilibrium and Kinetic Stability of a Protein Affect its Mechanical Resistance?

Table 1 shows the mechanical, thermodynamic, and kinetic stability of a series of proteins studied by both chemical denaturant perturbation and single molecule AFM methods. Surprisingly, the unfolding force and thermodynamic stability (ΔG_{UN}) show little correlation.

Table 1. Comparison of the thermodynamic, kinetic and mechanical unfolding properties of a selection of proteins ranked mechanical strength.

Protein	SCOP Class	SCOP Fold	ΔG_{UN} / kJmol^{-1}	$k_U^{0/s-1}$	Parallel terminal strands[†]	Force/pN (speed/ nms^{-1})	Reference
Im 9	all α	EF Hand-like	−27	1.2×10^{-2}	No	<15 (600)[#]	[122]
Spectrin R16	all α	Spectrin repeat-like	−26 to −27	1.7–3.2×10^{-3}	No	60 and 80 (3000)	[82] and [123]
Barnase	α+β	Microbial Ribonuclease	−43	3.3×10^{-5}	No	70 (300)	[40]
TNfn3	all β	Immunoglobulin-like β-sandwich	−28	7.2×10^{-5}	Yes	~120 pN (600)	[65] and [124]
PKD1	all β	Immunoglobulin-like β-sandwich	−4 to −8	5.6×10^{-3}	Yes	> 170 pN (1000)	[46]
I27 wt	all β	Immunoglobulin-like β-sandwich	−31	4.9×10^{-4}	Yes*	204	[47]
Ubiquitin	α+β	β-grasp	−34	1.1×10^{-3}	Yes*	203 (400) and 230 (1000)	[51], [61] and [125]

[#] A. Blake, E. Hann and D. Brockwell, unpublished data.
[†] proteins marked * have parallel terminal strands which are directly hydrogen bonded

The disparity between chemical and mechanical unfolding is clearly demonstrated by comparing the relative thermodynamic and mechanical stabilities of the first PKD domain of the membrane-associated protein polycystin-1 (PKD1)[46] and of I27. These proteins have a similar Ig-like fold, yet have vastly different thermodynamic stabilities ($\Delta G_{UN} = -4$ to $-8\,$kJmol^{-1} for PKD1[46] and $-31\,$kJmol^{-1} for I27[47]). However, despite these different thermodynamic stabilities, PKD1 is able to withstand forces as high as those measured for I27.

To understand why thermodynamic stability has little effect on the mechanical unfolding process, it is necessary to consider the timescale of the experiment relative to the rate constants of the transitions between the native and unfolded ground states. Figure 16 shows retract-approach cycles for myosin (a coilied coil) and titin. The force-extension profile for mysosin is very similar to that observed for the polysaccharide dextran[93]; the approach and retract traces superimpose and a kink or plateau indicating a structural transition occurs at intermediate force, which is independent of the retract-approach speed. By contrast, for the domains of titin, there is a large hysteresis between the approach and retract force-extension profiles and we have already seen that the unfolding force is dependent on the applied extension rate. However, despite these contrasting force-extension profiles, both proteins undergo a two-state transition when denatured by force. The difference in behaviour arises as a consequence of the rate of interconversion between each state. In the case of myosin (and dextran), interconversion is rapid relative to the rate of extension (the systems are either at or close to equilibrium), whereas unfolding of the domains from titin occurs far from equilibrium.

To date, myosin is the only natively folded protein studied that shows truly elastic behaviour when mechanically unfolded using the AFM[94]. This topologically simple protein is able to extend to more than twice its original contour length and yet refolds on a timescale of less than a second against an applied force of up to 30 pN with little dissipation of energy. By contrast, proteins with a more complex topology, such as I27, are more brittle with asymmetric energy landscapes. These proteins are able to resist very high forces over small extensions but can only refold slowly at low applied forces, producing a large hysteresis. A tandem array of such proteins can thus function as a force safety-catch in vivo, preventing damage to other structures; since unfolding one of these domains at a relatively high force dissipates large amounts of energy, as well as creating a large increase in contour length, which decreases the applied force, allowing the polypeptide chain to remain intact. This mechanism has been postulated to be important for the mechanical properties of biological materials such as shells, fibres, and

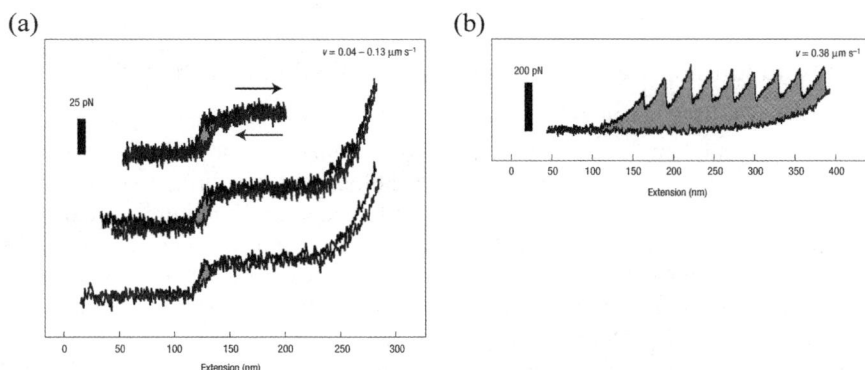

Figure 16. Force-extension profiles of (A) rabbit skeletal myosin and (B) human cardiac titin. Shaded areas represent the energy dissipated during a retract-approach cycle. Reprinted by permission of Macmillan Publishers Ltd[94].

adhesives[95]. Titin may also act as a shock absorber at large muscle extensions[96]. However, as is not the case with myosin, refolding the topologically complex Ig-like domain upon a reduction of force may lead to the formation of misfolded states. Such a deleterious situation could be avoided by allowing a part of the protein to show dynamic behaviour upon application of force. This would enable the mechanically strong protein to reduce the tension applied to it by undergoing a small structural reorganisation before the unfolding transition state is reached. As the high energy barrier has not been traversed, a reduction in force allows the native state to be reformed rapidly. In this way, mechanically robust proteins can act as effective shock absorbers but still maintain their native fold. This is exactly the mechanism proposed for the intermediate in the mechanical unfolding of I27[84], and this type of behaviour has also been observed in other proteins that have a force bearing or responsive function in vivo[97, 98].

From the discussion above it is clear that for most proteins, mechanical denaturation using the AFM is a far from equilibrium process and, as such, is kinetically controlled. As folding is effectively ablated upon application of even relatively small forces, the observed unfolding force would therefore be expected to be related to the height of the barrier to unfolding measured by standard chemical denaturant techniques. Examination of Table 1 demonstrates that this is not the case and suggests that chemical and mechanical denaturation occur via different processes that are limited by different transition states. Indeed, comparison of chemical and mechanical Φ-values obtained for the unfolding of I27 and TNfn3 shows that these proteins do unfold by distinct mechanisms when denatured by force and chemical denaturant[65, 80]. The unfolding mechanisms differ probably because chemical denaturants act by solubilising all parts of a protein (i.e., the backbone and hydrophobic side chains) and therefore may act globally; whereas force, by contrast, acts as a highly local peturbant. In this case, the height of the rate limiting transition state is governed by the strength of the interactions between the points which are being extended.

Most proteins studied to date by single molecule AFM experiments have been unfolded in the context of a synthetic tandem array of proteins in which the C-terminus of one domain is linked to the N-terminus of the next by an unstructured linker. Such polyproteins are ideal for these experiments and the resulting extension geometry is of functional relevance in naturally occurring, tandemly arrayed proteins such as those from cell adhesion molecules (CAMs)[99], fibronectin[41, 100], filamin[98, 101], myomesin[38, 39], spectrin[82, 83, 102], tenascin[36], and titin[96]. If the height of the mechanical unfolding barrier is related to the strength of interactions between the points being extended, then the mechanical resistance of tandemly arrayed proteins must be related to the type of secondary structure and its orientation (the topology) relative to the termini of the protein.

4.3   The Relationship Between Topology and Mechanical Stability

As more proteins are being studied by AFM and related methods, a correlation between topology and mechanical resistance is becoming clear. Initially, the nature of this relationship was investigated by Lu and Schulten who performed steered molecular dynamics on ten proteins of varied topology and secondary structure[67]. Analysis of the force-extension profiles revealed two main classes of response. Some proteins showed a dominant force peak at short extension (termed class I), whereas other proteins showed no dominant force peak, just a gradual increase during stretching (class II). Proteins in class I were found to be all-β-proteins with parallel terminal β-strands such as FNIII domains from fibronectin and the immunoglobulin-like fold of the cell adhesion proteins (cadherin and V-CAM), and I27 from titin. Class II comprised the all-α-helical B domain of protein A (a three-helix bundle), cytochrome c6 (a five-helix protein), and the C2 domain of synaptotagmin I (C2A), an eight-stranded β-sandwich domain with antiparallel terminal strands. This work suggested that all-α-helical proteins are mechanically weaker than most all-β-proteins; and subsequent experimental studies have shown that, at typical extension rates, α-helical proteins unfold at very low forces (<60 pN[52, 82]), or even below the thermal noise limit of the technique[52].

The significant difference in the mechanical strength of proteins containing different secondary structural elements is thought to arise because the array of hydrogen bonds between adjacent β-strands of β-sheet proteins provides more stability against local mechanical deformation than the hydrophobic contacts between helices in α-helical proteins. An ideal test of this theory would be to select two proteins that have the same sequence and structure, but whose termini are located in different regions of secondary structure. These criteria are of course very difficult to fulfil when only naturally occurring proteins are considered, but can be met by using a technique called circular permutation. A circular permutant is one where the topology of a variant of the protein is identical to its wild-type counterpart, but the location of the termini and, therefore, the order in which the secondary structural elements are linked, is different. At the DNA level, the position of the N- and C-termini can be changed by altering the point at which the start codon appears. Any residues previously 5' to the new start codon are simply added 3' to the terminal residue, along with some codons equivalent to glycine residues, which are added to link the original termini. Importantly, whilst the positions of the termini change, the structure remains unchanged, allowing the relationship between the topology local to the points of extension and mechanical strength to be measured in the context of the same protein. This method has recently been used to compare the mechanical strength of dihydrofolate reductase (DHFR) and a circularly permutated derivative[103]. When wild-type DHFR was extended by its natural N- and C-termini, which form two strands of a four-stranded β-sheet (Figure 17a), DHFR domains were found to unfold at a force similar to other β-sheet proteins (131 pN). By contrast, a circularly permutated DHFR domain (CP P25), whose termini occurred at the N-terminus of a helix and a loop region (Figure 17b), was found to be significantly weaker, unfolding at an average force of 56 pN.

Whilst the presence of an α-helical structure at the points of extension generally correlates with mechanical lability, the mechanical strength of all-β-proteins varies markedly in both experimental studies[52] and in simulations[67]. For example, I27, a protein with an Ig-like fold, unfolds at ~200 pN at 700 nms^{-1}(ref [47]); whereas FNIII domains, which have a similar topology to I27, typically unfold at much lower forces (TNfn3 unfolds at ~120 pN at 600 nms^{-1})[65]. Finally, C2A, a β-sandwich protein, unfolds at ~60 pN[52]. The differences in mechanical stability of these proteins can be partly rationalised by considering that the relative orientation of the N- and C-terminal strands affects how the protein is extended when concatenated. When proteins with parallel terminal β-strands are concatenated, the points of the protein that are pulled apart (the N- and C-terminal residues) are at opposite ends of each strand. To mechanically unfold these proteins by this geometry ('shearing'), it is necessary to simultaneously break all the interactions that form a 'mechanical clamp' between these two strands before the stable core is exposed to the applied force (see Figure 18a). By contrast, when β-stranded proteins whose termini are antiparallel are concatenated, the N- and C-terminal residues are at the same end of each strand. This results in a 'peeling' action that allows sequential breakage of the interactions which stabilise the protein and as such do not offer a high mechanical resistance (Figure 18c). This hypothesis is in accord with theoretical predictions[58] and molecular models[104] on simplified systems showing that the longitudinal shearing of interactions between parallel strands requires a significantly larger force than peeling apart the same system. This is because in the latter case, the orthogonal application of force allows each bond to break in turn, whereas in longitudinal shearing, strands bonded by n interactions require the consecutive breakage of all these bonds at a force of slightly less than n times the force required for a single rupture.

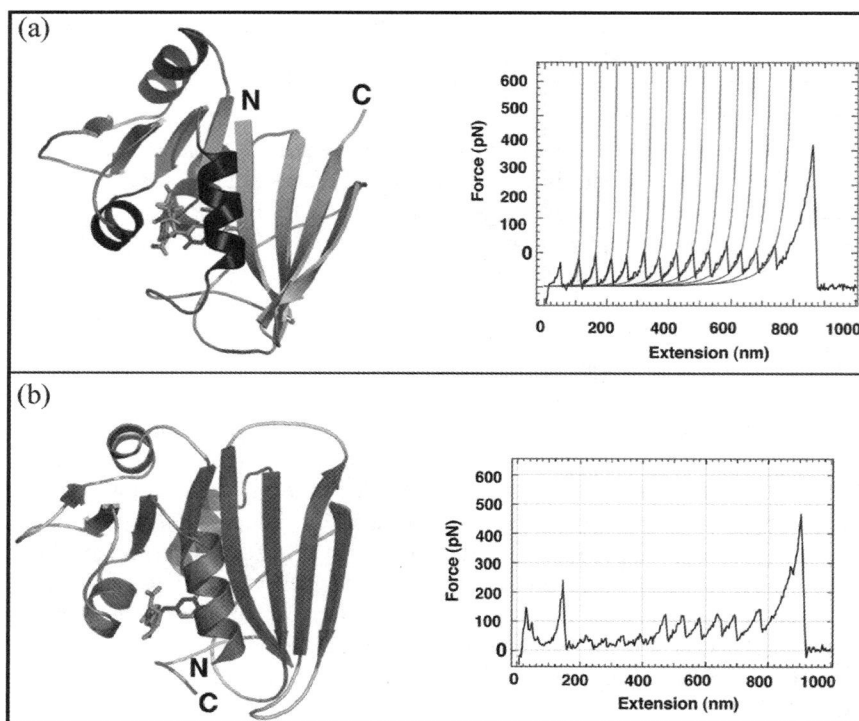

Figure 17. The mechanical resistance of a protein is affected by the type of secondary structure close to the extension points. (A) When a homopolymer of wild-type dihydrofolate reductase, (DHFR)$_{16}$, is extended by its termini, each monomer unfolds at ~131 pN. However, when a circularly permutated form (DHFR CP P25) is extended by its termini (which are at the N-terminus of an α-helix and a loop region) this protein unfolds at a force of ~56 pN (B). Note in this case that the construct used was the heteropolyprotein (DHFR)$_4$(DHFR CP P25)$_4$, and consequently the force-extension profile in (B) shows four unfolding events at high forces and four at low forces (in addition to the nonspecific events that occur at the start of this trace)[103]. Reprinted with permission of the National Academy of Science, U.S.A.

The orientation of terminal β-strands explains the huge difference in mechanical behaviour between I27 and C2A. However, simulations and experiments have shown that not all proteins with parallel terminal strands are mechanically strong. In their original study on the effects of topology on the mechanical strength of proteins, Lu and Schulten further divided their mechanically resistant class I proteins into class Ia, which contained the highly resistant proteins I27, V-CAMI, and the cadherins; and class Ib, the FNIII domains and V-CAMII, which showed weaker mechanical strength. The difference in mechanical strengths of these proteins is thought to arise because the terminal parallel β-strands in class Ia proteins are directly hydrogen-bonded to one another. To unfold these proteins it is necessary to simultaneously rupture all of these bonds, and this allows little deformation of the native structure before unfolding. By contrast, whilst FNIII domains have parallel terminal strands, they are not directly hydrogen-bonded to one another (see Figure 18b) and consequently can undergo a greater deformation with gradual breakage of native state interactions. The mechanical resistance of a protein therefore not only depends on the type and orientation of the secondary structural elements within a protein, but also on the geometry by which these elements are extended. However, determination of the effect of extension geometry upon mechanical

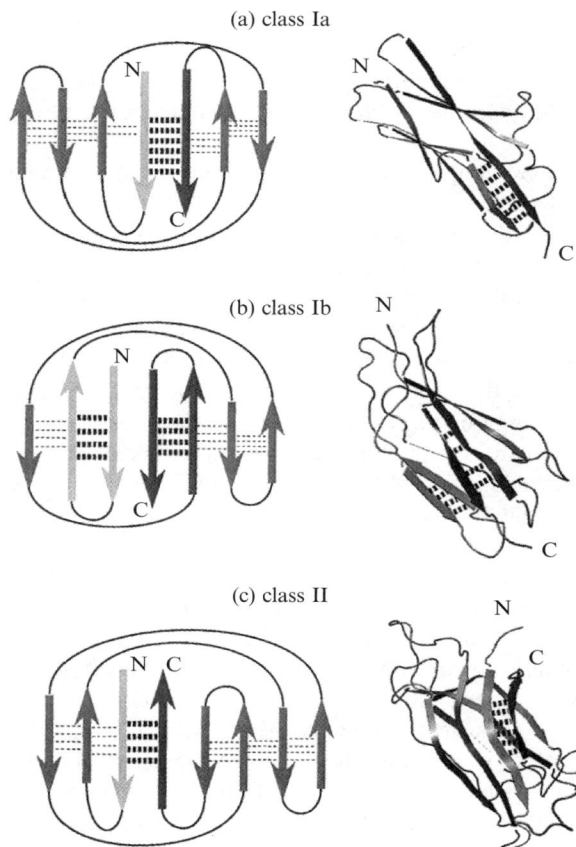

Figure 18. Classification of β-sheet proteins into three mechanical phenotypes. The topologies (left) and three-dimensional structures (right) are shown for (A) V-CAM1 (pdb accession code 1vsc), (B) [7]FNIII (1fnf) and (C) C2A (1rsy). Arrows represent β-strands, and hydrogen bonds (dashed lines) thought to be essential for the mechanical behaviour of each protein as discussed in the text are shown in bold in both representations[67]. Used by permission.

resistance by comparison of different proteins is difficult since, as well as having distinct topologies, they also vary in sequence and kinetic/thermodynamic stability. Direct determination of the effect of pulling geometry on mechanical resistance of a protein thus requires that the *same* domain be studied. Two studies where this has been achieved are discussed below.

4.4 Mechanical Stability of Proteins Pulled in Different Directions

In order to mechanically unfold the same protein by either the shearing or peeling apart of two β-strands, it is necessary to be able to control the points of extension. We have seen that this is possible by using circular permutants. However, this technique sometimes results in significant differences in thermodynamic stability and, importantly, the termini have to be close in space, obviating the possibility of performing a shearing geometry by extending distal points on each strand. To extend a protein in different geometries it is necessary to apply 'handles' onto the test protein at specific locations. This can be performed in vitro, usually by insertion of a unique cysteine residue that can be specifically derivitised by an appropriate tag, or in vivo, by utilising specific post-translational modifications that are performed on some proteins as part of their function. Whilst the former technique has recently been used to measure the mechanical properties of RNaseH close to equilibrium using laser tweezers[6], the latter technique has been used by two groups to explore the mechanical unfolding energy landscapes of E2lip3[105] and ubiquitin[61] by AFM.

The innermost lipoyl domain of the dihydrolipoyl transferase subunit (E2lip3) of the pyruvate dehydrogenase multi-enzyme complex from *Escherichia coli* is post-translationally modified by the covalent attachment of lipoic acid to the N6-amino group of Lys 41 (see Figure 19). As this modification enables the domain to be immobilised to a gold substrate via the dithiolane moiety of lipoic acid, it is possible to examine the mechanical properties of E2lip3 when extended in a shearing geometry (the N-terminus and Lys 41, denoted E2lip3(+)), when extended. In addition, by engineering two cysteine residues at the C-terminus of E2lip3, it is also possible to extend this protein by a peeling geometry (the N- and C-termini, denoted E2lip3(-)), as shown in Figure 19. When a shear-like force was applied, E2lip3 was found to be highly resistant to extension, unfolding at 177 pN at an extension rate of 700 nms^{-1})[105]. By contrast, when extended in a peeling geometry, this E2lip3(-) was found to be highly mechanically labile, unfolding at a force below the thermal noise limit of the AFM (<15 pN). Steered molecular dynamics simulations gave qualitatively similar results and demonstrated that E2lip3 unfolds by different pathways when extended by a shearing or peeling geometry. A second study, examining the effect of extension geometry on the mechanical resistance of ubiquitin, reported similar results. In this study, the authors utilised the variation in ubiquitin chains that occur naturally, owing both to the presence of different genes for this protein that encode monomeric or nonameric proteins[106] and to the formation of polymers of ubiquitin that are linked by their C-terminus and a variety of surface-exposed lysines (Lys 6, Lys 29, Lys 48. and Lys 63)[107]. As with E2lip3, ubiquitin showed distinct mechanical properties when extended from different points in the polypeptide chain. N-C–linked ubiquitin was highly resistant and unfolded at a force of 203 pN at a retraction speed of 300 nms^{-1}, whereas the Lys 48-C linked form unfolded at a mean force of 85 pN at the same speed. As both forms of ubiquitin showed measurable unfolding forces, it was possible to compare the speed-dependence of the unfolding force. N-C–linked ubiquitin was found to be more sensitive to the extension rate, suggesting that when unfolded in this geometry the unfolding transition state is more native-like than when

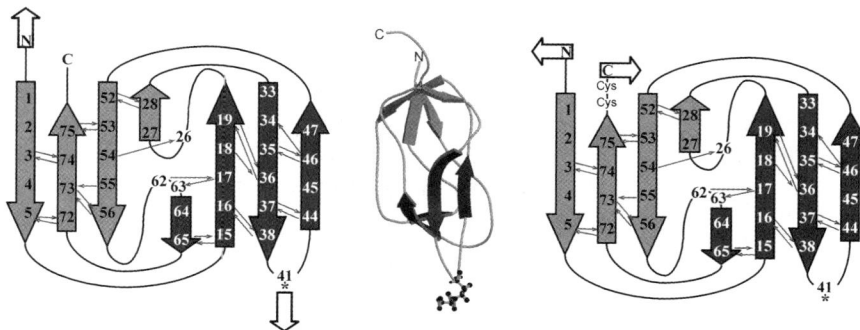

Figure 19. Three-dimensional structure (middle) and topology diagrams showing points of extension for E2lip3(+) (left), and E2lip3(-) (right). Note that whilst attachment of E2lip3(-) to the cantilever and substrate via the N- and C-termini results in a peeling geometry (open arrows), attachment via the N-terminus and lipoic acid at Lys 41 for E2lip3(+) results in a more complex situation (shear-like, open arrows). Hydrogen bonds are shown as thin arrows. Residues involved in elements of secondary structure or forming hydrogen bonds are numbered. Lipoylation of E2lip3(+) occurs at residue 41 (*) which is shown in ball and stick representation (middle).

unfolded by extension of Lys 48 and the C-terminus (x_u estimated to be 0.25 nm and 0.63 nm, respectively). This was in accord with MD simulations.

4.5 The Role of Side Chain Interactions in Force Response

We have seen that topology plays an important role in defining the mechanical properties of proteins. However, as proteins with similar topologies can display very different mechanical properties (for example ^{1}FNIII and ^{10}FNIII [see note 41], and I27 and I32 [see note 96]), it is clear that the type and arrangement of secondary structural elements are not the only factors involved in mechanical stability.

protein L has a mixed α/β topology in which a four-stranded β-sheet packs against a single α-helix. The topology of this domain is such that the terminal β-strands are directly hydrogen-bonded and are parallel with respect to each other. As discussed above, proteins with this arrangement of terminal strands would be expected to be mechanically strong; and in accord with this hypothesis, protein L was found to unfold at a force of 152 pN at 700 nms^{-1} (see note 54). The topology of protein L is similar to that of ubiquitin (both belong to the β-grasp fold SCOP family[108], see Figure 20). However, despite both proteins being mechanically resistant, ubiquitin unfolds at a force 70 pN higher than protein L at all speeds measured. The unfolding distance x_u for both proteins is similar (0.22 nm for protein L[54] and around 0.24 nm for ubiquitin[51, 61]), implying that the proteins unfold via similar pathways. The difference in the mechanical properties of the two proteins cannot be attributed simply to the number of hydrogen bonds between the terminal strands of each protein, as ubiquitin possesses fewer hydrogen-bonded pairs in this region compared with protein L (five and six, respectively), yet is mechanically more resistant. In agreement with the experimental results, molecular dynamics simulations suggest that both proteins unfold by the shearing apart of two separate units, one of which is comprised of the helix and strands I and II, and the other of strands III and IV. Comparison of contact maps constructed for each protein in its native state reveal that although the hydrophobic core of both proteins are similarly sized, the number of long-range contacts that span the two unfolding units in protein L was significantly smaller

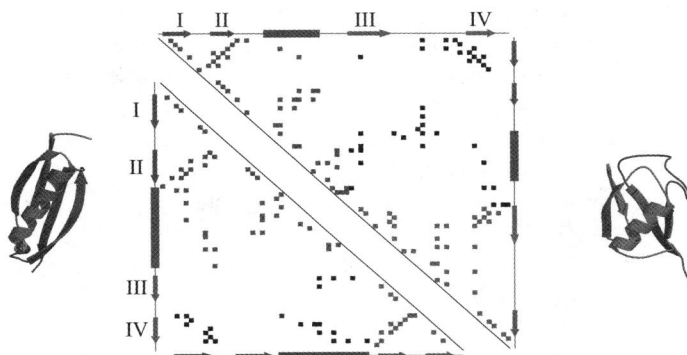

Figure 20. Contact map of protein L (bottom left) and ubiquitin (top right). Side chain contacts (the smallest distance between atoms of two residues < 5Å, calculated by CSU software[121]) made by pairs of amino acids within structural unit 1 (β-hairpin 1 and the helix) or within structural unit 2 (β-hairpin 2) are shown by green and red squares, respectively. Contacts made between these structural units are shown in black. β-strands (labelled I to IV) and α-helices were defined from the solution or X-ray structures using DSSP[108] and are shown as arrows and rectangles, respectively, alongside each contact map. The two structural units are coloured green (unit 1) and red (unit 2) in each protein and are also shown superimposed onto the three-dimensional structure of protein L (left) and ubiquitin (right)[54]. Used by permission. (*See Color Plates*)

(22 contacts) and in fewer clusters than those in ubiquitin (38 contacts). This suggests that a protein in which the side chains from each unit are enmeshed may have a higher mechanical resistance, as these residues must first be extracted, then pulled past other side chains to allow the protein to extend. Thus, although each protein has to be extended to a similar degree to reach the transition state to unfolding, in the case of ubiquitin a significantly greater force is required to reach the transition state. Hence, protein mechanical stability does not only depend on the extension geometry relative to the topology, but also on the extent to which the domain is globally and cooperatively stabilised across the surfaces that are to be sheared.

The degree of cooperativity within mechanical substructures relative to that across mechanical interfaces may also determine whether unfolding intermediates are populated over the lifetime of the experiment. For example, simulations suggest that ubiquitin unfolds via the shearing apart of two mechanical subunits in a similar mechanism to that of protein L. Experimental constant force techniques have shown that this protein sometimes unfolds in two steps (the usual protein end-to-end increase of 20 nm is broken up into two smaller steps of 8 nm and 12 nm[60]). Interestingly, these step sizes are in accord with the simulated unfolding pathway in which the first mechanical sub-unit remains largely native-like after traversing the transition state. These observations suggest that the mechanical sub-units of ubiquitin are metastable and that this protein may mechanically unfold via a high energy intermediate that is rarely populated. Protein L, by contrast, unfolds by a similar pathway but is always observed to unfold in a two-state manner owing to the high cooperativity between unfolding units.

Whilst the intermediate populated during the mechanical unfolding of ubiquitin is highly unstable, I27 unfolds via an intermediate state that is highly stable and native-like apart from the detachment of strand A. A variant lacking the A-strand is surprisingly stable (ΔG_{UN} = 20 kJmol^{-1}), presumably because key core mutations have not been disrupted[87]. The ability of this domain to significantly increase its end-to-end length in a rapid and reversible manner whilst still maintaining its complex native structure may have functional use. Two other proteins with an Ig-like fold similar to that of I27 (the tenth FNIII domain of fibronectin [^{10}FNIII][97] and domain four of filamin from *Dictyostelium discoideum* [ddFLN4][98]) have also been to shown to unfold via intermediates. Both of these domains are thought to have mechanical functions; filamin cross links actin molecules in *Dictyostelium discoideum* and is constantly subjected to mechanical forces during the complex life cycle of this organism, whilst fibronectin is an important component of the extracellular matrix which is proposed to act as a force-sensitive physiological trigger. However, unlike I27, these domains are far less mechanically stable (the native and intermediate states unfold at 63 and 53 pN, and 90 and 50 pN for ddFLN4, and ^{10}FNIII, respectively). The intermediate for each is also more extensively unfolded than the intermediate of I27. Site-directed mutagenesis and loop engineering have revealed that both ddFLN4 and ^{10}FNIII intermediates are formed by the detachment of strands A and B from the core of the protein, although a second intermediate state for ^{10}FNIII was identified in which either the F and G or the A and G strands are also detached from the core[97]. Interestingly, in both cases some mutations were found that destabilised the native state to such an extent that the intermediate was populated even under zero applied force, suggesting the presence of locally cooperative sub-structures in these domains. These proteins therefore have two significant mechanically resistant states, which may facilitate the refolding process when the proteins have been subjected to stress during their normal cellular functions.

Schwaiger et al.[109] investigated the refolding pathway of ddFLN4 in an elegant study in which the AFM was used to perform mechanical 'double jump' experiments. A construct of ddFLN domains was unfolded by extension to 100 nm, after which the cantilever was brought down to 70 nm above the surface, at which point the entropic force applied onto the polypeptide chain was still sufficient to ablate refolding. The polyprotein was then subjected to a series of approach-retract cycles in which the time spent on the surface was varied between 5 and 40 ms. As ddFLN4 had previously been shown to refold significantly faster than the other ddFLN domains[98], this approach allowed the unfolding and refolding of ddFLN4 to be studied in detail. A refolding intermediate similar in terms of size (a mechanically resistant structure of sixty amino acids) to the unfolding intermediate was found to form at a rate constant of $55 s^{-1}$ and fold to the native state with a rate constant of $179 s^{-1}$. The total refolding rate was found to be $42 s^{-1}$, an order of magnitude higher than homologous ddFLN domains that do not populate unfolding intermediates. The population of unfolding intermediates in which a significant fraction of the core is unfolded may expose 'cryptic' binding sites, allowing these domains to act as molecular strain sensors, or as force 'safety catches', as previously described for I27.

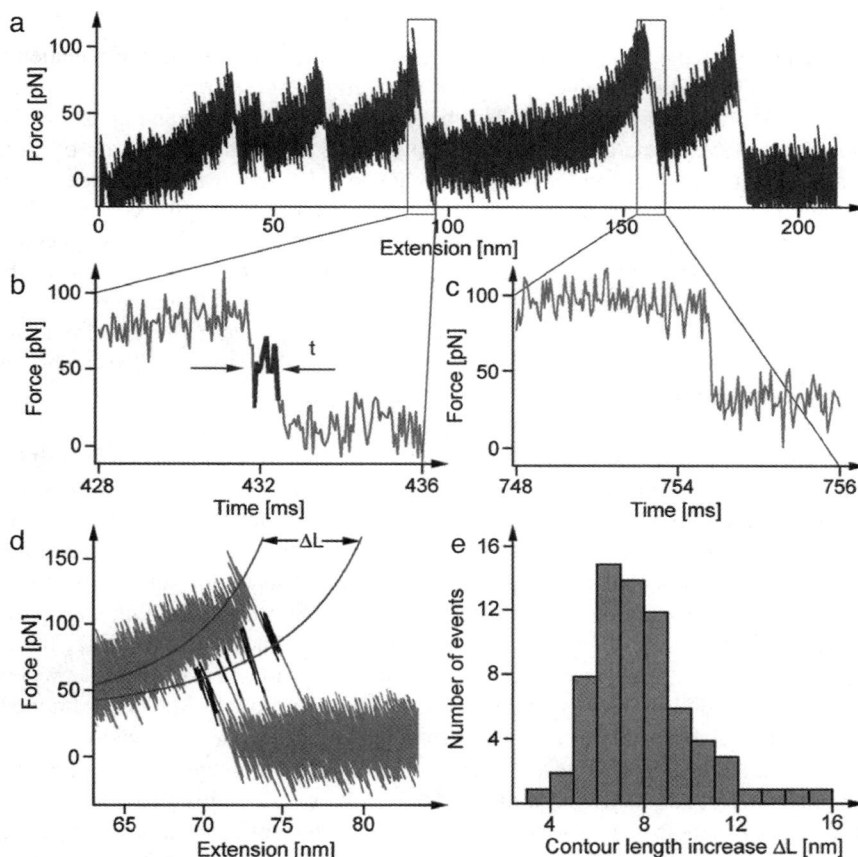

Figure 21. Identification of a short-lived intermediate in the mechanical unfolding of GFP. By measuring force-extension profiles at the full measurement bandwidth (A), deviations from the expected time course can be observed (B)—the area in black indicates the short lived intermediate state. These deviations are not observed for the ddFLN scaffold domains found in the same heteropolyprotein (C). By superimposing six force-extension profiles at the region of GFPΔα unfolding (D), it is possible to calculate the distribution of contour length increases during this transition (E). An increase in contour length of 6.8 nm determined in this manner was postulated to be a result of the abstraction of either the N- or C-terminal β-strand from the folded barrel[50]. Reprinted with permission of the National Academy of Sciences, U.S.A.

The ability of AFM to detect intermediates also depends on the lifetimes of these metastable species. An elegant example is the difference between two intermediates formed during the mechanical unfolding of the 238-residue green fluorescent protein (GFP)[50]. One of these intermediates is thought to be formed by the removal of the seven-residue N-terminal α-helix, which is a near equilibrium process and observed as a hump on the rising edge of the GFP unfolding peak. Although thermodynamically unstable, this intermediate (termed GFPΔα) is populated on the millisecond timescale. The second intermediate, formed by the abstraction of either the N-terminal or the C-terminal strand from the β-barrel of GFP, is highly unstable, and its presence was detected only by analysing the time course of the very fast cantilever relaxation phase after the unfolding peak over the full measurement bandwidth (see Figure 21).

5 Future Perspectives

Single protein molecule mechanical unfolding experiments were initially made feasible by developments in AFM instrumentation that allowed biomolecules to be manipulated and their mechanical properties characterised with sub-nanometre distance and picoNewton force resolution. The interest generated by these initial studies led to the application of molecular biological techniques to construct protein molecules ideally suited to these measurements[45, 47]. This interfacial technique is continually evolving, driven partly by the continual improvements in hardware components and partly by the needs of the researchers engaged in this field. In this section, we shall briefly discuss three relatively new applications of mechanical unfolding using the AFM and see what information may be gained from them.

Perhaps the most controversial novel technique to be reported in this area in recent years is the application of force-clamp spectroscopy to monitor protein folding. This technique was made feasible by the availability of soft cantilevers and the commercial availability of a state of the art piezoelectric translator which allowed rapid changes in length with subnanometer resolution. To date, just one study has been carried out, the measurement of the refolding of concatenated ubiquitin domains[110]. In this study, each domain in the polymer was unfolded by the application of 120 pN constant force that resulted in a stepped length versus time profile. Each 20 nm step marked the unfolding of a single domain and so the number of domains between the tip and substrate was known. When the force was rapidly quenched to ~ 35 pN, the unfolded polymer was found to spontaneously refold. Successful folding was verified by re-application of a high force and counting the number of unfolding events. Since the end-to-end length of the ubiquitin polymer could be monitored as a function of time, it was possible to measure the refolding trajectory of the immobilised polyprotein. The data obtained were not as simple to analyse as those obtained from unfolding experiments, because rather than a series of discrete stepwise folding events that shortened the protein length by 20 nm, refolding was characterised by four stages (regardless of the number of domains refolding) that showed distinct changes in length over variable time periods. Interruption of the folding process at different time points by application of a high force allowed the degree of folding at each stage to be assessed. Refolding appears to occur only during the final two stages of collapse (7% in stage III and 93% in stage IV). The authors suggested that this behaviour was similar to that observed for a coil-globule phase transition and, in the absence of stochastic step-like folding behaviour, concluded that these domains fold in a cooperative process in which most of the ubiquitin proteins in the chain follow similar folding changes at the same time. However, simulation of the folding process using Gō models showed that even if two-state folding of each domain did occur during the force quench, the entropic spring-like nature of the remaining unfolded polypeptide chain could cause large fluctuations in the end-to-end length of the chain, masking the discrete 20 nm folding steps[111]. Whilst this suggests that the experimental observations are consistent with an independent two-state folding mechanism for each domain in the polyprotein, the absence of any folding events early in the experimental trajectory does not support such a conclusion. This technique remains controversial and further experiments on different proteins and their variants, or under different refolding conditions, will be needed to elucidate the mechanism of protein folding when initiated by this method.

Surprisingly, the use of temperature to modulate the mechanical properties of proteins has only recently been addressed in detail. This is partly due to the need to construct vibration-free temperature controllers that can be mounted onto the sample plate[112] or in an isolated box around the AFM,[55] but also partly because the resulting data are difficult to interpret at the molecular level. In an extension of their previous work on ddFLN4, Schlierf and Rief measured the mechanical unfolding forces for the native-to-intermediate (NI) and intermediate-to-unfolded (IU) state transitions over a range of temperature from 5 °C to 37 °C[55]. Kinetic parameters were then obtained by directly fitting the force distributions for each transition at each temperature. Interestingly, the distance between each ground state and the unfolding barrier was found to increase significantly with increasing temperature (x_{ni} and x_{iu} were found to be 2.7±0.5 nm and 4.5±0.5 nm, and 7.8±0.5 nm and 7.6±0.5 nm, at 5 °C and 37 °C, respectively). The authors postulate that the movement of the transition state towards the denatured state with increasing temperature broadens the potential well of the native state and consequently softens the protein to mechanical deformation. Schlierf and Rief interpret this softening as a change in the relative importance of hydrogen bonding patterns and hydrophobic interactions to the mechanical stability of this protein.

However, temperature also affects protein dynamics on a shorter length scale ($\Delta x \ll$ dimensions of the protein) because at this scale, energy landscapes for proteins are not smooth but rough. Energy landscape roughness is a term used to describe the magnitude of the smaller barriers that are superimposed onto the smooth landscape (on a long length scale). A protein must cross these small barriers to explore configurational space and these barriers therefore control transitional dynamics and conformational heterogeneity. Since the relative roughness of the landscape controls the rate at which molecules diffuse through the energy landscape, and temperature affects the ability of proteins to traverse this landscape, it might be expected that a change in temperature would affect the measured unfolding forces, even if the potential of energy landscape remained constant (i.e., x_u remains constant). Conversely, by adopting a theory by Zwanzig[113], Hyeon and Thirumalai[114] showed that the roughness of a protein's energy landscape could be assessed by measuring the effects of temperature on the mechanical strength of proteins, since an increase in relative roughness (a decrease in temperature) slows down the 'diffusion' of a molecule, thereby increasing the unfolding force. Nevo et al. estimated the energy landscape roughness by measuring the force needed to break apart a protein complex (Ran.GTP:imp-β) over a range of temperatures between 7 °C and 32 °C using the AFM[112]. The relatively high value obtained ($\sim 6 k_B T$) was suggested to be of functional relevance since imp-β binds specifically to an array of ligands. In this case a rough landscape is advantageous because it would oppose thermally induced transitions between different conformations of similar free energy.

Whilst the effects of temperature on the observed mechanical properties of proteins are relatively easy to measure, deconvoluting the effects of energy landscape roughness and movement of the transition state is complex. For example, by assuming a temperature-independent barrier height, the data presented by Schlierf and Rief could be interpreted as the effects of temperature on an energy landscape with a roughness of $\sim 4 k_B T$[55]. Furthermore, in addition to changes in the unbinding rate, Nevo et al. report a significant dependence of the transition state placement on the temperature[112]. It is likely that many more of these studies will be carried out in the future, not only because a full understanding of these effects is important in order to describe the role that protein dynamics plays in their function, but also because changes in the position of the transition state strongly affect the predicted mechanical properties of proteins.

All of the mechanical unfolding data discussed in this chapter has been obtained by using constant force or constant velocity techniques. Whilst highly informative, such techniques can only provide information on the mechanical resistance and stiffness of a protein. However, since the dynamics of a protein's structure are very important to its function, there has been great interest in developing techniques that are capable of measuring the viscoelastic properties of proteins. By oscillating either the cantilever or the stage and monitoring the phase and amplitude of the driving oscillation and/or the cantilever as a function of extension or applied force, it is possible to extract the elastic (conservative) and viscous (dissipative) properties of the protein as it is being extended. Due to the infancy of this field, most work has been performed on simple test systems such as dextran[13, 115, 116]; however the dynamic response of bacteriorhodopsin during extraction from a native purple membrane[117] and the dynamic response of bovine carbonic anhydrase (BCA) during extension[118] have been reported. In the latter study, Okajima et al. used a thin disk piezo to oscillate the substrate at 40–60 Hz and collected a time series of cantilever deflection signals every few nanometers while extending a single BCA molecule at $20–30 \, nms^{-1}$. By doing this, the authors showed that the inactive and active forms of this protein had distinct viscoelastic properties. Interestingly, the active form showed an out-of-phase response to the external oscillation that was suggested to be due to the refolding of a partially unfolded region of the BCA. In these studies, as the protein is being extended and oscillated over relatively large amplitude, the system in not in equilibrium. Kawakami et al. have reported a novel quasi-equilibrium method whereby the power spectral density of the system is obtained by recording either the thermally[13] or magnetically[14] driven cantilever deflection signals at a series of set points along the force-extension profile of the biomolecule of choice (in these cases dextran). The viscoelastic properties of the system can then be extracted from the data by fitting with a suitable model. This allows relatively fast dynamical responses to be measured, although the ability of this technique to extract parameters from more complex biomolecules such as proteins is yet to be shown.

Overall, therefore, it is clear that the AFM now has a major role to play in determining the fundamental origins of protein mechanical strength and the nature of conformational fluctuations of functional relevance, in the design of proteins as biomaterials, and for the determination of new features of the unfolding free energy landscape. Further development in instrumentation and of simulations and theory is needed to extract the maximum information from these powerful experiments. Rather than just imaging a single molecule, mechanical unfolding by the AFM promises to become a powerful tool in improving our biophysical understanding of amino acid polymers in the forthcoming years.

Acknowledgments

We would like to thank our collaborators at Leeds without whom much of the work described in this chapter would not have been possible: Godfrey Beddard, Peter Olmsted, Emanuele Paci, and postgraduate students past (Anthony Blake, Rebecca Zinober-Moore) and present (Eleanore Hann, David Sadler, and Dan West). We would also like to acknowledge support from the BBSRC, EPSRC, the Wellcome Trust, and the University of Leeds. DJB is an EPSRC-funded White Rose doctoral training centre lecturer.

References

1. Janmey, P. A.; Weitz, D. A., Dealing with mechanics: mechanisms of force transduction in cells. *Trends Biochem. Sci.* **2004**, 29, (7), 364-370.
2. Hanson, P. I.; Whiteheart, S. W., AAA+ proteins: have engine, will work. *Nat. Rev. Mol. Cell. Biol.* **2005**, 6, (7), 519-529.
3. Clarke, J.; Cota, E.; Fowler, S. B.; Hamill, S. J., Folding studies of immunoglobulin-like beta-sandwich proteins suggest that they share a common folding pathway. *Struct. Fold. Des.* **1999**, 7, (9), 1145-1153.

4. Rief, M.; Fernandez, J. M.; Gaub, H. E., Elastically coupled two-level systems as a model for biopolymer extensibility. *Phys. Rev. Lett.* **1998**, 81, (21), 4764-4767.

5. Engel, A.; Gaub, H. E.; Muller, D. J., Atomic force microscopy: a forceful way with single molecules. *Curr. Biol.* **1999**, 9, (4), R133-R136.

6. Cecconi, R.; Shank, E.; Bustamante, C.; Marqusee, S., Direct observation of the three-state folding of a single protein molecule. *Science* **2005**, 309, 2057-2060.

7. Seidel, R.; van Noort, J.; van der Scheer, C.; Bloom, J. G. P.; Dekker, N. H.; Dutta, C. F.; Blundell, A.; Robinson, T.; Firman, K.; Dekker, C., Real-time observation of DNA translocation by the type I restriction modification enzyme EcoR124I. *Nat. Struct. Mol. Biol.* **2004**, 11, (9), 838-843.

8. Merkel, R.; Nassoy, P.; Leung, A.; Ritchie, K.; Evans, E., Energy landscapes of receptor-ligand bonds explored with dynamic force spectroscopy. *Nature* **1999**, 397, (6714), 50-53.

9. Moller, W.; Nemoto, I.; Matsuzaki, T.; Hofer, T.; Heyder, J., Magnetic phagosome motion in J774A.1 macrophages: Influence of cytoskeletal drugs. *Biophys. J.* **2000**, 79, (2), 720-730.

10. Evans, E.; Leung, A.; Heinrich, V.; Zhu, C., Mechanical switching and coupling between two dissociation pathways in a P-selectin adhesion bond. *Proc. Natl. Acad. Sci. USA* **2004**, 101, (31), 11281-11286.

11. Evans, E.; Leung, A.; Hammer, D.; Simon, S., Chemically distinct transition states govern rapid dissociation of single L-selectin bonds under force. *Proc. Natl. Acad. Sci. USA* **2001**, 98, (7), 3784-3789.

12. Marszalek, P. E.; Li, H. B.; Oberhauser, A. F.; Fernandez, J. M., Chair-boat transitions in single polysaccharide molecules observed with force-ramp AFM. *Proc. Natl. Acad. Sci. USA* **2002**, 99, (7), 4278-4283.

13. Kawakami, M.; Byrne, K.; Khatri, B.; McLeish, T. C. B.; Radford, S. E.; Smith, D. A., Viscoelastic properties of single polysaccharide molecules determined by analysis of thermally driven oscillations of an atomic force microscope cantilever. *Langmuir* **2004**, 20, (21), 9299-9303.

14. Kawakami, M.; Byrne, K.; Khatri, B. S.; McLeish, T. C. B.; Radford, S. E.; Smith, D. A., Viscoelastic measurements of single molecules on a millisecond time scale by magnetically driven oscillation of an atomic force microscope cantilever. *Langmuir* **2005**, 21, (10), 4765-4772.

15. Clausen-Schaumann, H.; Rief, M.; Tolksdorf, C.; Gaub, H. E., Mechanical stability of single DNA molecules. *Biophys. J.* **2000**, 78, (4), 1997-2007.

16. Florin, E. L.; Rief, M.; Lehmann, H.; Ludwig, M.; Dornmair, C.; Moy, V. T.; Gaub, H. E., Sensing specific molecular-interactions with the atomic-force microscope. *Biosens. Bioelectron.* **1995**, 10, (9-10), 895-901.

17. Allen, S.; Chen, X. Y.; Davies, J.; Davies, M. C.; Dawkes, A. C.; Edwards, J. C.; Roberts, C. J.; Sefton, J.; Tendler, S. J. B.; Williams, P. M., Detection of antigen-antibody binding events with the atomic force microscope. *Biochemistry* **1997**, 36, (24), 7457-7463.

18. Dammer, U.; Hegner, M.; Anselmetti, D.; Wagner, P.; Dreier, M.; Huber, W.; Guntherodt, H. J., Specific antigen/antibody interactions measured by force microscopy. *Biophys. J.* **1996**, 70, (5), 2437-2441.

19. Green, N. H.; Williams, P. M.; Wahab, O.; Davies, M. C.; Roberts, C. J.; Tendler, S. J. B.; Allen, S., Single-molecule investigations of RNA dissociation. *Biophys. J.* **2004**, 86, (6), 3811-3821.

20. Krautbauer, R.; Rief, M.; Gaub, H. E., Unzipping DNA oligomers. *Nano Letters* **2003**, 3, (4), 493-496.

21. Binnig, G.; Quate, C. F.; Gerber, C., Atomic force microscope. *Phys. Rev. Lett.* **1986**, 56, (9), 930-933.

22. Burnham, N. A.; Colton, R. J., Measuring The nanomechanical properties and surface forces of materials using an atomic force microscope. *J. Vac. Technol. A* **1989**, 7, (4), 2906-2913.

23. Hutter, J. L.; Bechhoefer, J., Calibration of atomic-force microscope tips. *Rev. Sci. Instrum.* **1993**, 64, (7), 1868-1873.

24. Kedrov, A.; Krieg, M.; Ziegler, C.; Kuhlbrandt, W.; Muller, D. J., Locating ligand binding and activation of a single antiporter. *EMBO Rep.* **2005**, 6, (7), 668-674.

25. Cisneros, D. A.; Oesterhelt, D.; Muller, D. J., Probing origins of molecular interactions stabilizing the membrane proteins halorhodopsin and bacteriorhodopsin. *Struct. Fold. Des.* **2005**, 13, (2), 235-242.

26. Janovjak, H.; Kessler, M.; Oesterhelt, D.; Gaub, H.; Muller, D. J., Unfolding pathways of native bacteriorhodopsin depend on temperature. *EMBO J.* **2003**, 22, (19), 5220-5229.

27. Oesterhelt, F.; Oesterhelt, D.; Pfeiffer, M.; Engel, A.; Gaub, H. E.; Muller, D. J., Unfolding pathways of individual bacteriorhodopsins. *Science* **2000**, 288, (5463), 143-146.

28. Kedrov, A.; Ziegler, C.; Janovjak, H.; Kuhlbrandt, W.; Muller, D. J., Controlled unfolding and refolding of a single sodium-proton antiporter using atomic force microscopy. *J. Mol. Biol.* **2004**, 340, (5), 1143-1152.

29. Kedrov, A.; Janovjak, H.; Ziegler, C.; Kuhlbrandt, W.; Muller, D. J., Observing folding pathways and kinetics of a single sodium-proton antiporter from *Escherichia coli*. *J. Mol. Biol.* **2006**, 355, (1), 2-8.

30. Muller, D. J.; Janovjak, H.; Lehto, T.; Kuerschner, L.; Anderson, K., Observing structure, function and assembly of single proteins by AFM. *Prog. Biophys. Mol. Biol.* **2002**, 79, (1-3), 1-43.

31. Frederix, P.; Akiyama, T.; Staufer, U.; Gerber, C.; Fotiadis, D.; Muller, D. J.; Engel, A., Atomic force bio-analytics. *Curr. Opin. Chem. Biol.* **2003**, 7, (5), 641-647.

32. Tokunaga, M.; Aoki, T.; Hiroshima, M.; Kitamura, K.; Yanagida, T., Subpiconewton intermolecular force microscopy. *Biochem. Biophys. Res. Commun.* **1997**, 231, (3), 566-569.

33. Ashby, P. D.; Chen, L. W.; Lieber, C. M., Probing intermolecular forces and potentials with magnetic feedback chemical force microscopy. *J. Am. Chem. Soc.* **2000**, 122, (39), 9467-9472.

34. Mitsui, K.; Hara, M.; Ikai, A., Mechanical unfolding of alpha(2)-macroglobulin molecules with atomic force microscope. *FEBS Lett.* **1996**, 385, (1-2), 29-33.

35. Rief, M.; Gautel, M.; Oesterhelt, F.; Fernandez, J. M.; Gaub, H. E., Reversible unfolding of individual titin immunoglobulin domains by AFM. *Science* **1997**, 276, (5315), 1109-1112.

36. Oberhauser, A. F.; Marszalek, P. E.; Erickson, H. P.; Fernandez, J. M., The molecular elasticity of the extracellular matrix protein tenascin. *Nature* **1998**, 393, (6681), 181-185.

37. Labeit, S.; Kolmerer, B., Titins - Giant Proteins In Charge Of Muscle Ultrastructure And Elasticity. *Science* **1995**, 270, (5234), 293-296.

38. Schoenauer, R.; Bertoncini, P.; Machaidze, G.; Aebi, U.; Perriard, J. C.; Hegner, M.; Agarkova, I., Myomesin is a molecular spring with adaptable elasticity. *J. Mol. Biol.* **2005**, 349, (2), 367-379.

39. Bertoncini, P.; Schoenauer, R.; Agarkova, I.; Hegner, M.; Perriard, J. C.; Guntherodt, H. J., Study of the mechanical properties of myomesin proteins using dynamic force spectroscopy. *J. Mol. Biol.* **2005**, 348, (5), 1127-1137.

40. Best, R. B.; Li, B.; Steward, A.; Daggett, V.; Clarke, J., Can non-mechanical proteins withstand force? Stretching barnase by atomic force microscopy and molecular dynamics simulation. *Biophys. J.* **2001**, 81, (4), 2344-2356.

41. Oberhauser, A. F.; Badilla-Fernandez, C.; Carrion-Vazquez, M.; Fernandez, J. M., The mechanical hierarchies of fibronectin observed with single- molecule AFM. *J. Mol. Biol.* **2002**, 319, (2), 433-447.

42. Rounsevell, R. W. S.; Steward, A.; Clarke, J., Biophysical investigations of engineered polyproteins: implications for force data. *Biophys. J.* **2005**, 88, (3), 2022-2029.

43. Best, R. B.; Brockwell, D. J.; Toca-Herrera, J. L.; Blake, A. W.; Smith, D. A.; Radford, S. E.; Clarke, J., Force mode atomic force microscopy as a tool for protein folding studies. *Anal. Chim. Acta* **2003**, 479, (1), 87-105.

44. Brockwell, D. J.; Beddard, G. S.; Clarkson, J.; Zinober, R. C.; Blake, A. W.; Trinick, J.; Olmsted, P. D.; Smith, D. A.; Radford, S. E., The effect of core destabilization on the mechanical resistance of I27. *Biophys. J.* **2002**, 83, (1), 458-472.

45. Steward, A.; Toca-Herrera, J. L.; Clarke, J., Versatile cloning system for construction of multimeric proteins for use in atomic force microscopy. *Protein Sci.* **2002**, 11, (9), 2179-2183.

46. Forman, J. R.; Qamar, S.; Paci, E.; Sandford, R. N.; Clarke, J., The remarkable mechanical strength of polycystin-1 supports a direct role in mechanotransduction. *J. Mol. Biol.* **2005**, 349, (4), 861-871.

47. Carrion-Vazquez, M.; Oberhauser, A. F.; Fowler, S. B.; Marszalek, P. E.; Broedel, S. E.; Clarke, J.; Fernandez, J. M., Mechanical and chemical unfolding of a single protein: A comparison. *Proc. Natl. Acad. Sci. USA* **1999**, 96, (7), 3694-3699.

48. Marko, J. F.; Siggia, E. D., Stretching DNA. *Macromolecules* 1995, 28, (26), 8759-8770.

49. Bustamante, C.; Marko, J. F.; Siggia, E. D.; Smith, S., Entropic elasticity of lambda-phage DNA. *Science* **1994**, 265, (5178), 1599-1600.

50. Dietz, H.; Rief, M., Exploring the energy landscape of GFP by single-molecule mechanical experiments. *Proc. Natl. Acad. Sci. USA* **2004**, 101, (46), 16192-16197.

51. Chyan, C. L.; Lin, F. C.; Peng, H.; Yuan, J. M.; Chang, C. H.; Lin, S. H.; Yang, G., Reversible mechanical unfolding of single ubiquitin molecules. *Biophys. J.* **2004**, 87, (6), 3995-4006.

52. Carrion-Vazquez, M.; Oberhauser, A. F.; Fisher, T. E.; Marszalek, P. E.; Li, H. B.; Fernandez, J. M., Mechanical design of proteins-studied by single-molecule force spectroscopy and protein engineering. *Prog. Biophys. Mol. Biol.* **2000**, 74, (1-2), 63-91.

53. Evans, E.; Ritchie, K., Dynamic strength of molecular adhesion bonds. *Biophys. J.* **1997**, 72, (4), 1541-1555.

54. Brockwell, D. J.; Beddard, G. S.; Paci, E.; West, D. K.; Olmsted, P. D.; Smith, D. A.; Radford, S. E., Mechanically unfolding the small, topologically simple protein L. *Biophys. J.* **2005**, 89, (1), 506-519.

55. Schlierf, M.; Rief, M., Temperature softening of a protein in single-molecule experiments. *J. Mol. Biol.* **2005**, 354, 497-503.

56. Rounsevell, R.; Forman, J. R.; Clarke, J., Atomic force microscopy: mechanical unfolding of proteins. *Methods* **2004**, 34, (1), 100-111.

57. Best, R. B.; Fowler, S. B.; Toca-Herrera, J. L.; Clarke, J., A simple method for probing the mechanical unfolding pathway of proteins in detail. *Proc. Natl. Acad. Sci. USA* **2002**, 99, (19), 12143-12148.

58. Williams, P. M.; Evans, E., Dynamic force spectroscopy II: multiple bonds. In *Les Houches-Ecole d'Ete de Physique Theorique.*, Flyvbjerg, H.; Julicher, F.; Ormos, P.; David, F., Eds. Springer-Verlag GmbH: Heidelberg, **2002**; Vol. 75.

59. Oberhauser, A. F.; Hansma, P. K.; Carrion-Vazquez, M.; Fernandez, J. M., Stepwise unfolding of titin under force-clamp atomic force microscopy. *Proc. Natl. Acad. Sci. USA* **2001**, 98, (2), 468-472.

60. Schlierf, M.; Li, H. B.; Fernandez, J. M., The unfolding kinetics of ubiquitin captured with single- molecule force-clamp techniques. *Proc. Natl. Acad. Sci. USA* **2004**, 101, (19), 7299-7304.

61. Carrion-Vazquez, M.; Li, H. B.; Lu, H.; Marszalek, P. E.; Oberhauser, A. F.; Fernandez, J. M., The mechanical stability of ubiquitin is linkage dependent. *Nat. Struct. Biol.* **2003**, 10, (9), 738-743.

62. Isralewitz, B.; Gao, M.; Schulten, K., Steered molecular dynamics and mechanical functions of proteins. *Curr. Opin. Struct. Biol.* **2001**, 11, (2), 224-230.

63. Ortiz, V.; Nielsen, S. O.; Klein, M. L.; Discher, D. E., Unfolding a linker between helical repeats. *J. Mol. Biol.* **2005**, 349, (3), 638-647.

64. Law, R.; Carl, P.; Harper, S.; Dalhaimer, P.; Speicher, D. W.; Discher, D. E., Cooperativity in forced unfolding of tandem spectrin repeats. *Biophys. J.* **2003**, 84, (1), 533-544.

65. Ng, S. P.; Rounsevell, R. W. S.; Steward, A.; Geierhaas, C. D.; Williams, P. M.; Paci, E.; Clarke, J., Mechanical unfolding of TNfn3: the unfolding pathway of a fnIII domain probed by protein engineering, AFM and MD simulation. *J. Mol. Biol.* **2005**, 350, (4), 776-789.

66. Lu, H.; Isralewitz, B.; Krammer, A.; Vogel, V.; Schulten, K., Unfolding of titin immunoglobulin domains by steered molecular dynamics simulation. *Biophys. J.* **1998**, 75, (2), 662-671.

67. Lu, H.; Schulten, K., Steered molecular dynamics simulations of force-induced protein domain unfolding. *Proteins: Struct., Funct., Genet.* **1999**, 35, (4), 453-463.

68. Lu, H.; Schulten, K., The key event in force-induced unfolding of titin's immunoglobulin domains. *Biophys. J.* **2000**, 79, (1), 51-65.

69. Alam, M. T.; Yamada, T.; Carlsson, U.; Ikai, A., The importance of being knotted: effects of the C-terminal knot structure on enzymatic and mechanical properties of bovine carbonic anhydrase II. *FEBS Lett.* **2002**, 519, (1-3), 35-40.

70. Ohta, S.; Alam, M. T.; Arakawa, H.; Ikai, A., Origin of mechanical strength of bovine carbonic anhydrase studied by molecular dynamics simulation. *Biophys. J.* **2004**, 87, (6), 4007-4020.

71. Afrin, R.; Okazaki, S.; Ikai, A., Force spectroscopy of covalent bond rupture versus protein extraction. *Appl. Surf. Sci.* **2004**, 238, (1-4), 47-50.

72. Krammer, A.; Lu, H.; Isralewitz, B.; Schulten, K.; Vogel, V., Forced unfolding of the fibronectin type III module reveals a tensile molecular recognition switch. *Proc. Natl. Acad. Sci. USA* **1999**, 96, (4), 1351-1356.

73. Li, P. C.; Makarov, D. E., Theoretical studies of the mechanical unfolding of the muscle protein titin: Bridging the time-scale gap between simulation and experiment. *J. Chem. Phys.* **2003**, 119, (17), 9260-9268.

74. Li, P. C.; Makarov, D. E., Simulation of the mechanical unfolding of ubiquitin: probing different unfolding reaction coordinates by changing the pulling geometry. *J. Chem. Phys.* **2004**, 121, (10), 4826-4832.

75. Izrailev, S.; Stepaniants, S.; Balsera, M.; Oono, Y.; Schulten, K., Molecular dynamics study of unbinding of the avidin-biotin complex. *Biophys. J.* **1997**, 72, (4), 1568-1581.

76. West, D. K.; Brockwell, D. J.; Olmsted, P. D.; Radford, S. E.; Paci, E., Mechanical resistance of proteins explained using simple molecular models. *Biophys. J.* **2006**, 90, (1), 287-297.

77. Cieplak, M.; Hoang, T. X.; Robbins, M. O., Thermal folding and mechanical unfolding pathways of protein secondary structures. *Proteins: Struct., Funct., Genet.* **2002**, 49, (1), 104-113.

78. Matouschek, A.; Fersht, A. R., Protein engineering in analysis of protein folding pathways and stability. *Methods Enzymol.* **1991**, 202, 82-112.

79. Serrano, L.; Matouschek, A.; Fersht, A. R., The folding of an enzyme.3. Structure of the transition-state for unfolding of barnase analyzed by a protein engineering procedure. *J. Mol. Biol.* **1992**, 224, (3), 805-818.

80. Best, R. B.; Fowler, S. B.; Herrera, J. L. T.; Steward, A.; Paci, E.; Clarke, J., Mechanical unfolding of a titin Ig domain: structure of transition state revealed by combining atomic force microscopy, protein engineering and molecular dynamics simulations. *J. Mol. Biol.* **2003**, 330, (4), 867-877.

81. Cecconi, C.; Shank, E. A.; Bustamante, C.; Marqusee, S., Direct observation of the three-state folding of a single protein molecule. *Science* **2005**, 309, (5743), 2057-2060.

82. Lenne, P. F.; Raae, A. J.; Altmann, S. M.; Saraste, M.; Horber, J. K. H., States and transitions during forced unfolding of a single spectrin repeat. *FEBS Lett.* **2000**, 476, (3), 124-128.

83. Altmann, S. M.; Grunberg, R. G.; Lenne, P. F.; Ylanne, J.; Raae, A.; Herbert, K.; Saraste, M.; Nilges, M.; Horber, J. K. H., Pathways and intermediates in forced unfolding of spectrin repeats. *Struct. Fold. Des.* **2002**, 10, (8), 1085-1096.

84. Marszalek, P. E.; Lu, H.; Li, H. B.; Carrion-Vazquez, M.; Oberhauser, A. F.; Schulten, K.; Fernandez, J. M., Mechanical unfolding intermediates in titin modules. *Nature* **1999**, 402, (6757), 100-103.

85. Li, H. B.; Carrion-Vazquez, M.; Oberhauser, A. F.; Marszalek, P. E.; Fernandez, J. M., Point mutations alter the mechanical stability of immunoglobulin modules. *Nat. Struct. Biol.* **2000**, 7, (12), 1117-1120.

86. Carrion-Vazquez, M.; Marszalek, P. E.; Oberhauser, A. F.; Fernandez, J. M., Atomic force microscopy captures length phenotypes in single proteins. *Proc. Natl. Acad. Sci. USA* **1999**, 96, (20), 11288-11292.

87. Fowler, S. B.; Best, R. B.; Herrera, J. L. T.; Rutherford, T. J.; Steward, A.; Paci, E.; Karplus, M.; Clarke, J., Mechanical unfolding of a titin Ig domain: structure of unfolding intermediate revealed by combining AFM, molecular dynamics simulations, NMR and protein engineering. *J. Mol. Biol.* **2002**, 322, (4), 841-849.

88. Klimov, D. K.; Thirumalai, D., Native topology determines force-induced unfolding pathways in globular proteins. *Proc. Natl. Acad. Sci. USA* **2000**, 97, (13), 7254-7259.

89. Zinober, R. C.; Brockwell, D. J.; Beddard, G. S.; Blake, A. W.; Olmsted, P. D.; Radford, S. E.; Smith, D. A., Mechanically unfolding proteins: The effect of unfolding history and the supramolecular scaffold. *Protein Sci.* **2002**, 11, (12), 2759-2765.

90. Li, H. B.; Oberhauser, A. F.; Redick, S. D.; Carrion-Vazquez, M.; Erickson, H. P.; Fernandez, J. M., Multiple conformations of PEVK proteins detected by single- molecule techniques. *Proc. Natl. Acad. Sci. USA* **2001**, 98, (19), 10682-10686.

91. Patel, A. B.; Allen, S.; Davies, M. C.; Roberts, C. J.; Tendler, S. J. B.; Williams, P. M., Influence of architecture on the kinetic stability of molecular assemblies. *J. Am. Chem. Soc.* **2004**, 126, (5), 1318-1319.

92. Williams, P. M.; Fowler, S. B.; Best, R. B.; Toca-Herrera, J. L.; Scott, K. A.; Steward, A.; Clarke, J., Hidden complexity in the mechanical properties of titin. *Nature* **2003**, 422, (6930), 446-449.

93. Rief, M.; Oesterhelt, F.; Heymann, B.; Gaub, H. E., Single molecule force spectroscopy on polysaccharides by atomic force microscopy. *Science* **1997**, 275, (5304), 1295-1297.

94. Schwaiger, I.; Sattler, C.; Hostetter, D. R.; Rief, M., The myosin coiled-coil is a truly elastic protein structure. *Nat. Mater.* **2002**, 1, (4), 232-235.

95. Smith, B. L.; Schaffer, T. E.; Viani, M.; Thompson, J. B.; Frederick, N. A.; Kindt, J.; Belcher, A.; Stucky, G. D.; Morse, D. E.; Hansma, P. K., Molecular mechanistic origin of the toughness of natural adhesives, fibres and composites. *Nature* **1999**, 399, (6738), 761-763.

96. Li, H. B.; Linke, W. A.; Oberhauser, A. F.; Carrion-Vazquez, M.; Kerkviliet, J. G.; Lu, H.; Marszalek, P. E.; Fernandez, J. M., Reverse engineering of the giant muscle protein titin. *Nature* **2002**, 418, (6901), 998-1002.

97. Li, L.; Huang, H. H.; Badilla, C. L.; Fernandez, J. M., Mechanical unfolding intermediates observed by single-molecule force spectroscopy in a fibronectin type III module. *J. Mol. Biol.* **2005**, 345, (4), 817-26.

98. Schwaiger, I.; Kardinal, A.; Schleicher, M.; Noegel, A. A.; Rief, M., A mechanical unfolding intermediate in an actin-crosslinking protein. *Nat. Struct. Mol. Biol.* **2004**, 11, (1), 81-85.

99. Carl, P.; Kwok, C. H.; Manderson, G.; Speicher, D. W.; Discher, D. E., Forced unfolding modulated by disulfide bonds in the Ig domains of a cell adhesion molecule. *Proc. Natl. Acad. Sci. USA* **2001**, 98, (4), 1565-1570.

100. Oberdorfer, Y.; Fuchs, H.; Janshoff, A., Conformational analysis of native fibronectin by means of force spectroscopy. *Langmuir* **2000**, 16, (26), 9955-9958.

101. Furuike, S.; Ito, T.; Yamazaki, M., Mechanical unfolding of single filamin A (ABP-280) molecules detected by atomic force microscopy. *FEBS Lett.* **2001**, 498, (1), 72-75.

102. Rief, M.; Pascual, J.; Saraste, M.; Gaub, H. E., Single molecule force spectroscopy of spectrin repeats: Low unfolding forces in helix bundles. *J. Mol. Biol.* **1999**, 286, (2), 553-561.

103. Wilcox, A. J.; Choy, J.; Bustamante, C.; Matouschek, A., Effect of protein structure on mitochondrial import. *Proc. Natl. Acad. Sci. USA* **2005**, 102, (43), 15435-15440.

104. Rohs, R.; Etchebest, C.; Lavery, R., Unraveling proteins: A molecular mechanics study. *Biophys. J.* **1999**, 76, (5), 2760-2768.

105. Brockwell, D. J.; Paci, E.; Zinober, R. C.; Beddard, G. S.; Olmsted, P. D.; Smith, D. A.; Perham, R. N.; Radford, S. E., Pulling geometry defines the mechanical resistance of a beta-sheet protein. *Nat. Struct. Biol.* **2003**, 10, (9), 731-737.

106. Wiborg, O.; Pedersen, M. S.; Wind, A.; Berglund, L. E.; Marcker, K. A.; Vuust, J., The Human Ubiquitin Multigene Family - Some Genes Contain Multiple Directly Repeated Ubiquitin Coding Sequences. *EMBO J.* **1985**, 4, (3), 755-759.

107. Pickart, C. M.; Fushman, D., Polyubiquitin chains: polymeric protein signals. *Curr. Opin. Chem. Biol.* **2004**, 8, (6), 610-616.

108. Murzin, A. G.; Brenner, S. E.; Hubbard, T.; Chothia, C., Scop - a Structural Classification of Proteins Database for the Investigation of Sequences and Structures. *J. Mol. Biol.* **1995**, 247, (4), 536-540.

109. Schwaiger, I.; Schleicher, M.; Noegel, A. A.; Rief, M., The folding pathway of a fast-folding immunoglobulin domain revealed by single-molecule mechanical experiments. *EMBO Rep.* **2005**, 6, (1), 46-51.

110. Fernandez, J. M.; Li, H. B., Force-clamp spectroscopy monitors the folding trajectory of a single protein. *Science* **2004**, 303, (5664), 1674-1678.

111. Best, R. B.; Hummer, G., Comment on "Force-clamp spectroscopy monitors the folding trajectory of a single protein". *Science* **2005**, 308, (5721).

112. Nevo, R.; Brumfeld, V.; Kapon, R.; Hinterdorfer, P.; Reich, Z., Direct measurement of protein energy landscape roughness. *EMBO Rep.* **2005**, 6, (5), 482-486.

113. Zwanzig, R., Diffusion In A Rough Potential. *Proc. Natl. Acad. Sci. USA* **1988**, 85, (7), 2029-2030.

114. Hyeon, C. B.; Thirumalai, D., Can energy landscape roughness of proteins and RNA be measured by using mechanical unfolding experiments? *Proc. Natl. Acad. Sci. USA* **2003**, 100, (18), 10249-10253.

115. Humphris, A. D. L.; Antognozzi, M.; McMaster, T. J.; Miles, M. J., Transverse dynamic force spectroscopy: A novel approach to determining the complex stiffness of a single molecule. *Langmuir* **2002**, 18, (5), 1729-1733.

116. Humphris, A. D. L.; Tamayo, J.; Miles, M. J., Active quality factor control in liquids for force spectroscopy. *Langmuir* **2000**, 16, (21), 7891-7894.

117. Janovjak, H.; Muller, D. J.; Humphris, A. D. L., Molecular force modulation spectroscopy revealing the dynamic response of single bacteriorhodopsins. *Biophys. J.* **2005**, 88, (2), 1423-1431.

118. Okajima, T.; Arakawa, H.; Alam, M. T.; Sekiguchi, H.; Ikai, A., Dynamics of a partially stretched protein molecule studied using an atomic force microscope. *Biophys. Chem.* **2004**, 107, (1), 51-61.

119. Best, R. B.; Clarke, J., What can atomic force microscopy tell us about protein folding? *Chem. Commun.* **2002**, (3), 183-192.

120. Zinober-Moore, R. Elasticity and mechanical unfolding of globular protein domains. PhD thesis **2005**, University of Leeds, Leeds,.

121. Sobolev, V.; Sorokine, A.; Prilusky, J.; Abola, E. E.; Edelman, M., Automated analysis of interatomic contacts in proteins. *Bioinformatics* **1999**, 15, (4), 327-332.

122. Friel, C. T.; Capaldi, A. P.; Radford, S. E., Structural analysis of the rate-limiting transition states in the folding of lm7 and lm9: similarities and differences in the folding of homologous proteins. *J. Mol. Biol.* **2003**, 326, 293-305.

123. Scott, K. A.; Clarke, J., Spectrin R16: broad energy barrier or sequential transition states? *Protein Sci.* **2005**, 14, 1617-1629.

124. Hamill, S. J.; Meekhof, A. E.; Clarke, J., The effect of boundary selection on the stability and folding of the third fibronectin type III domain from human tenascin. *Biochemistry* **1998**, 37, 8071-8079.

125. Maxwell, K. L.; Wildes, D.; Zarrine-Afsar, A.; De Los Rios, M. A.; Brown, A. G.; Friel, C. T.; Hedberg, L.; Horng, J. C.; Bona, D.; Miller, E. J.; Vallee-Belisle, A.; Main, E. R.; Bemporad, F.; Qiu, L.; Teilum, K.; Vu, N. D.; Edwards, A. M.; Ruczinski, I.; Poulsen, F. M.; Kragelund, B. B.; Michnick, S. W.; Chiti, F.; Bai, Y.; Hagen, S. J.; Serrano, L.; Oliveberg, M.; Raleigh, D. P.; Wittung-Stafshede, P.; Radford, S. E.; Jackson, S. E.; Sosnick, T. R.; Marqusee, S.; Davidson, A. R.; Plaxco, K. W., Protein folding: Defining a "standard" set of experimental conditions and a preliminary kinetic data set of two-state proteins. *Protein Sci.* **2005**, 14, 602-616.

Counting and Breaking Single Bonds

Dynamic Force Spectroscopy in Tethered Single Molecule Systems

Todd A. Sulchek, Raymond W. Friddle, and Aleksandr Noy

Overview

Tethered molecular systems, in which flexible polymer linkers connect the interacting molecules to the surfaces of the atomic force microscope probe and target sample, provide a particularly attractive platform for studying biological interactions using force spectroscopy. We describe the underlying nanomechanics of individual tether molecules. We provide a sampling from the literature that illustrates how knowledge and control of tether linkage can aid understanding of molecular interactions. We then describe the basic physical principles of force spectroscopy measurements of tethered biological interactions and show that well-defined mechanical properties of the tether linkages allow independent determination of the number of ruptured bonds. This approach allows us to show that forces between multiple biological bonds measured in a parallel configuration obey the predictions of a Markovian model of bond dissociation. Finally, we discuss the use of the dynamic force spectra of single and multiple protein-ligand bonds for determination of kinetic parameters for multivalent interactions. This ability to form, count, and dissociate biological bonds with nanomechanical forces provides a powerful method to study the physical laws governing the interactions of biological molecules.

1 Tethered Biomolecular Systems in Force Spectroscopy

Interactions between biological molecules drive the overwhelming majority of cellular processes and span a wide range of strength and complexity. Often biological interactions involve multivalent binding, where multiple individual ligand-receptor bonds combine to produce a stronger overall interaction. Multivalent binding events play an important role in adaptive immune response[1] and intercellular adhesion.[2] Complex bonds are also featured in the mechanism of action of many pharmaceuticals,[3] toxin inhibitors,[4] and serve as a generic affinity-enhancing approach[5, 6] in a variety of therapies and imaging techniques that target specific biological tissues.[7, 8]

The last decade saw an explosive growth in the number of experimental studies where researchers used mechanical force transducers to measure and apply stresses to individual molecules.[9–11] This ability provides an advantage over ensemble-based techniques as it eliminates spatial and temporal averaging that can obscure the details of the interaction.[12–15] In particular, atomic force microscopy (AFM) probes ligand-receptor interactions by simply applying a mechanical force to a ligand-receptor complex via a flexible cantilever spring.[16] The basics of the AFM force measurements have been well-covered in the previous chapters; therefore it is sufficient to note that in these measurements the bond strength is defined by the force at which bond rupture occurs most frequently during repeated tests of bond breakage on a given timescale, and that these measurements allow researchers to quantify kinetic off-rates and the distances to transition states.[17, 18]

The main conceptual difficulties in measuring and interpreting bond strength of individual and multivalent bonds are: (1) the necessity of separating the specific interactions from the nonspecific interactions of the probe tip and sample, and (2) the absence of a reliable way to determine the number of interacting molecules independent of the binding force values. In general, force spectroscopy does not provide an immediate way to discriminate between the receptor-ligand bond and nonspecific tip-surface interactions; therefore researchers have to address specificity either through careful experimental design that encourages[19] or discourages a particular interaction, or by using an independent reporter of specific interactions.[20]

A simple and powerful solution to the two problems discussed in the previous paragraph is linking the interacting molecules to the surfaces of an AFM tip and sample surface with long flexible polymer tethers (Figure 1). We now briefly enumerate the advantages of this experimental design. First, as shown for the antibody antigen binding system (Figure 2A), the specific interactions now occur at a fixed tip-sample distance given by the combined length of the two polymer tethers (Figure 2B). Second, the tethers spatially isolate nonspecific probe-sample interactions from the specific interactions of the tethered molecules (Figure 2C). Third, if the length of the tethers is much larger than the polymer persistence length,[21] they remain truly flexible, allowing the necessary conformational freedom to maximize the efficiency of the ligand-receptor docking. Fourth, this configuration separates the binding pair from the surfaces of the tip and sample, thus allowing the bond to rupture in an environment that closely resembles a natural solution environment. Fifth, as we will describe further in the chapter, tethered systems provide a convenient way to determine the number and manner of rupture of the interacting bonds independent of the magnitude of the rupture force. Finally,

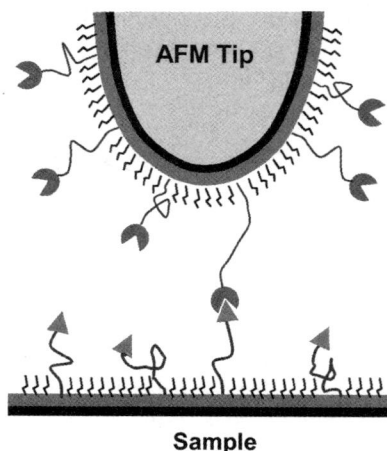

Figure 1. Schematic of force spectroscopy experiment in a tethered system. Both interacting molecules are connected to the surfaces of the AFM tip and sample with long flexible polymer chains.

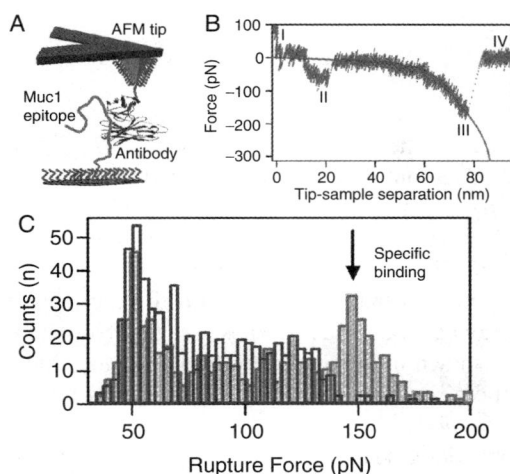

Figure 2. (A) Schematics of the tethered system for measuring interactions of peptide MUC1 with the antibody fragment recognizing MUC1. (B) Representative force vs. distance trace measured in this system. Solid line indicates a fit of the tether extension trace to the TLS-eFJC model (Eq. 3 and 7) (C). Histograms of the rupture forces measured in the one-linker length rupture region (filled bars) and two-linker lengths rupture region (unfilled bars). An arrow indicates the peak corresponding to the specific peptide-antibody interactions.

tethered systems provide a nearly-ideal experimental system for dynamic force spectroscopy measurements, as tether elasticity provides a means for very efficient rebinding suppression at moderate rates of force loading.

Tethering ligands and receptors to a solid substrate offers all the advantages that we just listed, albeit at the cost of making the experiment much more complex. In the next section, we introduce a physical description of polymer chains that account for many of the nonlinear elastic behaviors observed when a polymer tether is stretched under load. The nonlinear stiffness of polymer tethers will be a notable feature that we explicitly exploit to improve force spectroscopy measurements.

2 Entropic and Enthalpic Elasticity of A Polymer Chain

Solvent molecules moving in random directions constantly bombard a polymer chain in solution and supply the thermal energy that allows the polymer molecule to adopt one of its many coiled configurations. If we attempt to stretch the polymer beyond an entropically-favored configuration, the polymer chain will resist the stretching, producing a force that acts in the direction opposite to the external force. As we stretch the chain a distance that approaches its contour length, the polymer stiffness, or restoring force per displacement, rapidly increases as the number of configurations accessible to the system decreases: a polymer chain stretched beyond its natural conformation behaves as a non-Hookean "entropic" spring.

Although the state of a single polymer is constantly changing in time, we may assume Boltzmann's ergodic hypothesis and say that the average of a long-time measurement for a single polymer is equivalent to the instantaneous average of a statistical ensemble. In the following treatments we assume that the speed of stretching the polymer is slow enough to allow the thermal fluctuations to randomize the polymer chain configurations.

Most common models that describe the behavior of polymers can be derived from Kratky-Porod's[22] description of a chain of N rigid segments under a tension F along the z axis,

$$H = -\frac{B}{a}\sum_{i=2}^{N} \hat{a}_i \cdot \hat{a}_{i-1} - Fa\sum_{i=1}^{N} \hat{a}_i \cdot \hat{z} \tag{1}$$

where H is the Hamiltonian of the system, with segment lengths a, bending modulus B, and segment unit vectors $\hat{a}^i$. The configurational partition function Z is then calculated as the Boltzmann-weighted sum over the chain orientations $\hat{a}_i$. Under a constant force F the average extension of the molecule along the z axis, $\langle z \rangle$, is then given by the rate of change of the free energy with respect to applied force: $\langle z \rangle = d(k_B T \ln Z)/dF$.

The two prevailing models that are very useful in describing polymer mechanics are limiting cases of the energy equation (Eq. 1). These are the freely jointed chain (FJC) and the worm-like chain (WLC), shown in Figure 3. The freely jointed chain (FJC) is a particularly relevant description of polymer stretching for many tethered force spectroscopy measurements. The FJC model describes a polymer consisting of rigid segments ($B = 0$) that are free to rotate about their connections. The only parameter that governs the polymer's response to an external force is the Kuhn statistical segment length a. This length is related to the persistence length, P, of the polymer by $a = 2P$. The FJC model predicts that the mean extension $\langle z \rangle$ under constant force F follows a Langevin function:

$$\langle z \rangle = Na\left(\coth\frac{Fa}{k_B T} - \frac{k_B T}{Fa} \right) \tag{2}$$

Note that as the extension of the polymer approaches its contour length, the force approaches infinity. A more realistic extension of the FJC model allows for elastic stretching of the backbone in the high force regime:

$$\langle z \rangle = Na\left(\coth\frac{Fa}{k_B T} - \frac{k_B T}{Fa} \right)\left(1 + \frac{F}{K_s} \right) \tag{3}$$

where K_s is the stretching modulus of the polymer.

In some cases a polymer behaves more like a deformable rod than a series of loosely connected segments. The worm-like chain model is the limiting case of a continuous space curve with no discontinuities along its contour ($\lim a \rightarrow 0$). The WLC model describes an isotropic, homogeneous rod with no distinct segments as in the FJC. The amount of curvature the WLC

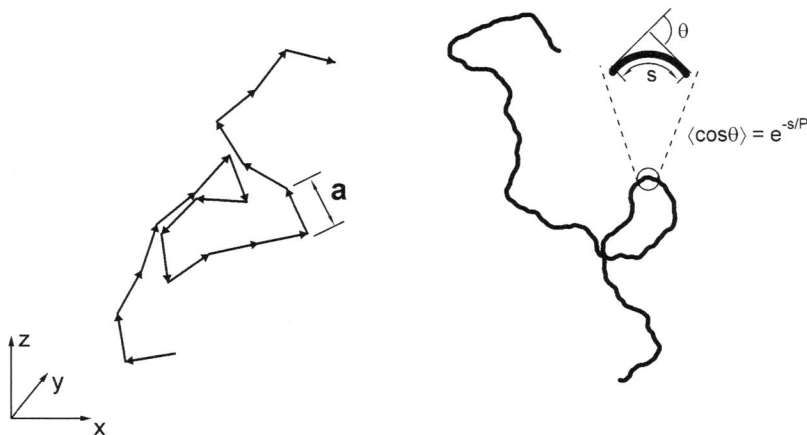

Figure 3. (Left) the freely jointed chain model; (Right) the wormlike chain.

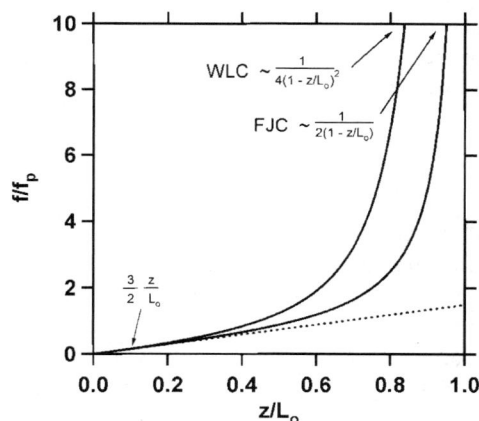

Figure 4. The Worm-Like Chain (WLC) and Freely Jointed Chain (FJC) force laws scaled by the characteristic force constant $f_p = K_B T/P$. Both models emerge from the Gaussian chain approximation at low extensions. At high extensions the WLC model is more resistant to streching due to fluctuations that are permitted to occur over distances less than the persistence length of the polymer.

will allow along a given length is determined by the bending modulus B, which is related to the persistence length by scaling with the thermal energy $P = B/k_B T$. However, the persistence length is strictly speaking a function of external force, and protein polymers show a persistence value that decreases at modest forces due to the shift from entropic to enthalpic sources of elasticity.[23]

While the WLC model cannot be solved analytically as a function of force, a very useful result that is asymptotically correct in both the low force Gaussian chain $f \approx \dfrac{k_B T}{P} \dfrac{3}{2} \dfrac{z}{L_o}$ and high force $f \approx \dfrac{k_B T}{P} \dfrac{1}{\left(1 - z/L_o\right)^2}$ regimes is given by Marko and Siggia.[24] Once again, most

$$f = \frac{k_B T}{P}\left(\frac{z}{L_o} - \frac{1}{4} + \frac{1}{4\left(1 - z/L_o\right)^2} \right) \tag{4}$$

physical molecules will be compliant when stretched at high forces. Therefore an extensible WLC based on a modification of Eq. 4 can be used when the stretching approaches the contour length,[25]

$$f = \frac{k_B T}{P}\left(\frac{z}{L_o} - \frac{1}{4} + \frac{1}{4\left(1 - z/L_o + f/K\right)^2} - \frac{f}{K} \right) \tag{5}$$

Notice that in principle the extensible-FJC and extensible-WLC models (eFJC and eWLC) can alone be used to encompass the phenomena of polymer stretching, since for an infinitely stiff stretching modulus they reduce to the standard FJC and WLC models. Notably, these two models begin to show significant differences as the extension begins to approach the polymer contour length (Figure 4).

Force-induced Transitions Within the Polymer Chain

Often the effect of an external load goes beyond simply straightening the polymer chain, and causes internal conformation transitions. Unfolding of the Ig domains in a titin molecule[26] and conformational transitions in monomer subunits of poly(ethylene-glycol) (PEG)[27] and poly(ethylene-oxide) (PEO) molecules[28] represent examples of such conformational changes. Boltzmann partitioning of the states can elegantly model a polymer chain with subunits that undergo an equilibrium transition from a shortened state to a lengthened state,[27]

$$\frac{N_f}{N_u} = \exp\frac{\Delta G - F\Delta z}{k_B T} \tag{6}$$

where $\Delta G = \Delta G^*_u - \Delta G^*_f$ is the energy difference between the folded and unfolded state at zero force, Δz is the extension length between the two states, and N_f, N_u are the number of domains in the folded and unfolded states. Note that in an analogous manner to pH or solution ionic strength change shifting the equilibrium in a chemical reaction, the applied force can tilt the energy landscape and change the relative equilibrium populations of the folded and unfolded states (5B). Considering that $N = N_u + N_f$, it is straightforward to show that the fractional contribution of domains in the folded and unfolded states to the average length of the polymer is

$$L_C(F) = N\left(\frac{N_u}{N} l_u + \frac{N_f}{N} l_f\right) = N\left(\frac{l_u}{1 + e^{\Delta G(F)/k_B T}} + \frac{l_f}{1 + e^{-\Delta G(F)/k_B T}}\right) \tag{7}$$

where $\Delta G(F) = \Delta G - F\Delta z$ is the force-tilted energy barrier, and l_u and l_f are the domain lengths when unfolded or folded, respectively. Substitution of this equation in place of the contour length in the eFJC model produces the force-extension relationship for a generic polymer chain with segments that can undergo structural transitions under load. Equation 7 provides the main framework for describing the elastic response of individual polymer linkers that are used extensively in force spectroscopy experiments. As evident from the comparison of the force extension curves for the three FJC models (Figure 5c), the most pronounced effect

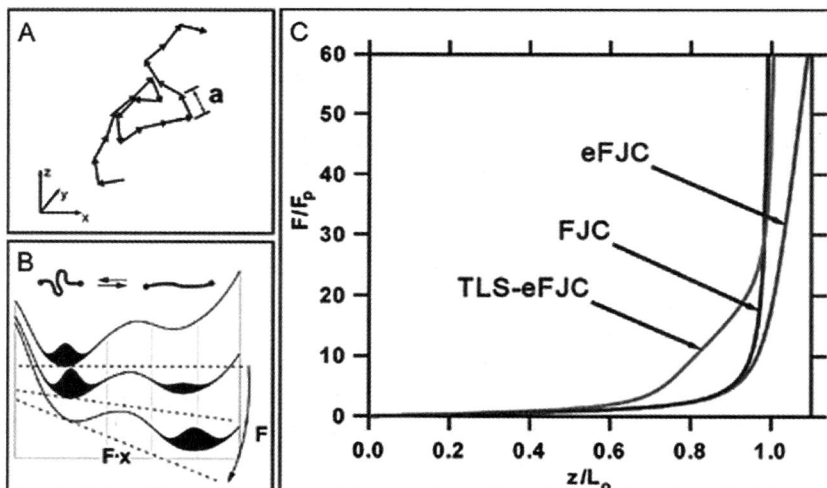

Figure 5. (A) Schematic representation of a freely-jointed chain. (B) Schematic representation and potential enegy profiles of a two-level system under an external loading force F. (C) Force-extension profiles for freely-jointed chain (FJC), extensible freely-jointed chain (eFJC), and freely-jointed chain with two-state conformation transition elements (TLS-eFJC). Forces are scaled by the characteristic force constant $F_P = k_B T/P$.

of the internal structural transitions on the shape of the elasticity curves is observed at the intermediate extensions in the range of 70–90% contour length extension.

3 Dynamics of Single Bond Rupture in Tethered Systems

The current description of the physics of non-covalent bond rupture originates from the framework proposed by Bell in the 1970s.[29] In this picture, external force lowers the energy barrier separating the bound state from the unbound state, while the thermal fluctuations continuously attempt to drive the molecular separation over that barrier. In the absence of force the intrinsic transition rate is small, but external loading can cause exponential amplification of this rate and eventually lead to the quick breakage of the bond.[17] Since the stability of the bond is affected by fluctuations in the system, the nature of the linkage between the interacting molecular pair and the force transducer can have profound consequences for the bond rupture kinetics. According to Kramers's description of thermally-activated transitions in an overdamped molecular system,[30] the intrinsic rate of bond rupture is inversely proportional to the damping coefficient of the system. Note that this observation can have immediate practical consequences for force spectroscopy experiments: if the interacting molecule is connected rigidly to an AFM tip, the effective spontaneous rate of dissociation in this system could differ from the dissociation rate of the free molecule by orders of magnitude. For example, an AFM tip within 1 nm of a surface can have a damping coefficient as high as 10^{-4} kg/s in solution.[31] When compared to the damping coefficients of small molecules, which are typically close to 10^{-11} kg/s, the difference in damping due to probe linkage can greatly decrease the spontaneous rate of dissociation (in this example by a factor of about 10^{-7}!). When a polymer linker separates the AFM tip and the molecule, the drag of the cantilever falls off with polymer length and is replaced with the drag of the polymer molecule itself. If the macromolecules that form the biochemical bond are substantially large in comparison to the linker, we can neglect the linker drag and assume that the dynamics will be dominated by the diffusion characteristics of the molecule.[32] For comparison, long tethers such as 25 nm-long PEG will have a molecular weight of approximately 4kD, whereas a typical antibody is over 100kD and a single chain variable fragment antibody (scFv) is approximately 25 kD. In this case the force spectroscopy measurement will determine the intrinsic dissociation rate accurately.

An additional important consideration arises from an often overlooked assumption in the Bell model (which has been carried over to most of the subsequent models of force spectroscopy experiments) that postulates that the forced rupture of a non-covalent bond is irreversible. In principle, this assumption is not always correct, as there is the possibility of rebinding at slow loading. However, the non-Hookean shape of the potential for stretching a typical polymer tether ensures that at almost any applied force the stretched polymer molecule contracts sufficiently upon bond rupture and thus inhibits rebinding events over all experimentally accessible loading rates. In this case the bond ruptures are truly irreversible and tethered systems represent a nearly ideal realization of the first order kinetic process that is described by most dynamic force spectroscopy models.

The inclusion of the nonlinear tether spring between the bond and the force transducer modifies the loading regime encountered in the AFM measurements. Typically, the AFM instrument moves the force transducer away from the sample at a constant rate, producing a constant increase in the loading force. In a tethered system, even if the force transducer moves at a constant rate, the bond is effectively stretched by the convolution of the tether spring and cantilever spring in series, and the loading force experienced by the bond will not increase linearly. Nonlinear loading introduces substantial complications into the analysis of the rupture kinetics, and in general precludes their analytical solution. Fortunately, Gaub and co-workers[33] showed that the loading rate can be replaced with an instantaneous loading rate at the moment of bond rupture without substantially altering the force spectrum for the bond, which should be considered a standard practice for analyzing tethered systems.

Examples From Literature

Implementing an independent readout of physical transitions that could be independently correlated with the measured forces transforms the force microscope into an extremely powerful tool for studying biological events on a single-molecule level. Such an independent readout can be implemented in a variety of ways that suit the individual measurement. Notably, it is possible to use the well -characterized extension properties of many common polymer linkers to study the specific mechanisms of bond rupture, or to distinguish multiple bond ruptures from single bond ruptures. Importantly, the knowledge and control of the linker characteristics provide a vital window into understanding the physical transitions underlying biological functionality.

P$_{EG}$ Extension

Careful studies of the force extension of the biologically inert polymer linker poly(ethylene glycol) (PEG) reveal that a simple entropic model does not adequately describe the polymer stiffness as a function of extension. In fact, researchers encountered three important regions of stiffness of individual PEG molecules as the force was ramped up to 600 pN.[34] This experimental work was then complemented by molecular dynamics stretching simulations.[35] At low forces, the polymer chain behaves like a Hookean spring, producing in a linear force versus distance, shown in the low force limit of Equation 4. At moderate forces, PEG produces a nonlinear, non-Gaussian force-extension curve. In the high force regime, the force rapidly increases as the extension approaches contour length and the chain has few conformation states available.

A quick comparison between Figures 5 and 6 reveals that in a nonpolar solvent such as hexadecane, PEG acts like an ideal eFJC (Figure 6). Polar solvents such as water produce

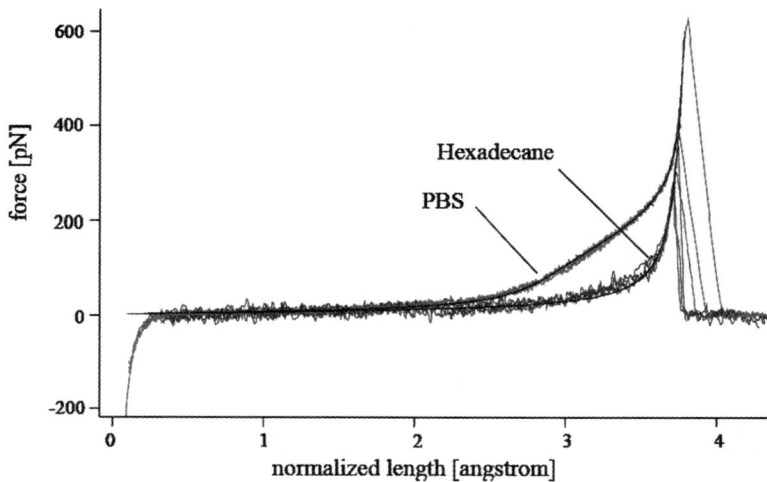

Figure 6. Force extension curves of PEG in a polar solvent (PBS) and a nonpolar solvent (hexadecane). The polar solvent stabilizes helical structures between PEG monomers that lead to distortion from the eFJC model. The freely-jointed chain with two-state conformation transition elements (TLS-eFJC) provides a more accurate fit (used with permission from Oesterhelt et al., Reference 26). (*See Color Plates*)

a marked deviation in the transition from entropic to enthalpic elasticity, as a secondary structure within the polymer becomes disrupted. This secondary structure of polymeric PEG results from the distribution of one of two isomeric states of the monomers, one state being a trans-trans-gauche (ttg) and the other a trans-trans-trans (ttt). The water bridge mediated state of the ttg conformation is energetically favorable to the elongated state of ttt by approximately $3k_BT$, as determined from quantum mechanical calculation. From Equation 6, the majority populated state under physiological conditions contains over $1-e^{-\Delta\Delta G/k_BT}$ or 95% of the total number of states at equilibrium. However, under force, that ratio skews towards the planar conformation, resulting in an increase in total contour length in accordance with Equation 7. The helical formation results in a significant deviation from the eFJC model. The net effect is that greater force is required to reach the same extension at moderate extensions.

Other Polymer Linkers

Much like PEG, poly(ethylene oxide), PEO, is a synthetic polymer that is biologically inert and can be used as a molecular tether. Rixman et al. studied the molecular elasticity of individual PEO chains and compared the results with several elasticity theories.[28] PEO stretching in polar solvent likewise conforms to the Markovian 2-state unfolding with extensible FJC. The solvent plays a large role in stabilizing ordered structures within a variety of polysaccharide molecules as well.[36]

Measuring Protein Stability Using Force-induced Unfolding

A major limitation of using single-molecule force spectroscopy to analyze intermolecular and intramolecular forces remains the difficulty in assigning the measured force of an interaction to a particular molecular rearrangement and transition. As a rule, it is difficult to determine which part of the molecule is stretched or how many molecules are pulled apart, and controlling these characteristics is even more difficult. Nevertheless, two techniques have begun to address these issues: the covalent attachment of the tip to unique positions on individual molecules, and utilizing linkers that allow a preferred application of the force vector. These approaches have already produced a significant improvement of measurement reproducibility.

A primary example in which single molecule force measurements can reveal the molecular mechanisms that govern biological interaction and function is protein conformational stability. Proteins define their unique function through spatial organization of polypeptide chains stabilized by specific interactions. Many proteins display a secondary structure, primarily of α-helices or β-sheets, that unfold as a unit upon mechanical destabilization.[37-40] The stability of a particular native protein conformation can be quantified by the resistance of the various unfolding pathways to an applied force. As a transducer increases the applied force, secondary structure unfolding manifests itself as a sharp increase in polymer length caused by the rapid increase of the polypeptide chain contour length. For example, Rief and colleagues recently engineered a globular protein with multiple sites of covalent attachment so that mechanical stability can be measured in up to five directional axes.[41] Such "mechanical denaturation" has a distinct advantage over the conventional thermal or chemical denaturation techniques, since it can probe intermediate states along various unfolding pathways. In each of these studies, knowledge or control of secondary structure arrangement, strength, and contour length was used to directly understand details of the intermediate state that underlie the overall protein stability.

AFM has long been used to resolve individual protein molecules and it is possible to combine molecular scale topographic imaging with force spectroscopy to confirm the unfolding of a single protein molecule. Bacteriorhodopsin, a light activated ion channel, contains seven transmembrane alpha helix domains. In the study by Muller et al., a two dimensional array of Bacteriorhodopsin protein (Br) was assembled upon a flat mica surface and imaged so as to resolve individual Br molecules.[42] A nonspecific adhesive bond then formed between the cantilever tip and a single Br protein. Retraction of the tip resulted in an unwinding of the helix domains (Figure 7A).

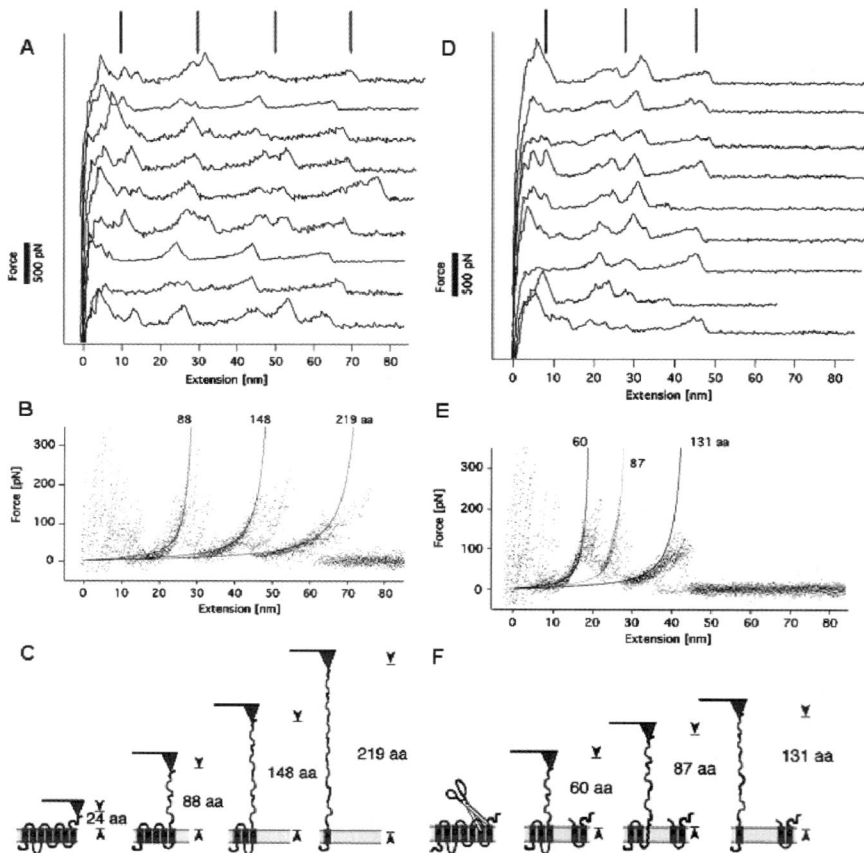

Figure 7. Single molecule unfolding of Bacteriorhodopsin. (A) Several force spectra taken on wild-type bR are shown with four peaks located around 10, 30, 50, and 70 nm. (B) Many spectra are superimposed where all but the location of the first peak is well defined. Fits are calculated from the WLC model. (C) This unfolding model describes the peaks in the force spectra as the sequential extraction from the bilayer and unfolding of a single Br. Helices F and G will be pulled out of the membrane and unfold together, followed by helices D and E. Peak 3 reflects unfolding of helices B and C followed by the last remaining helix A. Panels (D)-(F) repeat this measurement but using wild-type Br with the E-F loop cleaved enzymatically. (D) Only three main peaks are visible—around 5, 25, and 45 nm. (E) 17 spectra superimpose perfectly except for the first peak. (F) 45 nm length force curves can be recorded only when the free end of helix E is fixed to the tip. Thus, the first peak reflects extraction of helices D and E; the second reflects extraction and unfolding of helices B and C; the last peak shows extraction of the last remaining helix, A. (Figure used with permission from Muller et al., Reference 42).

Unfolding of the individual molecules of Br was verified by comparison with the known domain amino acid sequence length. A WLC fit (Kuhn length of 0.8 nm) of the contour length can be used to identify each unfolded poly-amino acid domain extension. Unfolding events repeatedly occurred at extension values of 10, 30, 50, and 70 nm due to the length of the transmembrane domains in Br. Assuming a value of peptide bond length of 0.36 nm, domain unfolding occurs with lengths corresponding to 88, 148, and 219 amino acid lengths which precisely corresponds to the sequential pair-wise extraction of domains F and G, D and E, and B and C, respectively, of a single Br protein. Interestingly, unfolding occurs primarily in order of domain sequence starting with the point of force application at the C terminal and does not occur sequentially from weakest to strongest domain as in the molecule titin.[37] Controls with engineering of a cleavage site between domains E and F support the sequential unfolding hypothesis (Figure 7b). Cleaving of the Br protein produces a shorter peptide unfolding measurement with a decreased contour length. Remarkably, repeating the single molecule imaging after a single protein unfolding event reveals a single Br protein gap in the topographic image.

Many proteins undergoing unfolding, such as bacteriorhodopsin or titin, display a reproducible pattern of rupture-distance events that result from structural transitions within intramolecular peptide domains. A feature unique to the single molecule approach is that a well-defined subset of data can be correlated with the unique force-extension profile of a particular unfolding pathway, without being obscured by the plurality of unfolding pathways that would exist in an ensemble measurement. Data outside this specific set can then be eliminated and only the specific pathway can be analyzed. This rational method for rejection of data is extremely useful in understanding thermodynamics and kinetics along unique pathways.[18, 43]

4 Surface Imaging Using Tether-mediated Receptor-ligand Interactions

Atomic force microscopy with a tethered biomolecular probe provides an ideal platform with which to map specific interactions on biologically active surfaces.[44–47] Compared with other techniques for probing the specific interaction forces between biologically active surfaces, namely surface force apparatus,[48] optical tweezers,[49] and biomembrane force probe,[12] the AFM provides the smallest probe and the highest spatial resolution. In addition to recording topography, which informs us of the volume and shape of particles immobilized on a surface, a molecular recognition imaging technique can record adhesion forces at each pixel and display a two-dimensional map where intensity corresponds to adhesion strength.[50, 51] Molecular mapping of specific proteins in living cells showed, for example, that adhesion molecules primarily exist in localized domains on the cell surface.[52]

A second use of adhesion mapping provides evidence for the biophysical mechanism of gating of gap junctions.[53] The adhesion and elasticity of a number of antibodies and peptides connected by PEG tethers to the AFM tip was measured upon binding to different regions of reconstituted connexin membrane protein. Researchers learned that the cytoplasmic tail of connexin protein is flexible and unavailable for binding in the presence of high Ca^{+2} concentration. The data suggest the cytoplasmic tail of the connexin protein might serve as a self-regulating steric gate to ions regulated by Ca^{+2} concentration.

The ability to map the spatial distribution of individual receptor sites of specific ligands on biologically mimetic and cell surfaces could also significantly aid the understanding of biological function. Hinterdorfer and colleagues have applied single-molecule tools for this purpose using a robust method in which topography and specific adhesion events are measured simultaneously over a two-dimensional surface.[54] Dynamic recognition force mapping forgoes quantification of adhesion force, but at the benefit of a dramatic increase in the imaging rate. In

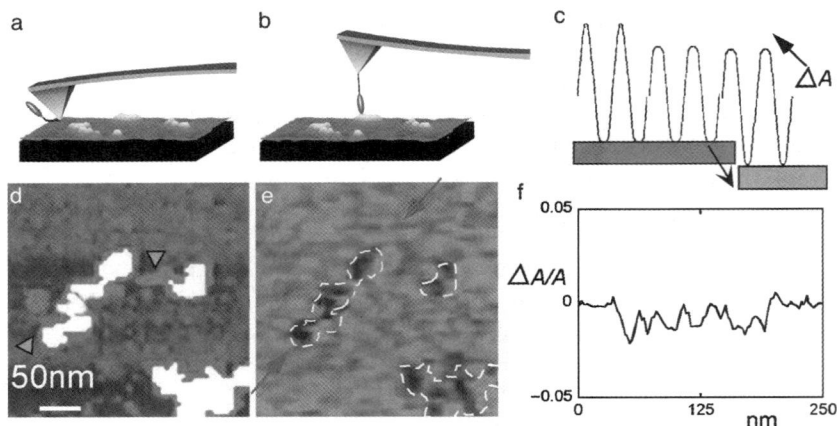

Figure 8. Dynamic recognition force mapping of an antibody-coated tip to an immobilized chromatin molecule. A schematic of topographic imaging (A) and recognition imaging (B). The resulting motion of the tip and servo (C) during no recognition binding (green and blue traces) and the molecular recognition event (red curve). A topographic image of chromatin (D) and the corresponding recognition image (E). (F) shows a plot of the peak oscillation signal between the arrows in the recognition map, where the dips in signal correspond well with the location of nucleosomes. (Reprinted with permission from Stroh et al., Reference 56.)

the method depicted in Figure 8, AFM tips functionalized with tethered ligands oscillate at high frequency, and topography is recorded by the impeding of the upward oscillation of the tip. At the point of specific recognition of a receptor by a tip-immobilized ligand, upward oscillation is impeded by the short linker.[55] Because of the low quality factor oscillation of the levers in aqueous environments, the amplitude of the oscillating lever rapidly conforms to the stretched tether, and the decrease in peak amplitude shows up as a dark region in the recognition map. This technique has been used to visualize a specific intramolecular modification of a histone protein by an antibody-functionalized tip in a complex sample of chromatin.[56]

One major drawback of this technique remains the slow image acquisition time. Mapping recognition events using dynamic recognition improves the image acquisition rate to 1–5 minutes per image.[57] This time resolution is still too limited to observe many dynamic processes in real time; therefore development of new imaging modes that could greatly increase imaging rates remains an important goal.[58, 59]

5   Force Spectroscopy of Multivalent Interactions: Counting Individual Tethered Biological Bonds

With traditional bulk biochemical analytical tools, it is extremely difficult to study molecules that can bind a target in more than one configuration. The problem with bulk techniques, such as calorimetry or surface plasmon resonance, stems from the fact that binding is measured over many individual molecules, which reveals an average binding measurement of all the individual modes of binding. To use the parlance of the single-molecule community, traditional methods always measure ensembles of interactions.[9] If a molecule can bind to a receptor in several modes that have different binding strength, then such averaging can introduce serious errors in estimating the strength of the interactions.

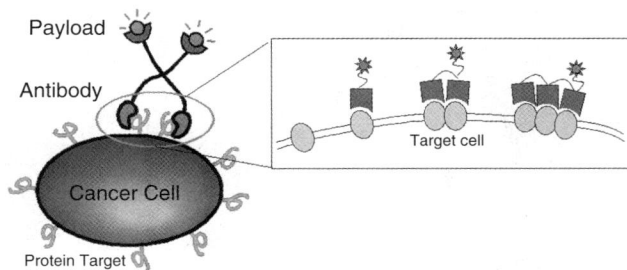

Figure 9. A schematic of a radioimmunotherapeutic system, in which antibodies target a protein in a cancer cell. A therapeutic payload such as a radioisotope is attached through a linker. In the inset, multivalent recognition elements are shown binding to a drug membrane target.

Multivalent binding which may include multiple protein-protein, ligand-receptor, protein-oligosaccharide, or antigen-antibody[60–62] bonds represents a special, but extremely important case of such multimodal binding. Multivalent binding is exploited in a new class of therapeutics in which monovalent antibody recognition elements are linked to form multivalent complexes with dramatically increased drug specificity and residency time (Figure 9). This approach offers the hope of controlling the quantity and binding time of a drug to its therapeutic target. As a step towards that goal, understanding how drug binding depends upon therapeutic valency would significantly improve the medical community's ability to fight disease.[63, 64]

Single-molecule force spectroscopy measurements offer several advantages for studying multivalent binding. First, these measurements can resolve the variation of binding strength resulting from changing valency. In addition, because a mechanical force greatly accelerates the bond dissociation rate, force spectroscopy can characterize extremely strong bonds that may have thermodynamic off-rates that are much longer than experimental time scales. Finally, force spectroscopy measurements can work on a wide variety of substrates, including live cell surfaces.

Measuring the multivalent binding strength in a parallel configuration, where the load is shared by all bonds equally, has been achieved in a variety of molecular systems.[15, 45, 52, 65] In some cases, multivalent bond interactions can be visualized simply with the force spectrum alone. The higher bond number interactions are ascribed to quantized force distributions. For example, Wong et al. utilized carbon nanotube tips functionalized with biotin to limit the contact area with a straptavidin functionalized surface (Figure 10). Such a small contact area provides a clean background, such that one and two parallel bonds can be distinguished and measured to reveal roughly an integer multiple of the single bond strength of 200pN. In the work by Dupres et al. force interactions that occur at approximately twice the single bond force of heparin and a complementary adhesion molecule are ascribed to two simultaneous interactions, though it is not possible to say whether the interaction of heparin is with two single adhesion molecules or rather to an adhesion molecule dimer.

Counting Individual Bonds

Creating the experimental conditions that bias the measurement towards the single bond formation requires using a dilute concentration of the interacting molecule that could decrease the probability of multiple bond interactions. Commonly a dilute concentration is obtained by mixing functional and spacer molecules in a defined ratio such that the covalent functionalization on the probe reflects the ratio of the molecules in solution. Conversely,

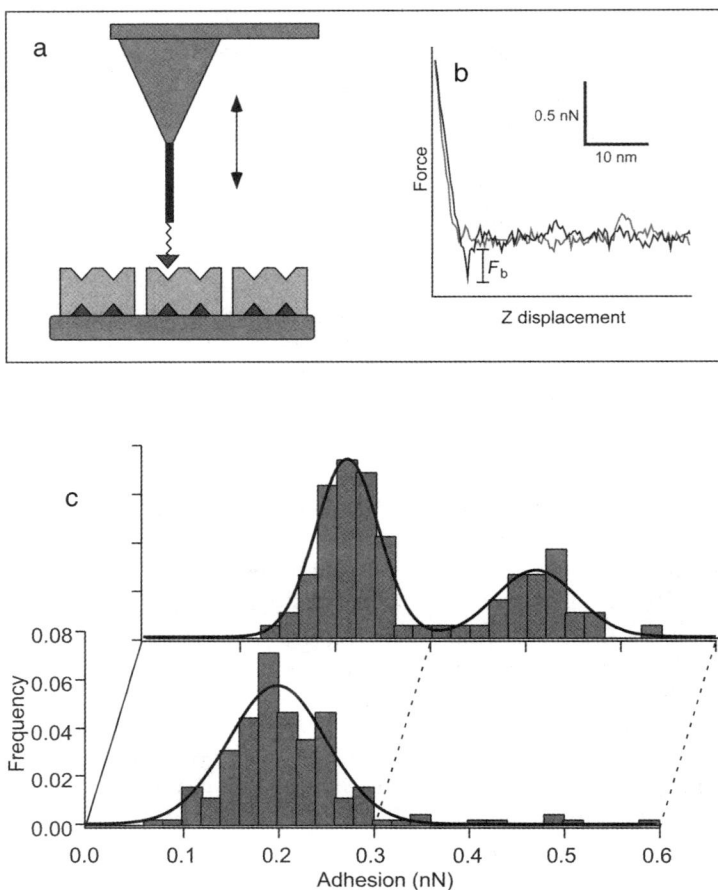

Figure 10. (A) A schematic of a biotin-functionalized single-walled carbon nanotube tip. (B) A representative force curve. (C) Two force histograms showing two biotin molecules per tip. (Used with permission from Wong, Lieber et al., Reference 65.)

a higher concentration of immobilized biomolecules is necessary if the experiment aims to study multiple bonds. For example, a 50:1 ratio of spacer molecules to functional molecules produces a 10% chance of interaction and a less than 1% chance of multiple bonds, assuming a Poisson distribution. However, increasing the functional density to a 30:1 dilution improves the probability of measuring an interaction to about 30% and biases those interactions to include more multiple bonds. Note that molecular binding still remains a stochastic process and therefore it is not possible to dictate the number of bonds, but only their probability in a distribution through changes in the surface coverage or time of probe contact with the surface. How can we then extract relevant information from the force spectroscopy measurements?

In general, it is challenging to determine the number of interacting bonds during each measurement solely based upon the rupture force. Bond rupture is described by a diffusive potential barrier crossing where the barrier is reduced by mechanical force.[66] As such, bond rupture is a stochastic process and individual rupture forces will show a spread around a most probable force. As the number of molecular interactions increase, the spread in rupture force will increase linearly with the valency of interaction.[47, 67] If the spread of rupture force is large compared to the

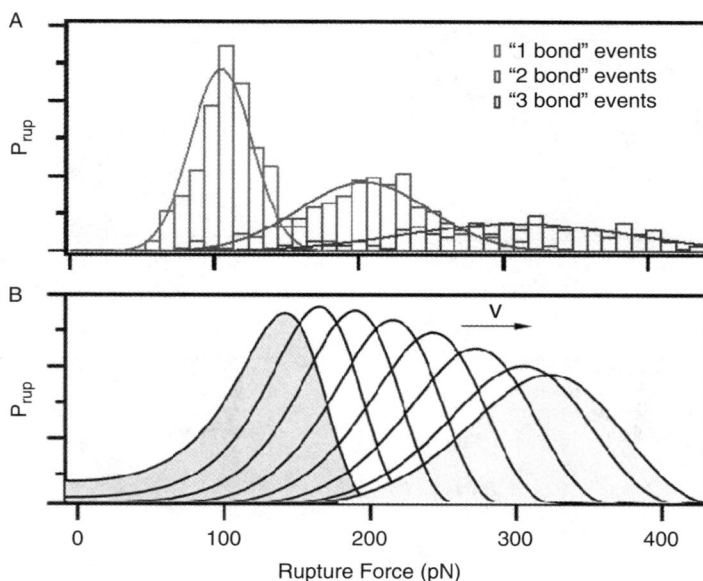

Figure 11. (A) hypothetical bond strength distribution assuming a single bond rupture force of 100pN, a Gaussian spread of 35pN, and a linear increase in rupture force and spread with valency. (B) A hypothetical bond strength distribution as a function of loading velocity (v). The leftmost distribution corresponds to a "slow" pulling rate, while the rightmost distribution represents a "fast" pulling rate.

average rupture force, the force distributions for multivalent interactions will overlap significantly. Therefore, a particular bond rupture will be "indeterminate" in regards to valency—it is impossible to know the number of molecules involved in the binding event (Figure 11A). Compounding the problem, rupture force is also load rate dependent.[12] Variations in measurement load rate resulting from interaction valency variations or from variations of the retraction rate of the probe will lead to a further spread in bond rupture force (Figure 11B). In fact, most single-molecule measurements specifically avoid measuring multivalent bonds by significantly diluting the receptor and ligand concentrations on the surface of the probe. Therefore to remove this ambiguity, we need to develop an independent "molecular counter" for these measurements.

In the remaining subsections, we describe how elasticity of the polymer linkers could be used to develop such a "counter". Specifically, we will show that the shape of the force-extension curve could be unambiguously correlated to the interaction valency. As a result, we could obtain both the valency and the composite bond strength for the multivalent interactions.

Force Extension of Multiple Identical Tethers

If we consider nt identical parallel PEG tethers (each containing n_m monomers) the force-extension relationship for n_t PEG tethers will be:

$$L(F, n_t) = L_C(F, n_t) \cdot \left(\coth\left(\frac{Fa}{n_t k_B T} \right) - \frac{n_t k_B T}{Fa} \right) + \frac{n_m F}{n_t K_S} \tag{8}$$

where L_c is the force-dependent contour length given by Equation 8, a is the Kuhn length, and K_s is the elastic constant. Typically, the tethers are much longer than any offset introduced by

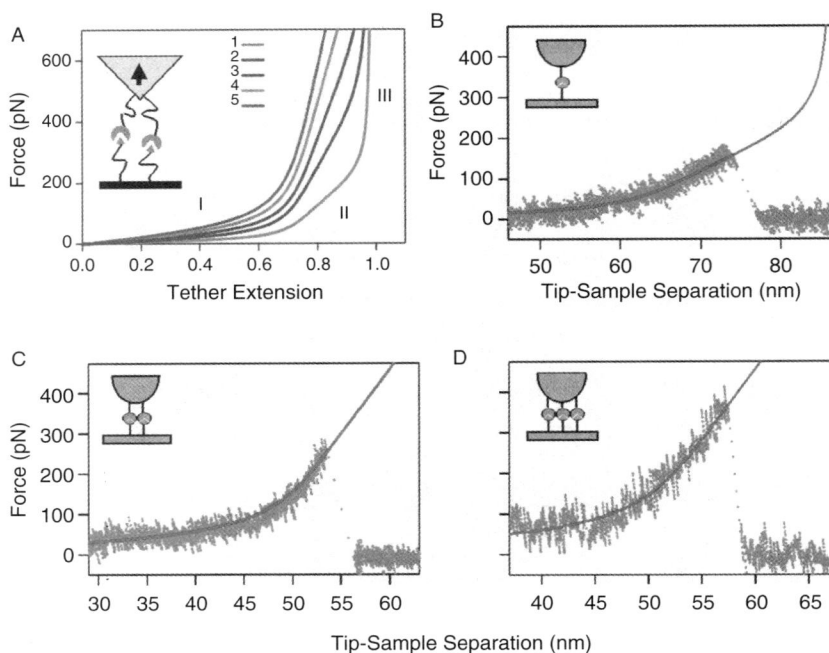

Figure 12. (A) Force-extension profiles calculated for several PEG linkers connected in parallel. (B–D) Stretching traces for one, two, and three MUC1-scFv interacting pairs connected by PEG tethers. Solid blue lines indicate fits to the TLS-eFJC model.

variations in the tether position on the probe tip; therefore we can consider all PEG linkers to have identical length. (The analysis of the rupture kinetics in cases where the tethers have different lengths is more complicated, and we refer the reader to the journal publication.[58]) The parameters for calculating this force-extension relationship for PEG in aqueous solution were found by Oesterhelt et al.[27] A comparison of the stretching profiles for multiple identical PEG tethers (Figure 12A) shows that an increase in the number of tethers produces a characteristic stiffness increase in the intermediate extension regime. This difference in the force-extension profiles provides a unique elastic signature that identifies the number of tethers that produce a particular rupture trace. Figures 12B–12D show that the measured single and multiple PEG tether extension traces fit this model very well, even if the model uses only a single parameter, n_m–, the number of PEG monomers in each tether.

Strength of Multiple Peptide-antibody Bonds

Significantly, this procedure allows us to identify the number of bonds corresponding to each of the large number of specific rupture events measured for interactions of the tethered peptide Mucin 1 (MUC1) and the single-chain antibody fragment (scFv) recognizing the MUC1 peptide.[47] Not surprisingly, these data (Figure 13) show that an increase in the number of bonds results in an increase in the measured rupture force. It is interesting to note that the measured force appears to increase nearly linearly with the number of bonds. Williams[68] has analyzed rupture of multiple identical bonds and showed that the measured rupture force, F^*, should scale with the number of bonds, N_B, and the measured loading rate, r_F, as:

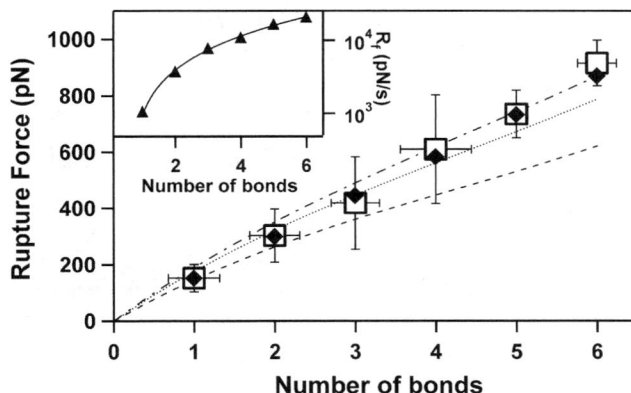

Figure 13. Plot of the measured rupture forces (open squares, □) as a function of the number of MUC1-scFv bonds between the AFM tip and sample. Solid diamond symbols (◆) correspond to the rupture forces calculated using the Markovian model for the strength of multiple bonds and the average loading rate determined for each particular number of bonds. Lines correspond to the predictions of the same model using a single value of the average measured loading rate for 1 (dashed), 3 (dotted), and 6 (dash-dotted) bonds over the whole range of bond numbers. Inset: The average measured instantaneous loading rates (solid triangles, ▲) for the full range of bond numbers used in our experiments. Solid line is provided as a guide to the eye.

$$r_F = k_{off} \frac{k_B T}{x_\beta} \left[\sum_{i=1}^{N_B} \frac{1}{i^2} \exp\left(-\frac{F^* x_\beta}{i k_B T} \right) \right]^{-1} \qquad (9)$$

where x_β is the characteristic bond width, k_{off} is the thermodynamic off-rate for a single bond, and $k_B T$ is the thermal energy scale. The rupture forces that we calculated using Equation 9 show reasonable agreement with the experimental data for the full range of loading rates used in the experiments (Figure 14); however, all calculated curves show a more pronounced curvature than the experimental data for both high and low loading rates.

This deviation disappears, however, when we consider the details of bond loading in a tethered system. As we discussed in the previous section, nonlinear stiffness of PEG tethers and fluctuations in the binding force cause every rupture event to occur at a different instantaneous loading rate. The increase in the number of bonds produces stronger connections that need to be loaded by higher forces, which then produce increased loading rates even if the AFM transducer retracts at a constant velocity. The net result is that the rupture events recorded in the experiments show a bias towards higher loading rates at higher bond numbers (Figure 13, inset). Indeed, if we calculate the expected rupture force using the average loading rate for each number of bonds, then the calculated forces match the experimental results over the entire range of the data.

Interestingly, the apparent linear dependence of the measured rupture forces on the number of bonds seen in Figure 13 can hint at the reasons why some AFM-based force measurements at first seem to match the predictions of the Poisson statistics model.[68] This method is based on the assumption that the rupture force for multiple parallel bonds scales linearly with the number of bonds; this assumption has since been shown to be incorrect due to the loading rate dependence on rupture force.[69] We note that in our experiments the "linear" scaling reflects the bias in loading rates specific to the AFM measurement with tethered ligands and not the fundamental scaling relationship for the strength of multiple bonds. Therefore, we stress that the Poisson statistics analysis will not produce a meaningful value for the single bond strength and researchers should use a full Markovian model[68] to analyze force spectroscopy measurements.

6   Dynamic Force Spectroscopy of Single and Multiple Tethered Bonds

The force spectroscopy measurements in tethered systems typically produce a continuous spectrum of loading rates due to coupling of the intentional variations in loading rate provided by changing the AFM cantilever retraction speed, and the natural spread in loading rate values resulting from the bond ruptures occurring at different points on the tether stretching curve. These variations provide the means to construct a plot of the measured rupture forces as a function of the logarithm of the instantaneous loading rate determined from the PEG elasticity fits. The dynamic force spectra obtained by this procedure (Figure 14) show the linear behavior predicted by Equation 9 and other phenomenological descriptions.[17] The dynamic force spectrum for the rupture of single bonds indicates that the unbinding events observed in these experiments correspond to a potential energy barrier located at 2.8±0.2 Å.[47] This value compares favorably with the bond distance of 2.6 Å, determined from the docking simulations of MUC1-antibody interactions. The kinetic off-rate of $2.6 \cdot 10^{-3}$ s^{-1} determined from the dynamic force spectrum (Table 1) also compares favorably with the $0.4 \cdot 10^{-3}$ s^{-1} off-rate determined from bulk surface plasmon resonance (SPR) measurements. We note that the values of these rates can rule out the possibility of protein unfolding in our experiments: kinetic off-rates for scFv unfolding measured by bulk spectroscopy techniques[70] are two orders of magnitude slower than the off-rates measured in our experiments. In addition, Hinterdorfer and colleagues showed that the forces required for breaking antibody-antigen interactions are much weaker than the forces required for antibody unfolding.[71] Our procedures for counting single bonds also allow construction of the dynamic force spectra for the rupture of different numbers of multiple bonds.

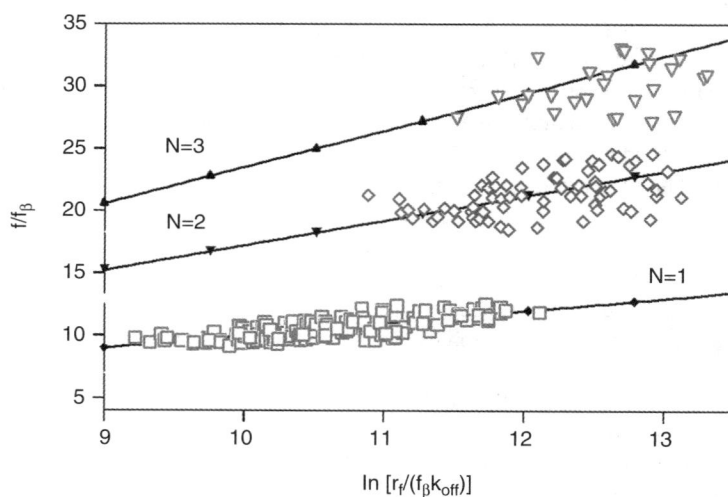

Figure 14. Comparison of the normalized dynamic force spectra for the rupture of one (□), two (◇), and three (▽) MUC1-antibody bonds with the prediction of the uncorrelated multiple bond rupture model. The experimental data were normalized according to Equation 10. Solid lines represent the results of the numerical solutions of Equation 11 for N = 1 (–(◆), 2 (–▼–), and 3 (–▲–)..

Table 1. The distances to the transition state, χ_β, kinetic off-rates, k_{off}, and the average bond lifetime $\tau_{off}=1/k_{off}$ values determined from the dynamic force spectroscopy measurements, where the force $k_{off} = \dfrac{\chi_B r_f\left(F=0\right)}{k_B T}$ and $\chi_\beta = \dfrac{k_B T}{dF/d\left(\ln r_f\right)}$.

	1 Bond	2 Bonds	3 Bonds
χ_β, Å	2.8±0.2	2.0±0.4	2.4±1.5
k_{off}, s^{-1}	$2.6 \cdot 10^{-3}$	$7.2 \cdot 10^{-5}$	$3.6 \cdot 10^{-8}$
τ_{off}	284 s	3.8 hrs	320 days

Remarkably, a comparison of the force spectra for rupture of one, two, and three bonds shows that they exhibit very similar distances to the transition state (Table 1). The effective kinetic off-rates for mono- and multivalent MUC1-antibody interactions, determined by fitting the spectra to the equations inset in Table 1, show the expected large drop in the kinetic off-rates with an increase in the number of bonds.

These data clearly illustrate that the main benefit of multivalent interactions is the reduction of the kinetic off-rate and the corresponding increase in the bond lifetime. Interestingly, these data point out a pathway for optimizing the design of multivalent radioimmunotherapeutics: researchers need to match the expected lifetime for the multivalent binding interaction with the time needed for radiation induced apoptosis. For instance, radiation-induced cell death in a significant fraction of cells requires over 8 hours;[72] therefore an ideal multivalent targeting molecule featuring MUC1-targeting antibodies should link three scFv units to achieve the necessary binding efficiency.

More generally, these experimental data also allow testing of the general predictions of the theory of uncorrelated rupture of multiple parallel bonds. The uncorrelated failure mode implies no particular mechanical coupling between individual bonds, and Williams showed that the force-induced rupture of such a connection could be described as a Markovian sequence.[69] To simplify the description we will use the normalization for force and loading rate:[73]

$$F = \frac{f}{f_\beta}; \qquad R = \frac{r_f}{f_\beta k_{off}}; \qquad (10)$$

where f_β is the thermal force scale defined as $k_B T/x_\beta$. The equivalent single bond approximation[69, 73] then produces the following expression relating the normalized loading rate, R, to the most probable rupture force, F^*, and the number of bonds, N:

$$R = \left[\sum_{n=1}^{N} \frac{1}{n^2} \exp\left(-\frac{F^*}{n}\right)\right]^{-1} \qquad (11)$$

Rupture forces calculated from Equation 11 (which is a normalized form of Equation 9) show extremely good agreement with the experimental data (Figure 14), demonstrating that the Markovian model[69] provides an accurate description of the dynamic failure of multiple uncorrelated parallel connections. In addition, analysis of the measured variations in the rupture forces also agrees with the predictions of the Markovian model.[47]

Conclusions

We showed that the use of long flexible tethers to connect interacting molecules to the surfaces of an AFM tip and sample provides several critical advantages for force spectroscopy experiments: the tethers identify specific binding interactions and separate them from the nonspecific interactions; the elastic signature of the tether identifies the number of bonds independently of the measured rupture forces; and finally, the tethers suppress the rebinding process at any loading rate. These advantages allow researchers to construct accurate dynamic force spectra for the rupture of mono- and multivalent interactions and quantify the advantages of the multivalent binding. Measured bond strength and the dynamic force spectra show excellent agreement with the Markovian model for the rupture of multiple uncorrelated molecular bonds, providing a solid experimental corroboration for the theoretical predictions.

Tethered ligand systems can serve as a flexible and versatile model for studying the fundamental dynamics of individual bond rupture in biological systems. Multivalent binding is a common tool for molecular targeting that enables extended and more accurate delivery of drugs and molecular labels to specific tissues. Dynamic force spectroscopy measurements can provide an accurate measurement of the kinetic off-rates in such systems. These off-rates are the main determinant of drug efficiency, and quantification of the advantages of multivalent binding can provide a valuable input into design efforts. Force spectroscopy techniques are especially useful for characterization of very strong interactions, which could be difficult to observe on reasonable experimental time scales by other methods. These results could accelerate the efforts to design the next generation of superior multivalent binders for disease treatment and tissue imaging.

References

1. Cochran, J. R.; Cameron, T. O.; Stone, J. D.; Lubetsky, J. B.; Stern, L. J. J. Biol. Chem. 2001, 276, 28068–28074.
2. van Kooyk, Y.; Figdor, C. G. Curr. Opin. Cell Biol. 2000, 12, 542–547.
3. Mammen, M.; Choi, S. K.; Whitesides, G. M. Angew. Chem. 1998, International Ed. in English. 37, 2755–2794.
4. Mourez, M.; Kane, R. S.; Mogridge, J.; Metallo, S.; Deschatelets, P.; Sellman, B. R.; Whitesides, G. M.; Collier, R. J. Nat Biotech 2001, 19, 958–961.
5. Holliger, P.; Prospero, T.; Winter, G. PNAS 1993, 90, 6444–6448.
6. Todorovska, A.; Roovers, R. C.; Dolezal, O.; Kortt, A. A.; Hoogenboom, H. R.; Hudson, P. J. J. Immunol. Meth. 2001, 248, 47–66.
7. Souriau, C.; Hudson, P. J. Expert Opin Biol Ther 2003, 3, 305–18.
8. Rowland, G.; O'Neill, G.; Davies, D. Nature 1975, 255, 487–488.
9. Bustamante, C.; Macosko, J. C.; Wuite, G. J. L. Nature Rev. Molec. Cell Biol. 2000, 1, 130–136.
10. Clausen-Schaumann, H.; Seitz, M.; Krautbauer, R.; Gaub, H. E. Curr. Opin. Chem. Biol. 2000, 4, 524–530.
11. Noy, A.; Vezenov, D.; Lieber, C. Ann. Rev. Mat. Sci. 1997, 27, 381–421.
12. Merkel, R.; Nassoy, P.; Leung, A.; Ritchie, K.; Evans, E. Nature 1999, 397, 50–52.
13. Alon, R.; Hammer, D. A.; Springer, T. A. 1995, 374, 539–542.
14. Thoumine, O.; Kocian, P.; Kottelat, A.; Meister, J.-J. Eur. Biophys. J.1 2000, 29, 398–408.
15. Florin, E. L.; Moy, V. T.; Gaub, H. E. Science 1994, 264, 415–417.
16. Moy, V. T.; Florin, E. L.; Gaub, H. E. Science 1994, 266, 257–9.
17. Evans, E.; Ritchie, K. Biophys. J. 1997, 72, 1541–1555.
18. Dietz, H.; Rief, M. Proc. Natl. Acad. Sci. U.S.A. 2004, 101, 16192–16197.
19. Riener, C. K.; Stroh, C. M.; Ebner, A.; Klampfl, C.; Gall, A. A.; Romanin, C.; Lyubchenko, Y. L.; Hinterdorfer, P.; Gruber, H. J. Analyt. Chim. Acta 2003, 479, 59–75.
20. Madl, J.; Rhode, S.; Stangl, H.; Stockinger, H.; Hinterdorfer, P.; Schutz, G. J.; Kada, G. Ultramicroscopy 2006, 106, 645–651.
21. Kienberger, F.; Pastushenko, V.; Kada, G.; Gruber, H.; Riener, C.; Schindler, H.; Hinterdorfer, P. Single Molecules 2000, 1, 123–128.
22. Kratky, O.; Porod, G. Recueil des Travaux Chimiques des Pays-Bas 1949, 68, 1106–1123.

23. Rief, M.; Fernandez, J. M.; Gaub, H. E. Physical Review Letters 1998, 81, 4764.

24. Marko, J. F.; Siggia, E. D. Macromolecules 1995, 28, 8759–8770.

25. Odijk, T. Macromolecules 1995, 28, 7016–7018.

26. Tskhovrebova, L.; Trinick, J.; Sleep, J.; Simmons, R. Nature 1997, 387, 308–312.

27. Oesterhelt, F.; Rief, M.; Gaub, H. E. New J. Phys. 1999, 1, 1–11.

28. Rixman, M. A.; Dean, D.; Ortiz, C. Langmuir 2003, 19, 9357–9372.

29. Bell, G. I. Science 1978, 200, 618–27.

30. Kramers, H. A. Physica 1940, 7, 284–304.

31. Jeffery, S.; Hoffmann, P. M.; Pethica, J. B.; Ramanujan, C.; Ozer, H. O.; Oral, A. Phys. Rev. B 2004, 70, 054114–8.

32. Jeppesen, C.; Wong, J. Y.; Kuhl, T. L.; Israelachvili, J. N.; Mullah, N.; Zalipsky, S.; Marques, C. M. Science 2001, 293, 465–468.

33. Friedsam, C.; Wehle, A. K.; Kuhner, F.; Gaub, H. E. J. Phys.-Cond. Matt. 2003, 15, S1709-S1723.

34. Oesterhelt, F.; Rief, M.; Gaub, H. E. New J. Phys. 1999, 1, 1–11.

35. Heymann, B.; Grubmuller, H. Chem. Phys. Lett. 1999, 307, 425–432.

36. Zhang, Q. M.; Marszalek, P. E. Polymer 2006, 47, 2526–2532.

37. Rief, M.; Gautel, M.; Oesterhelt, F.; Fernandez, J. M.; Gaub, H. E. Science 1997, 276, 1109–1112.

38. Marszalek, P. E.; Lu, H.; Li, H. B.; Carrion-Vazquez, M.; Oberhauser, A. F.; Schulten, K.; Fernandez, J. M. Nature 1999, 402, 100–103.

39. Carrion-Vazquez, M.; Li, H. B.; Lu, H.; Marszalek, P. E.; Oberhauser, A. F.; Fernandez, J. M. Nature Struct. Biol. 2003, 10, 738–743.

40. Fernandez, J. M.; Li, H. Science 2004, 303, 1674–1678.

41. Dietz, H.; Berkemeier, F.; Bertz, M.; Rief, M. Proc. Natl. Acad. Sci. U.S.A. 2006, 103, 12724–12728.

42. Muller, D. J.; Kessler, M.; Oesterhelt, F.; Moller, C.; Oesterhelt, D.; Gaub, H. Biophys. J. 2002, 83, 3578–3588.

43. Dietz, H.; Rief, M. PNAS 2006, 103, 1244–1247.

44. Holliger, P.; Prospero, T.; Winter, G. Proc. Natl. Acad. Sci. U.S.A. 1993, 6444–6448.

45. Baumgartner, W.; Hinterdorfer, P.; Ness, W.; Raab, A.; Vestweber, D.; Schindler, H.; Drenckhahn, D. Proc. Natl. Acad. Sci. U.S.A. 2000, 97, 4005–4010.

46. Bonanni, B.; Kamruzzahan, A. S. M.; Bizzarri, A. R.; Rankl, C.; Gruber, H. J.; Hinterdorfer, P.; Cannistraro, S. Biophys. J. 2005, 89, 2783–2791.

47. Sulchek, T. A.; Friddle, R. W.; Langry, K.; Lau, E. Y.; Albrecht, H.; Ratto, T. V.; DeNardo, S. J.; Colvin, M. E.; Noy, A. Proc. Natl. Acad. Sci. U.S.A. 2005, 102, 16638–16643.

48. Leckband, D. E.; Israelachvili, J. N.; Schmitt, F. J.; Knoll, W. Science 1992, 255, 1419–1421.

49. Ashkin, A. Proc. Natl. Acad. Sci. U.S.A. 1997, 94, 4853–4860.

50. Ludwig, M.; Dettmann, W.; Gaub, H. E. Biophys. J. 1997, 72, 445–448.

51. Grandbois, M.; Dettmann, W.; Benoit, M.; Gaub, H. E. J. Histochem. Cytochem. 2000, 48, 719–724.

52. Dupres, V.; Menozzi, F. D.; Locht, C.; Clare, B. H.; Abbott, N. L.; Cuenot, S.; Bompard, C.; Raze, D.; Dufrene, Y. F. Nature Methods 2005, 2, 515–520.

53. Liu, F.; Arce, F. T.; Ramachandran, S.; Lal, R. J. Biol. Chem. 2006, 281, 23207–23217.

54. Hinterdorfer, P.; Baumgartner, W.; Gruber, H. J.; Schilcher, K.; Schindler, H. PNAS 1996, 93, 3477–3481.

55. Ebner, A.; Kienberger, F.; Kada, G.; Stroh, C. M.; Geretschlaeger, M.; Kamruzzahan, A. S. M.; Wildling, L.; Johnson, W. T.; Ashcroft, B.; Nelson, J. ChemPhysChem 2005, 6, 897–900.

56. Stroh, C.; Wang, H.; Bash, R.; Ashcroft, B.; Nelson, J.; Gruber, H.; Lohr, D.; Lindsay, S. M.; Hinterdorfer, P. Proc. Natl. Acad. Sci. U.S.A. 2004, 101, 12503–12507.

57. Hinterdorfer, P.; Dufrene, Y. F. Nature Methods 2006, 3, 347–355.

58. Sulchek, T.; Hsieh, R.; Adams, J. D.; Yaralioglu, G. G.; Minne, S. C.; Quate, C. F.; Cleveland, J. P.; Atalar, A.; Adderton, D. M. Appl. Phys. Lett. 2000, 76, 1473–1475.

59. Hobbs, J. K.; Vasilev, C.; Humphris, A. D. L. Analyst 2006, 131, 251–256.

60. Mammen, M., Helmerson, K., Kishore, R., Choi, S. K., Phillips, W. D., and Whitesides G. M. Chem. & Biol. 1996, 3, 757–763.

61. Wells, J. A. Proc. Natl. Acad. Sci. U.S.A. 1996, 93, 1–6.

62. Dower, S. K.; DeLisi, C.; Titus, J. A.; Segal, D. M. Biochemistry 1981, 20, 6326–6334.

63. Kipriyanov, S. M.; Little, M.; Kropshofer, H.; Breitling, F.; Gotter, S.; Dubel, S. Protein Eng. 1996, 9, 203–211.

64. Kortt, A. A.; Dolezal, O.; Power, B. E.; Hudson, P. J. Biomolec. Engin. 2001, 18, 95–108.

65. Wong, S. S.; Joselevich, E.; Woolley, A. T.; Cheung, C. L.; Lieber, C. M. Nature 1998, 394, 52–55.

66. Evans, E.; Ritchie, K. Biophys. J. 1997, 72, 1541–1555.

67. Evans, E.; Williams, P. Dynamic Force Spectroscopy: II. Multiple Bonds, in Physics of Bio-Molecules and Cell; Springer and EDP Sciences: Heidelberg, 2002.

68. Han, T.; Williams, J. M.; Beebe, T. P. Analyt. Chim. Acta 1995, 307, 365–376.

69. Williams, P. M. Analyt. Chim. Acta 2003, 479, 107–115.
70. Jager, M.; Pluckthun, A. FEBS Letters 1999, 462, 307–312.
71. Kienberger, F.; Kada, G.; Mueller, H.; Hinterdorfer, P. J. Molec. Biology 2005, 347, 597.
72. Marcel, V.; Harry, B. Cell and Tissue Research 2000, V301, 133–142.
73. Evans, E.; Williams, P. In Physics of Bio-Molecules and Cells; Flyvbjerg, H., Jülicher, F., Ormos, P., David, F., Eds.; Springer and EDP Sciences: Heidelberg, 2002; Vol. 75, p 187–203.

Direct Mapping of Intermolecular Interaction Potentials

Paul D. Ashby

Introduction

Atoms and molecules are the building blocks of nature's vast variety of materials, and short range intermolecular interaction potentials govern their structure and movement as they form materials or undergo reactions. Characterizing these interaction potentials and elucidating their scientific principles will not only facilitate a greater understanding of the natural world but also aid development of new technology, from computers built with nanowires[1] to less invasive medical devices.[2, 3]

For centuries, scientists have pondered the nature of materials and forces. In the 18th century, Clairaut suggested that the capillary rise of water is due to stronger attractions between glass and water than between water and itself. Later, van der Waals incorporated an intermolecular attractions term in his famous gas equation of state to describe deviations from ideal behavior. Concurrently, it was deduced that if the forces followed an inverse power law relationship the exponent must be greater than three, since the forces were not extensible.[4] In the twentieth century, many indirect techniques for investigating intermolecular potentials developed as part of the burgeoning field of colloid and surface science, but the creation of the surface forces apparatus (SFA) ushered in the era of direct molecular force measurement. Precise displacement of surfaces and measurement of interfacial distances allowed direct evaluation of force as a function of distance.[4-6] This capability yielded information about many different types of interactions such as van der Waals, double layer, steric, and solvation forces along with dissipative interactions associated with friction and wear.

While the SFA has led the field of direct force measurement over the last few decades, the atomic force microscope (AFM)[7] promises a new revolution in direct measurement of intermolecular interaction potentials. Its nanoscale probe reduces contact areas to square nanometers, providing the opportunity for single molecule studies[8-11] and discrimination of chemical and material properties spatially at length scales on the order of molecules.[12-14] Position detection is orders of magnitude more precise with the availability of high bandwidth. In addition, its simplicity and relatively low cost enables its ubiquitous presence in scientific laboratories across all disciplines.

The core of an atomic force microscope consists of a sharp tip attached to a spring positioned near a surface using piezo-electric translators. Microfabricated silicon or silicon nitride cantilevers are used most frequently as springs for AFM experiments. Tip-sample interaction forces deflect the cantilever and are measured as position changes of a laser beam

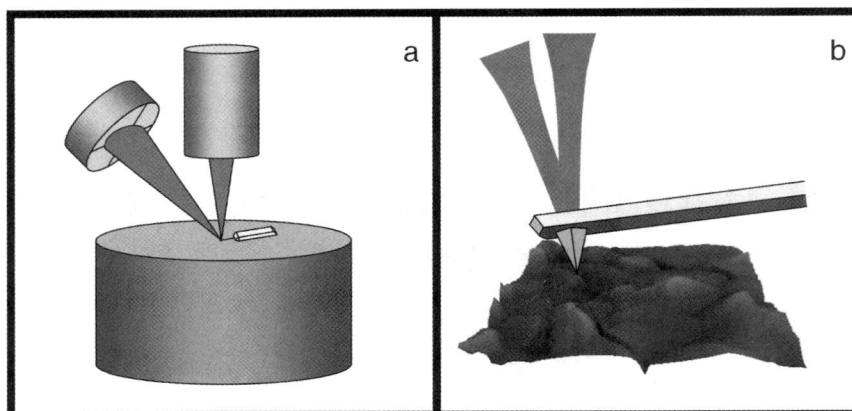

Figure 1. Cartoon of atomic force microscope. (A) A laser and quadrant photodiode are used to measure precisely the deflection of a microscopic cantilever. The sample is displaced using piezo-electric translators. (B) The sharp tip of the cantilever pyramid probes intermolecular interaction potentials which are sensed with piconewton precision by the deflection detection mechanism.

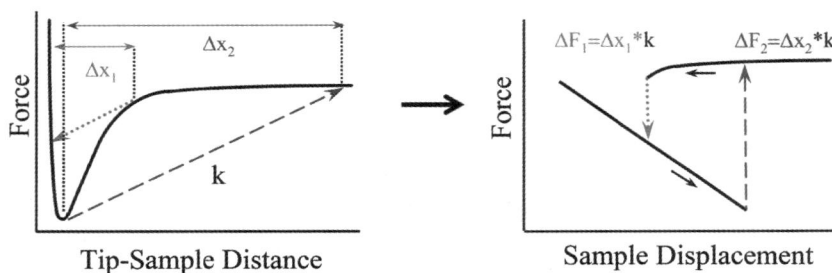

Figure 2. Schematics of force profile (left) and measured force vs. Z-piezo displacement (right). The gray arrows indicate the trajectory of the cantilever in the region where the gradient in the profile exceeds the fixed spring constant k of the cantilever leading to snap in and snap out.

reflected from the back of the cantilever at a photodiode. Figure 1a depicts macroscopic parts of the microscope such as the laser, photodiode, piezo-electric translators, and chip used to support the cantilever. Panel b depicts a pyramidal tip on the cantilever over a sample with exaggerated topography.

Idealized treatments of cantilever deflection approximate the lever as a simple harmonic oscillator where all mass and damping is localized at the tip and the spring follows Hooke's law, $F = k \cdot x$, where F is the applied force, k is the spring constant, and x is the bending of the cantilever from its equilibrium position. Thus, measuring deflection is a direct measure of force. A force profile is the derivative of a one-dimensional projection of the potential energy surface, where the reaction coordinate is defined by the tip-sample geometry and the distance traversed during the experiment. The potential energy surface is subsequently reconstructed by integration.

Traditional force profile measurements record cantilever bending while tip and sample are brought in and out of contact. The resulting force curve plots force as a function of sample displacement. Figure 2 depicts the relationship between a force profile and the corresponding force curve measured with a weak spring. For most direct force experiments, a region in the force profile exists where the derivative exceeds the cantilever stiffness. This condition causes the tip to snap to contact during approach and snap out during separation, depicted as Δx_1 and Δx_2 in figure 2. These instabilities, associated with the nonequilibrium crossing of energy barriers, preclude measurement of attractive portions of force profiles at small separations and similarly prevent integration to obtain intermolecular potentials.

Using stiffer springs such that cantilever stiffness is greater than interaction stiffness removes these characteristic instabilities but correspondingly decreases cantilever deflection, such that noise sources often obscure force profile information. However, through careful instrument design and implementation of a few data collection and analysis techniques, precise direct force profile measurement is possible for extremely stiff intermolecular interaction potentials. This chapter is intended to be a practical guide, by providing instrument design techniques and experimental procedures, for enabling direct mapping of intermolecular interaction potentials.

Noise Considerations

Instrument and thermal noise sources limit the precision of direct measurements of intermolecular interaction potentials with the AFM. This becomes more evident as stiff cantilevers are used to avoid instabilities and cantilever deflections become exceedingly small. Since cantilever deflection and noise signals are dynamic with many frequency components, noise spectra are a useful tool for performing noise analysis. The power spectral density, $P(\omega)$, describes how the power of a time series, $f(t)$, is distributed with frequency. Mathematically, it is the square of the magnitude of the Fourier transform scaled per unit frequency, $P(\omega) = F(\omega)^* F(\omega)/Hz$, where $F(\omega) = \dfrac{1}{\sqrt{2\pi}} \displaystyle\int_{-\infty}^{\infty} f(t) e^{-i\omega t} dt$.

Thus, the RMS noise of certain spectral components, $\omega_1 < \omega < \omega_2$, of a time signal are related to the power spectral density by $\sqrt{\left\langle f(t)^2 \right\rangle} = \sqrt{\displaystyle\int_{\omega_1}^{\omega_2} P(\omega) d\omega}$. For a time signal in volts, V, the corresponding power spectrum has units of V^2/Hz, but frequently, power spectra are represented in $V/\sqrt{Hz}$ for ease of scaling by other factors. When calculating sums of noise powers, the sum is performed in units of V^2/Hz. A characteristic noise power spectrum depicting the three major types of noise sources, flicker, white, and thermal, is shown in Figure 3. Direct force measurements collect low frequency components near zero hertz (DC) up to the filter cutoff frequency. A lower frequency cutoff removes more high frequency noise components and increases precision, but it may also remove high frequency components of cantilever deflection associated with tip-sample interaction and distort force data. Flicker noise usually dominates at frequencies associated with force curve collection, 0 to 1 kHz, while white noise dominates at higher frequencies. Thermal noise is a function of cantilever properties and environment. Frequently, only thermal noise near the resonance frequency is observed due to mechanical gain of the thermal excitations by the oscillator, bringing the motion above the high frequency instrument noise. However, under special circumstances low frequency thermal noise components are detected in the force curve data collection window.

Thermal noise arises due to buffeting of the cantilever by its surrounding medium. The magnitude of this constrained Brownian motion is related to the damping of the cantilever by its medium and can be calculated using the wave equation of motion and equipartition theorem. Within the simple harmonic oscillator approximation, the wave equation of motion for the cantilever is $k \cdot x + b \cdot \dot{x} + m \cdot \ddot{x} = F_t$, where x, $\dot{x}$, and $\ddot{x}$ are the displacement of the cantilever

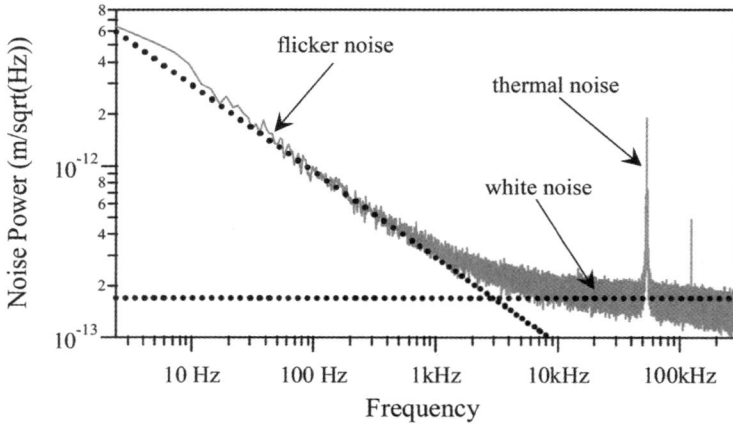

Figure 3. Noise spectrum for commercial instrument showing flicker, white, and thermal noise components. Flicker noise dominates most direct mapping measurements working in a 1 kHz bandwidth.

from equilibrium and its first and second time derivatives respectively, m is the effective mass, b is the damping, k is the spring constant, and F_t is thermal force due to buffeting. Solving for the amplitude spectrum using the wave equation and Ansatz, $x = A(\omega)e^{i(\omega t - \varphi)}$, with amplitude A, time t, arbitrary phase φ, and angular frequency ω, produces $|A(\omega)|^2 = \dfrac{F_t^2}{\left(k - m\omega^2\right)^2 + b^2\omega^2}$.

More intuitive and experimentally accessible cantilever parameters are the resonant frequency, $\omega_0 = \sqrt{k/m}$, and quality factor, $Q = \sqrt{km}/b$. Thus the amplitude function can be expressed as $|A(\omega)|^2 = \dfrac{\left(\dfrac{F_t}{k}\right)^2}{\left(1 - \left(\dfrac{\omega}{\omega_0}\right)^2\right)^2 + \left(\dfrac{\omega}{\omega_0 Q}\right)^2}$, which consists of two parts: transfer function,

$\dfrac{1}{\left(1 - \left(\dfrac{\omega}{\omega_0}\right)^2\right)^2 + \left(\dfrac{\omega}{\omega_0 Q}\right)^2}$, and scaler, $\left(\dfrac{F_t}{k}\right)^2$. The transfer function determines the spectral

shape of the resonance curve with values of one near DC and of Q near resonance. The scalar is a multiplicative factor whose value is computed using the equipartition theorem. From the equipartition theorem, each degree of freedom has thermal energy distributed in its motion such that the average motion, $\langle x^2 \rangle$, follows the relationship $\dfrac{k_B T}{2} = \dfrac{k \langle x^2 \rangle}{2}$. The integral of the amplitude function $\int_0^\infty |A(\omega)|^2 d\omega$ equals the average value of the square of the deflection, $\langle x^2 \rangle$, which leads to the scalar being a function of only temperature, damping, and spring constant when calculating noise power in units of deflection, $F_t = \sqrt{4 k_B T b} = \sqrt{\dfrac{4 k_B T k}{\omega_0 Q}}$.

Values of thermal noise near DC and at resonance for three different cantilever and environment combinations are organized in Table 1. Only weak springs in high damping environments have significant contributions from thermal noise near DC. For force spectroscopy experiments intent on avoiding instabilities by using stiff cantilevers, thermal noise is unlikely to

Table 1. Thermal noise power near DC and near resonance for a weak spring in water (highly damped), a medium spring in water, and stiff spring in air (lightly damped).

	Weak spring in water	Medium spring in water	Stiff spring in air
k (N/m)	0.030	2.0	40
Q 3.0	5.0	150	
$f_0 = \omega_0/2\pi$ (Hz)	3,500	30,000	250,000
Thermal noise near DC $(m/\sqrt{Hz})$	2.9×10^{-12}	9.4×10^{-14}	1.3×10^{-15}
Thermal noise near resonance $(m/\sqrt{Hz})$	1.1×10^{-11}	4.7×10^{-13}	2.0×10^{-13}

be a major source. Thus the focus for increasing the signal to noise ratio should be on decreasing the relative contribution of flicker instrument noise in the bandwidth near DC.

Many systems in nature exhibit flicker noise. Although universally observed, the mechanism for flicker noise in semiconductor devices is poorly understood. Presently, the best theory focuses on contaminants and defects in the crystal lattice. Devices made from III-IV materials have significantly higher flicker noise than silicon FET or bipolar devices, while bipolar in many instances are better than FET. [15] FET transimpedance amplifiers are frequently used for photodiode sensing of low light levels for their extremely low input current noise. However, AFM applications have enormous light levels and small feedback resistors, making input voltage noise the dominant flicker noise source. Thus, using bipolar op amps for this application reduces flicker noise by up to an order of magnitude. Flicker noise for the AFM also originates from mode hopping of the laser light source. Mode hopping manifests itself in power and pointing fluctuations, which can be minimized by using a feedback stabilized single mode fiber coupled laser.

Photocurrent shot noise is the most significant source of white noise for AFM. Shot noise derives from individual charge carriers crossing a potential barrier stochastically although the DC current is steady. Its relation to current is $i_n = \sqrt{2qI}$, where q is the charge of an electron and I is the measured current. Most AFMs produce tens of microamps per photodiode segment and together with the feedback resistor produce shot noise levels on order of 1 $\mu V/\sqrt{Hz}$.

While reducing noise sources is important, the most effective means of reducing the relative contribution of flicker noise is to increase the sensitivity to cantilever deflections. Optical lever sensitivity is a measure of light power transferred between photodiode segments for a specified cantilever deflection measured in V/nm. The inverse of this value, InVOLS, is used during experiments to calculate deflection in meters from the photodiode signal. The sensitivity is directly related to the photon flux at the boundaries of the photodiode segments so methods to increase photon flux are desirable.

A simple technique for increasing sensitivity is to use reflective coatings on the cantilever. Most cantilevers are made of materials such as silicon or silicon nitride with high transmittance. A metal coating increases the reflected power appreciably. Micrometer scale surface roughness also affects the quality of the reflection. Nonspecular reflection of the laser produces a diffuse spot on the detector which decreases the photon flux. Special care should be taken to preserve or reapply reflective coatings if functionalization procedures expose cantilever surfaces to harsh conditions, which may roughen the surface. Along with improving the reflection of the laser light, optimization of the other optical components to produce a low numerical aperture increases sensitivity.

A low numerical aperture incident beam has little divergence upon reflection, which maintains a high photon flux at the detector. Diffraction establishes the relationship between the focal beam waist and beam diameter at a specified distance. Spot radius, w(z), a distance z from the focal point, follows $w(z) = w_0 \sqrt{1 + \left(\dfrac{\lambda z}{\pi w_0^2}\right)^2}$, where w_0 is the beam waist radius at the focal point and λ is the wavelength of light. Numerical aperture is a measure of a beam's convergence and divergence, $N.A. = \sin(\theta / 2)$, where θ is the angle of the converging rays. The beam waist is related to numerical aperture by $w_0 = \dfrac{\lambda}{\pi \cdot N.A.}$. A shorter focal length produces a smaller focus on the cantilever but also causes the beam to diverge more before hitting the photodiode thus reducing sensitivity. Optimal precision is achieved by reducing numerical aperture such that the focus has similar dimensions as the cantilever. Although the angle is not constant over the length of the cantilever, its effects on the reflected laser light are averaged together and do not produce significant nonlinearities or inaccuracies. While designing the optics with the appropriate lens is best for retaining laser power, inserting an aperture in the beam path may be an easy and effective method to reduce the numerical aperture. Reflected laser power should be maintained by increasing the incident laser power so that the signal is not compromised by having too few photons. Using a low numerical aperture while maintaining a high reflected laser power maximizes the photon flux at the detector for high sensitivity deflection detection.

A noise power spectrum of an instrument which incorporates the above noise reduction techniques is shown in Figure 4. The same 1.3 N/m, 220 μm long cantilever used for Figure 3 is used for this spectrum. Although instrument noise still contributes the major noise components, the integrated noise is 0.83 pm in a 1 mHz-1 kHz bandwidth, which is a factor of 32 less than the commercial instrument depicted in Figure 3. Further gains are possible by using nonstandard spring designs such as a torsion spring, where the whole reflector is rigid and relative tip motion is leveraged.[16, 17]

Having outstanding acoustic and vibration isolation becomes imperative as high sensitivity cantilever deflection detection detects subtle external noise sources coupling into

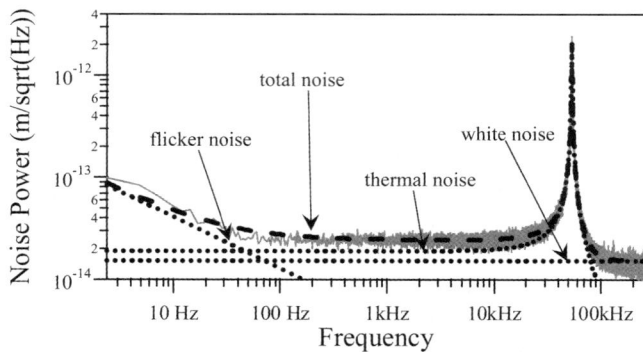

Figure 4. Noise spectrum for instrument fabricated using design principles presented in this chapter. Instrument noise is significantly reduced compared to the commercial instrument.

the instrument. Building vibrations and acoustic noise are in the measurement bandwidth for force profile experiments. When designing a work space, isolate acoustic enclosures from the microscope foundation and avoid rooms with noisy air ducts and ventilation. Acoustic enclosures also reduce the influence of turbulent drafts which can add significant flicker noise. Bungee cords provide good vibration isolation across the whole measurement bandwidth but their implementation can be inconvenient and require thoughtful design to accommodate height changes as mass is added and removed from the instrument. Active vibration isolation units are effective and convenient, but may require some passive damping such as a silicon rubber pad for high frequency (>1 kHz) components. Good isolation from external noise sources and high precision measurement of deflection detection are the foundation of direct measurements of intermolecular interaction potentials.

Cantilever Parameter Optimization

Choosing an appropriate cantilever becomes important after optimizing the instrumentation for low noise and high sensitivity deflection detection. The primary focus for cantilever choice is spring constant. Weak springs suffer from data loss due to instabilities. The cantilever should be stiff enough to avoid instabilities, but not overly stiff and thus needlessly lose force resolution. The scalar for the thermal noise spectrum is $F_t = \sqrt{4k_B T b}$ in units of force $\left(N/\sqrt{Hz} \right)$, which is a function only of damping and temperature and independent of spring constant. When instrument noise sources do not contribute, large spring constants can be used to avoid instabilities without compromising force resolution. However, most instruments have large flicker noise components in the measurement bandwidth which are accentuated as the spring constant is increased. For the instrument shown in Figure 3, avoiding the deleterious effects of flicker noise requires using a spring as weak as 0.05 N/m cantilever in water. Thus the conventional wisdom of decreasing the spring constant to increase force sensitivity is often true. Yet an instrument with performance similar to that of Figure 4 can use significantly stiffer cantilevers and retain an excellent signal to noise ratio.

Second to the spring constant in importance is damping. Recently, cantilever damping came into focus as weak cantilevers were used for non-equilibrium single molecule pulling experiments in solution.[18] The high damping of the solution raised the DC thermal noise levels above the instrument noise levels and limited the resolution. By making short, less damped cantilevers, lower force noise and higher scanning rates were achieved. Cantilever damping is similarly important for those interested in equilibrium mapping of interaction potentials; however, competing goals must be balanced. Smaller cantilevers reduce damping, which increases force resolution. But small cantilevers limit the reflected laser spot size, which may increase numerical aperture, lower detection sensitivity, increase the relative contribution of flicker noise, and effectually decrease force resolution. Results similar to those of Figure 4 require a spot diameter near 75 μm for large commercial cantilevers. A bending beam with force applied to the end has the shape $x_{end}(s) = \dfrac{3Ls^2 - s^3}{2L^3}$,

where L is the cantilever length and s is the distance along the cantilever.[19] Most beam curvature is located within L/2 from the base such that sensitivity is leveraged, since the angle at the end is greater than a rigid beam pivoting at its base. Because the last half of the cantilever is relatively free from bending, the spot should fill this region so a cantilever that is roughly twice the laser beam diameter is ideal, 150 μm for the example spot above. Despite hundreds of AFM cantilever products, little selection for those interested in measurements with stiff springs is available. Most cantilevers come in roughly three lengths, 125 μm, 225 μm, and

450 μm, and only one width, 30 μm. For 125 μm and 225 μm long cantilevers, different thicknesses afford only 3 stiffness options, ~2 N/m, ~25 N/m, and ~40 N/m. As interest grows for non-standard cantilever designs, manufacturers may produce more suitable products. A torsion spring design is more complicated to produce but it has a number of advantages, such as 25% less damping than a cantilever of the same plan dimensions; more usable area for laser reflection, which further increases sensitivity; and the whole body is rigid, reducing higher order bending mode contributions to cantilever motion. Although there is presently little selection, scientific progress is not limited by what is available. A 1.3 N/m 220 μm long cantilever in water has only 5 pN of RMS noise in a 1 mHz–1 kHz bandwidth. Using smaller cantilevers in lower damping environments will improve on this spectacular result.

Impressive precision and high stiffness are available using passive cantilevers and specifically designed instrumentation. However, some investigators have attempted to improve performance by using feedback to modulate cantilever parameters. Cantilever feedback techniques apply force to the cantilever derived from the deflection signal. Magnetic actuation is used most frequently, along with electrostatics and piezoelectrics. A general equation of motion for feedback is $k \cdot x + b \cdot \dot{x} + m \cdot \ddot{x} = F_t - G \cdot x \cdot e^{i\theta}$, where G is feedback loop gain and θ is a phase shift in the feedback loop. This equation leads to the amplitude spectrum,

$$|A(\omega)|^2 = \frac{\left(\dfrac{F_t}{k}\right)^2}{\left(1-\left(\dfrac{\omega}{\omega_0}\right)^2\right)^2 - \dfrac{2G}{k}\left(1-\left(\dfrac{\omega}{\omega_0}\right)^2\right)\cos(\omega) + \dfrac{2G\omega}{kQ\omega_0}\sin(\omega) + \dfrac{G^2}{k^2} + \left(\dfrac{\omega}{Q\omega_0}\right)^2}.$$

Q-control is a special case of cantilever feedback where $\theta = \pi/2$ and the amplitude spectrum becomes $|A(\omega)|^2 = \dfrac{\left(\dfrac{F_t}{k}\right)^2}{\left(1-\left(\dfrac{\omega}{\omega_0}\right)^2\right)^2 + \left(\dfrac{\omega}{Q\omega_0}+\dfrac{G}{k}\right)^2}$. Feedback loop gain effectively changes

damping such that higher gain increases damping and decreases Q. However, thermal force noise experienced by the lever is determined by passive damping. These excitations experience the higher feedback damping leading to reduced motion and suppression of thermal noise.[20] Another π of phase shift, $\theta = 3\pi/2$, reverses the effect and increases Q but consequently amplifies the thermal noise sometimes to the point of self oscillation and instability. Using Q-control to suppress residual thermal noise after proper cantilever selection is appealing. With enough gain, thermal motion is reduced at all frequencies including the low bandwidth used for direct force profile measurements without affecting the spring constant. Unfortunately, a feedback loop with constant phase lag and gain over a large bandwidth is not realizable with analog circuitry. Digital signal processing (DSP), with its ability to implement non-casual algorithms, can produce the appropriate feedback loop properties; but instability may result, as complicated computations are performed on truncated information due to digitization of exceedingly small signals. Analog approximations are very effective in small bandwidths near the cantilever resonance, which is useful for AC mode operation where one is interested in scanning faster, noise squeezing in the lock-in bandwidth, or gentle tapping. Yet for direct interaction potential measurements, Q-control feedback is not helpful.

A cantilever feedback technique more appealing than Q-control for direct force measurements uses a π phase shift in the loop to modulate the spring constant instead of damping. A π phase shift has the resulting amplitude function of $|A(\omega)|^2 = \dfrac{F_t^2}{\left(k+G(w)-m\omega^2\right)^2 + b^2\omega^2}$.

In these experiments, the feedback loop not only functions as a method for stiffening the cantilever but also is a force measurement device. The voltage applied to the force actuators is proportional to the resulting force. Some experiments increased the feedback loop gain enough that most force information is passed from cantilever deflection to the feedback loop, making the voltage at the actuator the measure of tip-sample force. Calibration of the interaction forces is accomplished easily by applying a low frequency modulation signal to the force actuators and measuring the response of the passive cantilever. With this scheme, the best technique is to use a very weak cantilever with high force resolution and transfer force information to the feedback loop while increasing stiffness.

An inverter has near perfect gain and phase characteristics for implementation in the feedback loop as a π phase shift. Unfortunately, as with Q-control, any phase shift greater than π with a loop gain greater than unity leads to self oscillation. Thus, extra phase lag in the feedback loop greatly limits the useable gain range and spring constants that can be achieved. Extra phase lag often results from the limited bandwidth of high power op amps and reactive impedances associated with force actuators. One method of stabilizing the feedback loop uses a low pass filter in series within the loop.[21, 22] The low pass filter allows large loop gain near DC, but rolls off the gain at higher frequencies such that phase shifts do not total π before the unity gain frequency. This method produces a frequency dependent value for the spring constant. At high frequencies (fast velocities), the cantilever behaves as an intrinsic cantilever which allows non-equilibrium jumps to the surface. The feedback loop subsequently pushes the tip away from the surface and the system establishes a steady state of jump to contact and being pushed out. Unfortunately, the motion spans a large range of tip-sample distances and the average of this motion is an inaccurate measure of the local force gradient. A different implementation which uses a larger bandwidth for the feedback circuit is required for doing direct force measurements.

A superior method for stabilizing the feedback loop adds differential gain in parallel to reduce phase shifts to below π in the bandwidth of cantilever motion. Having the whole cantilever resonance within the feedback loop controls the cantilever motion. However, implementation of this technique may still be quite difficult. With magnetic actuation, it may not be possible to add enough phase compensation with the differential circuit to offset phase lag produced by a highly inductive solenoid. Also, as spring constant is increased, resonant frequency and Q increase. Higher Q reduces the tolerance of the system for phase lag and increases loop gain at the resonance frequency, increasing the chance of oscillation. Similarly, phase compensation of the differential gain is reduced at higher frequencies while phase lag from the electronics increases, limiting the ability of the circuit to remain stable as the resonant frequency is increased. Moreover, tip-sample contact is a very stiff interaction which further pushes the resonant frequency higher, possibly causing instability. These phenomena conspire to make the feedback loop unstable and spring constant control through feedback challenging.

The effects of instrument noise introduce a further challenge beyond stability for using feedback to modify cantilever parameters. The feedback process shifts information from the deflection signal to the actuator. When the measurement has instrument noise components, the feedback loop responds to signals from the photodiode that are not cantilever deflections. As a result, the actuator may induce motion in the cantilever, which cancels this falsely perceived motion at the detector. This process increases the range of tip-sample distances probed and reduces the accuracy of the local force gradient measurement. Both instrument noise and stability considerations make enhancement of cantilever properties with active feedback very challenging to use effectively.

Direct force profile measurement requires choosing the proper cantilever properties. The spring constant must be large enough to avoid information loss due to instabilities, even if it compromises force resolution. Smaller cantilevers may reduce thermal noise contributions in the measurement bandwidth, but short cantilevers may also reduce sensitivity for low numerical aperture instruments. Active feedback methods open the possibility of modifying the spring constant and damping; but for direct force profile measurement the advantage may not be worth the difficulty, especially when excellent instrument design more easily affords the same advantages. Thus most direct force measurements will be performed where the interaction stiffness and the cantilever stiffness are similar, because it maximizes the signal to noise ratio while avoiding non-equilibrium jumps to and from contact with the sample. In this regime, Brownian force profile reconstruction becomes a necessary tool for gleaning accurate equilibrium force profile information from the force curve.

Brownian Force Profile Reconstruction

When spring constant and interaction stiffness are similar, then cantilever motion due to thermal excitation begins to influence the average value of the deflection more significantly. Normal force curve techniques filter the deflection signal, which produces the arithmetic mean of the cantilever thermal motion. When thermal excitation allows the cantilever to hop over shallow barriers and probe regions of the energy landscape far from the center of the harmonic well, then the mean deflection value inaccurately describes the local changes in the tip-sample potential. However, information about the equilibrium potential surface is contained in the motion of the cantilever and can be obtained by analyzing the thermal motion using Brownian force profile reconstruction.

Brownian force profile reconstruction (BFPR) is a data collection and analysis technique that utilizes information content in cantilever thermal motion to probe interaction potentials more accurately.[23] During a force curve, if the probe is able to cross the barrier freely in both directions within the timescale of the measurement, then the system has reached quasi-equilibrium. As a result, the inverse of Boltzmann's distribution equation, $U(x) = -k_b T \ln(p(x)) + C$, governs the probability of finding the cantilever at a specific location. In the equation, x is the tip-sample distance, $U(x)$ is the combined potential of tip-sample interaction with the spring harmonic well, $p(x)$ is the probability distribution, $k_B T$ is thermal energy, and C is an arbitrary constant. Thus, the probability distribution of the deflection is a measure of the local potential energy landscape. Brownian force profile reconstruction samples the deflection signal at four times the resonant frequency, f_0. This makes the highest frequency component or Nyquist frequency two times f_0, which includes all thermal motion in the data collection window to accurately capture the deflection probability distribution. Next, the force curve is parsed into many small sections, on the order of 10,000 samples each. Deflection data is scaled to account for a larger variance due to instrument noise and binned into a histogram to calculate cantilever deflection probability density. The inverse of Boltzmann's distribution equation converts the cantilever deflection probability density into relative energy. A quadratic curve is fit to an energy section without tip-surface interaction to calculate the spring harmonic well. This value can be confirmed with independently determined values for the spring constant. The resulting fit is subtracted from each energy section and the sections are subsequently scaled for tip-surface distance, derived from the overall deflection of the cantilever and the position of the cantilever support relative to the sample. Unfortunately, the energy sections do not overlap because the arbitrary constant, C, in Boltzmann's distribution equation is unknown. Calculating force by taking the derivative removes the necessity of obtaining C and automatically calculates the proper scaling. Lastly, all force sections are averaged to produce the reconstructed Brownian force profile. Brownian force profile reconstruction augments low frequency deflection information with information about the force profile from the high frequency thermal motion. This added information not only corrects errors in accuracy but also increases precision. Figure 5 shows the steps of Brownian force profile reconstruction: high frequency sampling (a), binning into histograms (b and c), calculating force sections (d and e), and computing the average (f). The inaccurate force profile calculated using normal force curve techniques is included in Figure 5f.

Brownian force profile reconstruction leverages the convolution of the cantilever spring and $3–4 k_B T$ excursions of the thermal noise to probe deep potentials. In addition, the added energy from the thermal noise allows interactions that are stiffer than the intrinsic cantilever stiffness to be accurately measured. In Figure 5, the interaction is 4.0 N/m while the cantilever is 2 N/m. Furthermore, the cantilever may only need to hop over the barrier separating the two wells 5 to 10 times to obtain an accurate reconstruction. Yet energy barriers between wells increase rapidly as the distance between the wells increases. As the length scale of the interaction increases, the cantilever spring constant must be closer to the true interaction stiffness to achieve quasi-equilibrium. Thus, researchers seeking to probe complex tip-surface interactions composed of numerous force components should estimate the total interaction stiffness and length scale, then choose a cantilever stiffness as close as possible to the interaction stiffness, with the freedom to have a less stiff spring for short interactions. It is better to err on the side of having too stiff a cantilever, since too weak a cantilever could require

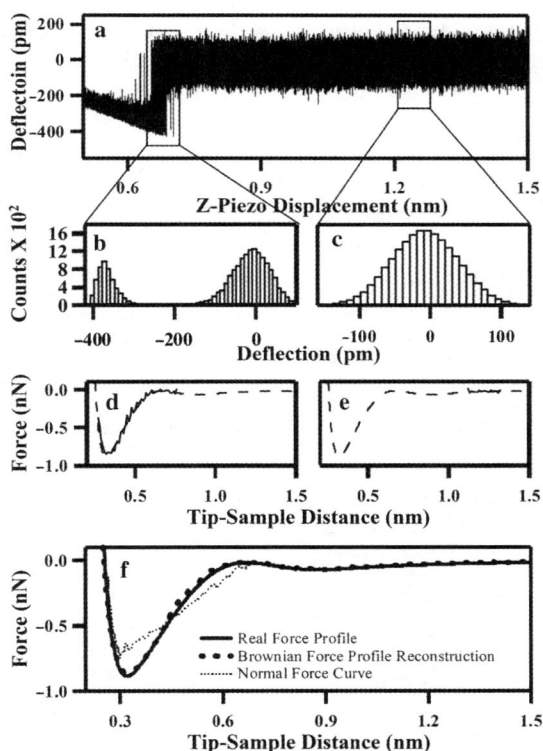

Figure 5. Brownian force profile reconstruction. (a) Force curve sampled at $4^* f_0$ to include thermal motion. (b and c) Histograms of deflection data for regions with strong and weak tip-sample interactions, respectively. (d and e) Force sections derived from histograms in b and c shown with force profile used in simulation. (f) Brownian force profile with force curve obtained using a low pass filter on deflection signal. Brownian force profile reconstruction accurately reconstructs the force profile, while the normal force curve is inaccurate in the high stiffness region.

unacceptably long data collection times to reach quasi-equilibrium. Since a large gap exists in spring constants that are commercially available between 5 and 20 N/m, the utility of BFPR can be further increased by artificially augmenting the Brownian motion such that it probes more of the energy landscape while maintaining sample temperature. The Brownian motion can be increased by driving the cantilever using an external source such as white noise, or using Q-control to reduce the effective damping. Externally exciting the cantilever motion effectively increases the cantilever temperature, which will have to be quantified and used in place of ambient temperature in Botlzmann's distribution equation during the conversion of probability to energy.

Brownian force profile reconstruction is a powerful tool for mapping interaction potentials. While passive techniques with outstanding instruments work well, errors may still result from using a cantilever with a stiffness similar to the interaction stiffness; and outstanding instruments such as the one shown in Figure 4 are not generally available. However, BFPR provides the means for researchers to investigate interactions stiffer than the spring constant with noisy instruments.

Outside the Approximation

Throughout this chapter, the simple harmonic oscillator and point mass approximation are used to describe cantilever dynamics. However, the cantilever is neither a point mass nor a true harmonic oscillator, causing inaccuracies that may become important during experiments. A more thorough understanding of cantilever dynamics provides the basis for estimating sources of error and determining if the error is significant.

As mentioned in the noise reduction section, laser diffraction determines the shape and spot size of the focused beam. This relatively large laser spot reflects from a bent cantilever, where the angle is a function of the position along the length of the beam. Above, it was recommended to have a cantilever twice the length of the laser spot, based on simple approximations.

However, an analytical expression for sensitivity is derived for spot size and position on the cantilever in Schaffer et. al.[24] Similarly, Proksch et. al. build on this work to describe the relationship between sensitivity measured at a specific location along the cantilever, and the tip deflection due to interaction forces.[25] They provide excellent graphs depicting sensitivity changes as a function of spot size and position and their relationship to tip motion. This work is extremely important because sensitivity is the initial calibration step to force from a photodiode signal and it plays in an important part in calibrating the spring constant.

Calibration of the spring constant is commonly the largest source of error. The thermal noise method is the gold standard and quite simple to implement with a spectrum analyzer and analysis software.[26] However, it relies heavily on an accurate sensitivity and the simple harmonic oscillator approximation. Sensitivity determination by contacting the sample may cause tip damage for hard samples or may be difficult to obtain on soft samples. Also, AC mode and contact mode produce different sensitivities due to the cantilever's response to static load, compared to that of an oscillatory drive.[19, 25] Furthermore, the thermal method is especially difficult to perform accurately in viscous fluids, since damping becomes a function of frequency.[27] Another method for cantilever calibration is the added mass method.[28] It is accurate and nondestructive, but it can be tedious and time consuming. Sader published an outstanding method that is easy and only requires Q, f_0 and the plan dimensions of the cantilever.[29] This method is fast becoming the new standard for spring constant estimation for its simplicity and accuracy; only cantilevers with uneven mass distributions, such as a cantilever with a magnetic particle glued to it for performing feedback, will require other techniques.

Conclusion

Direct mapping of intermolecular interaction potentials provides great insight into the properties of materials and the energetics of their interactions. Force spectroscopy with the AFM is ideal for directly measuring these properties because the spring stiffness is very high relative to the contact area, enabling equilibrium mapping of the interaction potential without instabilities. However, instrumental considerations become important, since using stiff cantilevers may compromise the signal to noise ratio. Instrument noise in the bandwidth of interest is drastically reduced by using low numerical aperture optics and a single mode fiber feedback stabilized light source. Attempts to use electronic force feedback to enhance cantilever properties for force measurement usually prove to be more challenging to effectively implement than using cantilevers with the desired properties intrinsically. Due to instrument noise considerations, most measurements will use a cantilever stiffness near the interaction stiffness. Under such conditions, thermal excitations may cause the mean value of the deflection signal to inaccurately represent the local interaction potential. Brownian force profile reconstruction gleans information about the local interaction potential from the distribution of thermal excitations and accurately maps interaction potentials.

References

1. Law, M.; Goldberger, J.; Yang, P. D., *Annual Review of Materials Research* **2004,** *34*, 83–122.
2. Jain, K. K., *Clinica Chimica Acta* **2005,** *358*, 37–54.
3. Kubik, T.; Bogunia-Kubik, K.; Sugisaka, M., *Current Pharmaceutical Biotechnology* **2005,** *6*, 17–33.
4. Israelachvili, J., *Intermolecular and Surface Forces.* second ed.; Academic Press: San Diego, CA, 1992.
5. Tabor, D.; Winterton, R. H. S., *Proceedings of the Royal Society of London* **1969,** *A312*, 435–450.
6. Israelachvili, J.; Tabor, D., *Proceedings of the Royal Society of London Series a-Mathematical and Physical Sciences* **1972,** *331*, 19–38.
7. Binnig, G.; Quate, C. F.; Gerber, C., *Physical Review Letters* **1986,** *56*, 930–933.
8. Wong, S. S.; Joselevich, E.; Woolley, A. T.; Cheung, C. L.; Lieber, C. M., *Nature* **1998,** *394*, 52–55.
9. Oberhauser, A. F.; Marszalek, P. E.; Erikson, H. P.; Fernandez, J. M., *Nature* **1998,** *393*, 181–185.
10. Rief, M.; Gautel, M.; Oesterhelt, F.; Fernandez, J. M.; Gaub, H. E., *Science* **1997,** *276*, 1109–1112.
11. Williams, P. M.; Fowler, S. B.; Best, R. B.; Toca-Herrera, J. L.; Scott, K. A.; Steward, A.; Clarke, J., *Nature* **2003,** *422*, 446–449.
12. Ashby, P. D.; Lieber, C. M., *Journal of the American Chemical Society* **2005,** *127*, 6814–6818.
13. Frisbie, C. D.; Rozsnyai, L. F.; Noy, A.; Wrighton, M. S.; Lieber, C. M., *Science* **1994,** *265*, 2071–2074.
14. Noy, A.; Sanders, C. H.; Vezenov, D. V.; Wong, S. S.; Lieber, C. M., *Langmuir* **1998,** *14*, 1508–1511.
15. Deen, M. J.; Pascal, F., *Iee Proceedings-Circuits Devices and Systems* **2004,** *151*, 125–137.
16. Gustafsson, M. G. L.; Clarke, J., *Journal of Applied Physics* **1994,** *76*, 172–181.
17. Beyder, A.; Sachs, F., *Ultramicroscopy* **2006,** *106*, 838–846.
18. Viani, M. B.; Schaffer, T. E.; Chand, A.; Rief, M.; Gaub, H. E.; Hansma, P. K., *Journal of Applied Physics* **1999,** *86*, 2258–2262.
19. Sarid, D., *Scanning Force Microscopy: With Applications to electric, magnetic and Atomic Forces.* 2nd ed.; Oxford University Press: New York, 1994.
20. Liang, S.; Medich, D.; Czajkowsky, D. M.; Sheng, S.; Yuan, J.-Y.; Shao, Z., *Ultramicroscopy* **2000,** *84*, 119–125.
21. Ashby, P. D.; Chen, L. W.; Lieber, C. M., *Journal of the American Chemical Society* **2000,** *122*, 9467–9472.
22. Jarvis, S. P.; Yamada, H.; Yamamoto, S.-I.; Tokumoto, H.; Pethica, J. B., *Nature* **1996,** *384*, 247–249.
23. Ashby, P. D.; Lieber, C. M., *Journal of the American Chemical Society* **2004,** *126*, 16973–16980.
24. Schaffer, T. E.; Hansma, P. K., *Journal of Applied Physics* **1998,** *84*, 4661–4666.
25. Proksch, R.; Schaffer, T. E.; Cleveland, J. P.; Callahan, R. C.; Viani, M. B., *Nanotechnology* **2004,** *15*, 1344–1350.
26. Hutter, J. L.; Bechhoefer, J., *Review of Scientific Instruments* **1993,** *64*, 1868–1873.
27. Sader, J. E., *Journal of Applied Physics* **1998,** *84*, 64–76.
28. Cleveland, J. P.; Manne, S.; Bocek, D.; Hansma, P. K., *Review of Scientific Instruments* **1993,** *64*, 403–405.
29. Sader, J. E.; Chon, J. W. M.; Mulvaney, P., *Review of Scientific Instruments* **1999,** *70*, 3967–3969.

Index

Printed in the United States of America.